Germany's Ancient Pasts

Germany's Ancient Pasts

Archaeology and Historical Interpretation since 1700

BRENT MANER

THE UNIVERSITY OF CHICAGO PRESS   CHICAGO AND LONDON

The University of Chicago Press, Chicago 60637
The University of Chicago Press, Ltd., London
© 2018 by The University of Chicago
Published 2018
Printed in the United States of America

27 26 25 24 23 22 21 20 19 18   1 2 3 4 5

ISBN-13: 978-0-226-59291-6 (cloth)
ISBN-13: 978-0-226-59307-4 (paper)
ISBN-13: 978-0-226-59310-4 (e-book)
DOI: https://doi.org/10.7208/chicago/9780226593104.001.0001

Library of Congress Cataloging-in-Publication Data

Names: Maner, Brent, author.
Title: Germany's ancient pasts : archaeology and historical interpretation since 1700 / Brent Maner.
Description: Chicago ; London : The University of Chicago Press, 2018. | Includes bibliographical references and index.
Identifiers: LCCN 2018019131 | ISBN 9780226592916 (cloth : alk. paper) | ISBN 9780226593074 (pbk. : alk. paper) | ISBN 9780226593104 (e-book)
Subjects: LCSH: Archaeology—Germany—History. | Archaeology and history—Germany.
Classification: LCC CC101.G35 M36 2018 | DDC 930.10943—dc23
LC record available at https://lccn.loc.gov/2018019131

Contents

Illustrations

Abbreviations

BGfAEU   Berliner Gesellschaft für Anthropologie, Ethnologie und Urgeschichte

BNM   Bayerisches Nationalmuseum

DAG   Deutsche anthropologische Gesellschaft

GNM   Germanisches Nationalmuseum

GStA PK   Geheimes Staatsarchiv, Preußischer Kulturbesitz

KMVA   Königliches Museum vaterländischer Alterthümer

OA   *Oberbayerisches Archiv für vaterländische Geschichte*

RGZM   Römisch-Germanisches Zentralmuseum

SSM MM   Stiftung Stadtmuseum (Berlin), Märkisches Museum, Abteilung Ur- und Frühgeschichte

Introduction

Nation-states present their stories in different ways. In the United States, for example, textbooks have traditionally explained American history as a new beginning. An ocean separated European explorers and settlers, African slaves, and later waves of immigrants from their places of origin. These arrivals carved a new nation out of the wilderness.

Stories of origins are fundamentally different in Europe. European states have modern founding moments (1688 in England and 1801 for the United Kingdom; 1789 in France; 1871 and 1990 in Germany; the 1860s in Italy), but they look back upon long histories of settlement that extend into the premedieval era. These histories differ from the American case because many Europeans partly view the earlier inhabitants of their lands as their ancestors and the development of modern populations as a long-term process that took place on the soil they still inhabit.

The differences between these national stories have shaped the concept of prehistory and the practice of archaeology over the past three centuries. In the United States, the study of the distant past has represented the investigation of a culture that is distinct from the dominant culture. Native peoples lived in North America for millennia before Thomas Jefferson began to study Indian burial mounds in Virginia in the 1780s. Yet, as anthropologist David Hurst Thomas has argued, European Americans treated Native Americans more as natural history specimens than as the creators of North American culture, and in the young nation, American history became a relatively recent story about the birth of the Republic and how the West was "won." Research into the oldest settlements and cultural sites of North America remained fully distinct from this national narrative.[1]

In Europe, however, the archaeological investigation of indigenous peoples has been a form of self-investigation. The connection between people and territory became especially prominent in the nineteenth and early twentieth centuries, when European stories of national origins were tied directly to the study of ancient burial grounds and settlements. This approach treated pre- and early history as the critical era when languages, customs, political values, and even racial characteristics developed.

This book focuses on domestic archaeology in Germany since the eighteenth century. Domestic archaeology refers to excavations that took place in central Europe and were used to write the prehistory of this territory. This practice explored prehistoric, Roman, and barbarian sites, as opposed to the well-known projects carried out by classical archaeologists in Italy, Greece, and Turkey, and it includes both the activities known in German as *vaterländische Altertumskunde* (the regional antiquarianism of the nineteenth century) and the field of *Vor- und Frühgeschichte* (pre- and early history), which became a scientific discipline around 1900. I have intentionally chosen the general modifier "domestic" over "German" or "Germanic" archaeology to preserve the variety of meanings associated with archaeological interpretation in the period covered by this book, and I will reserve the use of ethnic labels for references to excavators and authors who related their work to a national story.

Germany offers an extreme and troubling example of the relationship between archaeology and stories about national origins. In the early twentieth century, racist ideologues drew firm lines of continuity between the distant past and their present day. Gustaf Kossinna (1858–1931), the Berlin philologist-turned-archaeologist, claimed that a "Germanic" people had continuously inhabited central and northern Europe from the Bronze Age to his present day. And Hans Günther (1891–1968), the notorious racial scientist of the Weimar Republic and Third Reich, popularized the idea that Germans carried forward the achievements of a superior prehistoric Nordic race. This glorified view of the ancient past became a fundamental component of the Nazi regime's vision of history. Its conception of race stressed Germanic heritage and biological purity; its "blood and soil" ideology bolstered the idea of an eternal bond between Germans and the land; and the Nazi state's foreign policy and wartime goals were predicated on the idea that eastern Europe belonged to Germany because "ancient Germans" had settled and developed the area in earlier centuries. The Nazi regime featured Germanic motifs in propaganda, taught their version of prehistory in schools, and supported the academic study of archaeology with new university positions and generous support for excavations.[2] The connection between the Germanic past and the German present produced not only strong bonds of inclusion but also exclusionary and racist identities that yielded hypernationalism, violence, and genocide.

To varying degrees, German humanists, nineteenth-century antiquarians, nationalist scholars, and Nazi ideologues took ethnic labels as givens and believed in clear lines of continuity from "ancient Germans" to themselves. In the wake of two world wars fueled by aggressive nationalism, though, scholars have cast a skeptical, even reproachful eye on attempts to treat nations as clearly defined, primordial communities. In a pioneering essay, historian Eric Hobsbawm explained that nations are relatively recent creations, but they "generally claim to be the opposite of novel, namely rooted in the remotest antiquity." Holidays, festivals, and folk customs are often modern inventions that lend political communities the feeling of being timeless and unchanging.[3] A growing literature applies this core idea to the development of archaeology, showing how states have funded excavations, museums, and eventually university positions in part to legitimate stories of ethnic descent and make claims on disputed territories.[4]

Studies based on archaeological and early historical sources have also challenged the basic assumptions behind nationalist interpretations of the distant past. Walter Pohl, a leading voice in questions of ethnicity in the early Middle Ages, suggests that the Germanic peoples (*die Germanen*) never existed as a clearly defined group. The use of *die Germanen* as a collective term comes from Roman sources. It did not emerge as a self-description of an ancient people. And, as Pohl explains, attempts to establish ethnic identities in the premedieval past face severe limits, whether they are based on linguistics, archaeological material, notions of lineage, political history, or classical sources.[5]

Historians who work with evidence from the early Middle Ages emphasize the fluid nature of group identity. The concept of ethnogenesis, for example, regards ethnic groups and tribal affiliations not as static communities but as the outcomes of political and cultural processes. Andrew Gillett explains that tribal identities like the Goths, Franks, or Langobards were not "fixed and simply hereditary." Rather, they "had to be generated and reified by the efforts of elites—the nobility of the barbarian groups, and particularly the royalty—in order for the diverse individuals who constituted their followings to accept that they were members of one group, and that they owed loyalty to that group's leaders."[6] Another assumption that has undergone tremendous revision relates to the border between Germania and the Roman Empire. Nationalist interpretations stressed the nobility of tribes who defended their freedom from the grip of Roman imperialism. Historians today generally challenge this dichotomy and underscore the vast amount of interaction with the Roman Empire (trade, military service, tributes, alliances, etc.) that contributed to group formation. As historian Florin Curta has argued, we should view ancient "political frontiers as key elements in the *creation*, as opposed to *separation*, of ethnic groups."[7]

In the main, this scholarship has exposed the connection between ethnic groups in the distant past and modern national communities as exceedingly simplistic or even completely wrong. And it is tempting to view the political use of archaeology as a product of a nationalist age that peaked during the first half of the twentieth century and has been debunked by the science of archaeology during the postwar period. Yet the idea of ethnic roots has persisted, and the identification with ancient peoples remains a powerful and alluring way to explain national belonging. Surveying his field in the early 1960s, the eminent British archaeologist Glyn Daniel criticized the rhetoric of politicians and propagandists that drew unfounded connections between ancient and modern peoples, and he reminded his readers that archaeological sites can tell us about broad transformations in earlier times and the spread of new technologies, but "we must alas, for the most part, keep the builders and bearers of our prehistoric cultures speechless and physically neutral."[8] And historian Patrick Geary, distressed by the revival of nationalism in Europe after 1989, has called on scholars to address the persistence of naive ethnic thinking: "As a tool of nationalist ideology, the history of Europe's nations was a great success, but it has turned our understanding of the past into a toxic waste dump, filled with the poison of ethnic nationalism, and the poison has seeped deep into popular consciousness. Cleaning up this waste is the most daunting challenge facing historians today."[9] Notions of primordial descent underpin nationalist movements around the globe, and they continue to haunt Europe today in heated debates about immigration and citizenship. Given the extreme power of these ideas, it is imperative that we better understand how political communities came to seek their origins in prehistory.

The two approaches outlined above (studies of nationalism's influence on archaeology and an emphasis on the fluid nature of group identity in the early Middle Ages) are important contributions to the cleanup called for by Geary. They identify the nationalist interpretations that must be avoided, and they replace misguided assumptions about the ancient past with more accurate knowledge. Yet, when it comes to the history of archaeology, they are somewhat limited because they are trapped beneath the long shadow cast by nationalism (and, for Germany, National Socialism) over the study of the ancient past. They are so focused on demonstrating the ways that pre- and early history research facilitated nationalist and racist thinking that they obscure other meanings that archaeology has held for its practitioners. They also leave several key questions unanswered.

How, for example, should we understand the relationship between archaeological activity in the eighteenth and nineteenth centuries and nationalist interpretations that emerged around 1900? Can this earlier activity be conceived primarily as a pathological obsession with national origins and racial purity, especially when antiquarians in nation-states that did not become fascist dictatorships or racialist dystopias were also consumed with the excavation of "their ancestors" in the nineteenth century? Also, is it proper to view archaeology primarily as a nationalist practice even though most excavations were carried out by regional associations and exhibited in local or regional museums? Finally, if nationalism was the driving force behind archaeological investigation, why did archaeology continue to flourish in Europe after 1945, when nationalism largely became taboo?

To provide a fuller understanding of these issues, this book takes a step back from the conclusion that nationalist thought drove archaeological interpretation and places the development of pre- and early history research in Germany in a longer chronological framework and a broader political and cultural context.[10] This perspective uncovers several motivations that were not always tied to the concept of the nation, and it allows us to see that many historical actors developed aspects of archaeology that we still value today, including a strong commitment to historic preservation, a desire to teach others about the history of their surroundings, and methodologies that advance our knowledge but also check the tendency to romanticize or simplify the past.

This broader approach to the history of domestic archaeology yields three main conclusions. First, antiquarians and archaeologists throughout the time period covered by this book often challenged nationalist readings of excavated material for a variety of reasons. Many noted that archaeology, by its very nature, focused on a specific site and spoke most clearly to local circumstances. They therefore felt that archaeological finds were closely tied to local conditions and regional identities, not national categories.

Second, it is not enough to show that archaeology produced nationalist interpretations or to replace these interpretations with more accurate descriptions of the ancient past. We also need to understand what lent nationalist interpretations their political power and cultural influence. It is my contention that this power arose from the *narrative* nature of nationalist interpretations. "Archaeology and Historical Interpretation," my subtitle, points to the main interpretive challenge that faced antiquarians in the eighteenth and nineteenth centuries and was fatefully overcome by literary figures and archaeologists themselves in the late nineteenth and early twentieth centuries. Beginning in the 1820s, antiquarians in German-speaking Europe spoke about their desire to shed light on the *Dunkel* or *Grau der Vorzeit* (literally the "darkness or gray of pre-time"). Classical authors mentioned the conditions and peoples of central Europe in the distant past, but a critical mass of chronicles and religious documents that related the progression of events did not appear until the medieval period. Thus, for antiquarians interested in the history of their lands, the *Dunkel der Vorzeit* lasted a millennium longer than it did for ancient Greece or the Roman Empire. It shrouded the entire premedieval period.

Antiquarians hoped to solve prehistory's interpretive challenge through archaeological investigation. They wanted to transform excavated objects into historical sources and thereby annex earlier eras to the realm of history. Yet they acknowledged the limitations of the archaeological record. This inaugurated the great tension described in this book between the questions archaeology can answer and the fundamental questions about history and identity that people wish to have answered. Antiquarians did not want local, regional, and national histories to remain vague and timeless. They wished to know what had happened in their local areas in ancient times and who "their ancestors" were. They hoped that archaeology would clarify early history and that this narrative would then offer clear points of orientation for future developments. In short, they wanted to turn the disparate objects from the archaeological record into the elements of a story.[11]

Most of the authors described in this book respected the incomplete nature of the archaeological record. Beginning in the early nineteenth century, antiquarians described archaeology as a promising but under-developed way to study the ancient past. They believed they could use finds to document local events, but they did not think it was possible to explain finds in ethnographic terms until they had more material and a better grasp of comparative typologies. They knew they were a long way from writing sweeping narratives that connected the ancient past to the Middle Ages. Even during the nationalist euphoria of the early German Empire, leading scholars argued that empiricism simply did not allow connections between ancient peoples and modern nations. Other authors, though, were frustrated by this hesitancy and took the leap of turning the archaeological past into the prehistory of the German nation.

My third conclusion is that greater attention to multiple trends in the development of archaeological thought makes more sense of what actually happened in Germany after 1700. On the one hand, Germany did produce the prime example of the ideological abuse of prehistory with the rise of *völkisch* nationalism, racial studies, and Nazi ideology. On the other hand, archaeology in central Europe has been a pervasive but de-centralized endeavor that has explored mainly regional and local areas. It is difficult to find a German town that does not have a display of local artifacts. *Heimatmuseen* (local museums) that document the traditions of towns or regions almost always include a section devoted to archaeology. Larger city museums often begin with a display about pre- and early history before telling the story of urban expansion and civic culture. And major regional museums in Trier, Stuttgart, Bonn, Hannover, Kiel, Halle an der Saale, Weimar, Munich, and Berlin contain impressive archaeological collections that represent over one hundred years of local and regional efforts to collect and interpret the distant past. The largest and most prestigious exhibit for domestic archaeology in Germany, the Museum für Vor- und Frühgeschichte (Museum of pre- and early history) housed in the Neues Museum (New museum) in Berlin, began as a royal Prussian museum and expanded in the late nineteenth century not as an institution for national archaeology but as a scientific institution dedicated to European prehistory.

This study is divided into three chronological parts. Part 1 considers the attention antiquarians accorded to archaeological material before the creation of a unified German nation-state in 1871. Since the Renaissance, scholars had studied classical texts as sources for European prehistory. But, as chapter 1 shows, the eighteenth century marked an important turning point. Antiquarians began to view the physical remnants of the distant past as important sources, and they combined their knowledge of Greek and Roman texts with the study of antiquities to sort out the early territorial and religious history of central Europe. The number of people engaged in the archaeological investigation of central Europe's primeval past increased dramatically in the first half of the nineteenth century. Chapter 2 relates the excitement felt in hundreds of local settings where antiquarians excavated ancient sites. Individual antiquarians and the members of newly founded historical associations considered their work a patriotic endeavor that contributed new knowledge about the regional monarchies they inhabited. In this reading of archaeological material, the ancient past was inseparable from the landscape itself. Archaeology became a way of knowing one's environs intimately, and it generated local pride as much as it sought to explain national ties or ethnic descent.[12]

Archaeological activity during the first half of the nineteenth century brought a wealth of material to light, and chapter 3 explores initial attempts by antiquarians to connect their excavations to the early history of their local areas. These interpretations from the middle decades of the nineteenth century transformed excavation sites into historical places, as authors imagined the events that may have taken place nearby. These local stories added a deeper temporal dimension to historical thinking, and it became natural to view the undocumented era of premedieval migrations as the formative moment for regional and national identities that continued to exist in later historical epochs.

One might reason that the unification of Germany in 1871 produced the rise of a national archaeology, but the development of domestic archaeology was more complicated. Part 2 of the book examines three ways of viewing the distant past that coexisted in the German Empire. Chapter 4 elucidates the rise of an anthropological orientation in prehistory research. New discoveries during the 1850s and 1860s exploded both the chronological sweep and the geographical scope of earlier antiquarian practices. Investigators in the young field of geology discovered that the earth was much older than biblical chronology allowed, and many asked where the appearance of human beings fit into this longer time frame. Archaeology became an auxiliary science to anthropology, and it addressed new questions about daily life, social organization, and cultural development in the ancient past. In this context, most practitioners of domestic archaeology posed their research questions in terms of the general development of humankind, not the investigation of early national communities.

Regionalism flourished in the German Empire despite the celebration of a new national culture, and chapter 5 shows that the traditions of local archaeology and regional historical institutions became even stronger in the last third of the nineteenth century. The 1860s and 1870s produced another wave of popular engagement with the distant past, and the rise of a wealthier and more urbanized middle class created a larger audience for knowledge about science and history. States within the German Empire created historical commissions and built impressive regional museums (*Provinzialmuseen*) that included large displays devoted to archaeological material. These institutions contributed to the professionalization of domestic archaeology by combining the traditional focus on locales and regions with the scientific questions of anthropology. The result was the rise of cultural history narratives that explained the processes and traditions that had produced the conditions of the present day. The main framework for these studies was the region, not the nation. The connection between localities and artifacts was so strong, in fact, that many local groups resisted requests to hand objects over to larger regional or national collections. This in part explains one of the great ironies in the story of domestic archaeology in Germany: in the state that produced a hypernationalism that claimed connections to the "Germanic" past, there was no national museum for German archaeology.

A nationalist interpretation of the distant past arose in the German Empire, but it was a third perspective that emerged most clearly outside the meetings of historical and anthropological societies and regional museums. Chapter 6 documents how literary authors explored the ancient past through historical fiction. The incredible popularity of these stories grew out of the enthusiasm for local archaeology, but the authors narrated the ancient past in ways that were not possible for professional archaeologists. Museum officials and archaeologists understood the complexities of the distant past, including the waves of migrations that swept across the European continent and the shifting borders of cultural groups and political units over the past two millennia. Literary narratives, on the other hand, crafted a coherent national story that traced cultural traits out of the distant past and into the present day. This connection presented ancient peoples as the ancestors of modern Europeans, and it was strengthened by the assertion of other historical continuities that stretched back into the distant past, most notably the continuous possession of the land and the idea of biological purity.

Part 3 explores prehistory's achievement of greater scientific status in the early twentieth century and the concurrent rise of radical nationalism. Regional museums became key institutions for professional prehistory research after 1900, and chapter 7 documents the efforts by some archaeologists to defend the objectivity of their science in the turbulent political climate during and especially after World War I as others abandoned the caution required by the fragmentary nature of archaeological evidence and crossed the interpretive threshold to embrace seamless narratives that connected the ancient past to the present day. As archaeologist Sebastian Brather has argued, nationalists created an ethnic paradigm that treated *Volk*, nation, and race as interchangeable or even equivalent concepts, and archaeology during these years gave these ideas scientific authority.[13] The idea of a German racial identity rooted in the distant past became dogma in the Third Reich, and chapter 8 shows how archaeologists under this regime confirmed and spread this view. The epilogue considers the efforts to reestablish the field of prehistory as a credible scientific and public-outreach enterprise after the Nazi exploitation of the field.

The nationalist interpretation of the past (or the transformation of domestic archaeology into Germanic archaeology) relied on the popularization of prehistory that occurred over the course of the nineteenth century. But the archaeology performed by local historical associations during the eighteenth and most of the nineteenth century was not national chauvinism in the making. Germany has had many ancient pasts, and this study uncovers a variety of narratives, including ones that emphasized the local character of archaeological finds, a vision of prehistory as an international science removed from questions of ethnicity and nationality, and museum exhibits that primarily documented regional cultures.[14] This attention to the multiple forces at play in the history of archaeology certainly does not break the connection between archaeology and politics or underestimate the strength of nationalism. Instead, it reveals that a self-understanding based on prehistory relied on specific cultural efforts that responded to political contexts and changed over time.

The antiquarians who collected and preserved ancient objects in the eighteenth and nineteenth centuries shed light on an unknown and distant past. They unlocked the mystery of cultural change and directed feelings of belonging toward their localities, regions, and nation-state. Nationalist narratives combined these strands of thought with the political goal of legitimating the nation. An analysis of the interaction between these perspectives shows that domestic archaeology is a very deep-seated phenomenon that allows people to interact with their environment and relate to the past. Ultimately, this broader view deepens our understanding of the allure of nationalist interpretations without posing a rigid continuity between early archaeology and Nazi ideology. Nonnational approaches to archaeology endured alongside the racist and nationalistic perversion of prehistory in the Weimar Republic and even during the Third Reich. These other perspectives offered positive traditions for the field of domestic archaeology in the aftermath of World War II.

Today, archaeology in Germany receives the support of state offices for historic preservation that operate within the same geographical scope as the historical associations and regional museums of the nineteenth century. In this sense, the strongest continuity in the history of archaeology appears to be its local resonance, not a radical form of ethnic nationalism. The nonnational language in archaeology today and the emphasis on the connection between excavation and natural science methods hark back to earlier words of caution against overtly nationalistic readings of the past.

PART I

The Discovery of Germany's Ancient Pasts

The Sources for Prehistory

Texts and Objects in the Eighteenth Century

In 1754, an article in a Hannover gazette discussed how one might seek out antiquities "most successfully" and describe them "most usefully."[1] The author did not suggest searching texts by Greek and Roman authors. Instead, he advised readers to pay more attention to their surroundings and especially to rural villages as places where "the ancient" (*das Alte*) had maintained itself the longest. Local actors should investigate the physical remains left by earlier peoples and study legends passed down as oral traditions. Even the names of rivers, valleys, and caves could contain clues to past events. The author opened with two specific questions that this approach might address: "Where did Charlemagne encounter our ancestors, and where did they live the longest?"[2]

Questions about Charlemagne and "our ancestors" represented two new ideas about the ancient past. First, the author did not equate *das Alte* with the classical past of Greece and Rome. He ascribed value to the ancient events that transpired north of the Mediterranean world and viewed the premedieval inhabitants of the area around Hannover as the distant relatives of his eighteenth-century readers. Second, the author proposed a methodological break from the practices of classical philology. The article advised readers to look for sources other than texts, concluding that "the entirety of German history would become clearer and many of its gaps would be filled if one would describe the old artifacts [*die alten Denkwür-digkeiten*] of our villages more diligently and more often."[3]

History has not been particularly kind to the individuals who answered the call to collect and describe old artifacts. Nineteenth-century novels, like Walter Scott's *The Antiquary* (1816) and Charles Dickens's *Pickwick Papers* (1836), poked fun of antiquarians obsessed with the miscellany of history. Their efforts appeared pointless and comical because they did not contribute to a deeper understanding of history. More authoritatively, *The Cambridge Illustrated History of Archaeology* claims that many amateurs caused more harm than good. It reports that "while the aristocrats of northern Europe travelled to the Mediterranean to develop a refined appreciation of classical civilization, at home less exalted antiquarian interests continued to flourish. . . . A few of these early excavations were well conducted, many much less so. With hindsight, terrible damage was done to the prehistoric monuments of western Europe by people who thought nothing of digging half a dozen burial mounds before breakfast."[4]

Classicist Arnaldo Momigliano has offered a much greater appreciation for the antiquarian. Writing primarily about the study of ancient Italy, he noted that antiquarians contributed to the new, more skeptical history of the Enlightenment era. Scholars before the eighteenth century commented on ancient texts, but they generally assumed the validity of their sources. They read Livy, for example, as the history of ancient Rome without subjecting his work to historical criticism. Eighteenth-century antiquarians, on the other hand, initiated new questions by documenting inscriptions and monuments, and scholars came to use this nonliterary evidence to test their interpretations of ancient authors. Momigliano also noted that antiquarians could suggest entirely new fields of research. The study of pre-Roman peoples, and especially Etruscan culture, arose in the eighteenth century because of the antiquarian attention to archaeological sources. In these ways, the antiquarian "could turn himself into a historian or could help historians to write histories of a new kind."[5]

In central Europe, antiquarians provided new primary-source material to a scholarly world focused on texts. They understood burial grounds and ancient ruins as the physical evidence of peoples described by classical authors, and they posed new questions about cultural changes in the distant past. Their work also contributed to a new kind of territorial history during the eighteenth century. The modernization projects of enlightened absolutism, like roadbuilding, canal dredging, and land reclamation, brought state administrators into greater contact with archaeological material. This awareness fed a growing desire to investigate the cultural and historical forces that held princely territories and kingdoms together. With increasing frequency, states commissioned historical works that went beyond the story of rulers and their deeds to describe the resources, customs, and people of the ruler's lands. Archaeology fit neatly into this new engagement with territorial history. Princes and monarchs increased the number of artifacts in their royal collections, and they supported the work of scholars who discussed these artifacts in cultural, historical, and scientific publications like Hannover's *Gelehrte Anzeigen*.[6] This activity established a new relationship between texts and objects and represented a critical first step toward seeing archaeological artifacts as valuable historical sources.

The Divide between Texts and Objects before the Eighteenth Century

During the Middle Ages, historical writing focused on the classical world and the spread of Christianity. Chronicles placed events in relation to key religious turning points, and barbarian history received little attention. In the wake of the Reformation, however, new national historiographies broke away from this older paradigm and replaced points of origins associated with the classical world and the Christian church with genealogical links to the barbarian past.[7] This transition was further encouraged by the rediscovery of works by Greek, Hellenistic, and Roman historians during the Renaissance. These writings inspired great admiration for the achievements of the classical world, but they also provided glimpses of life north of the Mediterranean. Scholars became familiar with the vivid details of the military encounters in central Europe between the Romans and barbarians in Julius Caesar's *The Gallic War*, Livy's *History of Rome*, and Cornelius Tacitus's *Annals*. Renaissance scholars also appreciated the geographical descriptions of ancient Europe provided by Strabo and Ptolemy.

The text that provided the most detail about central Europe's ancient past, though, was Tacitus's *Germania*. Around 98 CE, the great classical author composed an ethnographic description of the peoples who lived to the north and east of the Roman Empire. The book's first twenty-seven chapters describe the cultural practices and social structure of "the Germans." Tacitus included comments on the languages of these people, their pride in their common descent, and their relative strength. He also provided colorful commentaries that associated general "Germanic" char-

acteristics with specific groups.[8] Tacitus explained, for example, that the
Batavi are "foremost in valour" (88); that the Chauci are "a people of
great renown among the Germani. . . . Devoid of greed and recklessness,
orderly and aloof" (91); that the Chatti grow their hair and beards long
until they have slain an enemy (89–90); and that the Semnones established the Germanic religious rituals centered on the forest grove (93).
Throughout the text, Tacitus highlighted the Germans' readiness for
battle, their respect for nature, and above all their commitment to the defense of family and community. In the second half of the *Germania*, Tacitus, moving along the Rhine from south to north and then around the
North Sea, reviewed the names and locations of twenty-three Germanic
tribes and several subgroups among the Suebi.

Contemporary historians of the Roman Empire and the early history
of Europe note major problems with reading the *Germania* as a straightforward ethnography of the peoples who lived east of the Rhine River in
the first century of the Common Era. Tacitus probably did not travel in
these lands, so his coverage is not based on direct observation. His text
gave "the Germans" a collective name that the peoples themselves did
not use, and it suggested a unified culture by describing character traits
and rituals and declaring what "the Germans" eat, drink, do, believe, and
value. As many historians have noted, it is critical to handle this cultural
portrait carefully. Tacitus was a master rhetorician, and he was writing for
a Roman audience. He used his description of the Germanic peoples as
pure, vital, and close to nature to chastise what he saw as the decadence
of the empire.[9] Yet these features did not prevent Renaissance scholars
and many subsequent readers from deriving a clear portrait of the ancient
Germans from this text.

Tacitus's *Germania* had been lost for several centuries when, in the
mid-fifteenth century, Italian manuscript hunters confirmed that a copy
of the text existed in the Hersfeld monastery (near Fulda, Germany).
Aeneas Silvius Piccolomini (who later became Pope Pius II) offered the
first influential interpretation of the text in 1457 when he published passages that recounted the barbaric nature of life in pre-Christian northern Europe. He used the *Germania* to argue that cultural development in
Germany had occurred through contact with the Catholic Church.[10]

Other commentators read the *Germania* in dramatically different
ways. German-speaking humanists treated Tacitus's ethnographic and
geographic information as hard facts, and they devoured the vivid details about the vigor and nobility of the "ancient Germans." Conrad Celtis

(1459–1508), a widely traveled scholar and poet and one of the most influential German humanists, set out to create a native history that contrasted the liberty and purity of ancient northern Europe with the sensuality and greed of Roman civilization. He planned a massive topographical project called *Germania illustrata* that would connect the ancient locations of the peoples named in the *Germania* to the forests, villages, and towns of his day. This would be accompanied by a genealogical section that would trace the descent of fifteenth-century rulers, including the current Habsburg emperor, from early tribal chieftains. After the *Germania* was translated into German in 1526, several other historical works, inspired by Celtis's cartographic idea, suggested connections between the ancient locations of tribes and present-day villages.[11]

This wave of scholarship offered a new interpretation of the German landscape. Tacitus had described the land north of the Roman Empire as cold and barren. As he famously pronounced, the area was so desolate that only a native would choose to live there (77). The patriotic humanists of the sixteenth century, however, reimagined the forests "as domesticated woodlands, intersected by arable land and orchards."[12] They viewed this territory, not as the home of rough barbarians, but as a natural setting that nurtured the virtues of the ancient Germans.

Tacitus also provided important details about the military event that would become so influential for later interpretations of early German history. In 9 CE, Publius Quinctilius Varus, the Roman governor of the Rhineland, and three legions of soldiers were ambushed by a coalition of Germanic peoples under the leadership of Arminius, the Cheruscan chief. Roman historians recorded this major defeat and offered commentaries on Varus's competency and imperial policies along the Rhine.[13] In the *Annals*, Tacitus's history of the Roman Empire from the reign of Tiberius to Nero's suicide, Tacitus referred to the *saltus Teutoburgiensis* as the place where "the remains of Varus and his legions were said to lie unburied."[14] In accounts of the event from the seventeenth century onward, this location was referred to as the Teutoburger Wald (Teutoburg Forest).[15] Furthermore, the *Annals* provided a portrait of Arminius's career after the battle, and Tacitus closed his assessment by describing Arminius as "[t]he liberator of Germany without doubt, and one who challenged not the formative stages of the Roman people, like other kings and leaders, but the empire at its most flourishing . . . , and [he] is still sung among barbarian races, though unknown to the annals of the Greeks, who marvel only at their own, and not celebrated duly in the Roman, since we extol the distant past, indifferent to the recent."[16] In later interpretations, the moniker "liberator of Germany" gained great significance, as Roman-German relations were depicted as a fight against imperial expansion and for the preservation of freedom that was won because of Arminius's ability to unite the Germanic peoples against a common enemy. As the comments about Greek and Roman annals suggest, Tacitus was conveying a more complicated point about the lessons of history and the kinds of events that enter into the collective memory. For many subsequent readers in Germany, though, the battle in the Teutoburg Forest offered a clear message that strong leadership and unity ensured liberty.

The search for ancestors among the ancient peoples of northern Europe was not limited to German-speaking lands. The London antiquarian William Camden, for example, replaced legends of Trojan origins and the Roman tradition with a new narrative about the Anglo-Saxon heritage of the British people in his *Britannia* (1586). Around the same time, French historiography disputed stories of Trojan origins and increasingly placed the beginning of French history during the migrations of the early Middle Ages. Polish and Czech scholars followed suit, referring to ancient Slavs and the early medieval origins of their territories.[17] The key sources for these native stories of origins were the sixth-century author Jordanes, who told of the great deeds of the Goths and traced their heroic lines; Gregory of Tours, also from the sixth century, whose *History of the Franks* related Clovis's military victories and conversion to Christianity; and the Venerable Bede, who provided a similar plotline for Great Britain and Ireland from the eighth century.[18]

As with the interpretation of Tacitus's *Germania*, modern scholars have raised serious doubts about interpretations that use early medieval sources to connect ancient peoples to later cultural groups. These treatments transform ancient peoples with no clear sense of a group identity into clearly demarcated groups or tribes. They also ignore the context of the authors, who often had goals in mind other than supplying the beginning of a national narrative.[19] Yet this reinterpretation of Europe's early past established the general belief that European nations were created in the early medieval period and that sixteenth-century populations were related to the earlier inhabitants of their lands.

Humanist scholars exhibited an interest in the early history of Europe, but only a few engaged the physical evidence available from this distant epoch. In *Britannia*, William Camden used the stone formations, Roman walls, and ruins on the British Isles, along with finds of coins and place-names, to verify Roman geography.[20] And Ole Worm (1588–1654) prepared a six-volume work on Danish monuments based in part on his own excavations and his personal collection of antiquities and curiosities.[21] These individuals were exceptional, though, and most of the knowledge about the prehistory of central Europe was based on texts, not objects.

From the Middle Ages to the eighteenth century, the most common (but least documented) encounter with prehistoric objects came not in scholarly works but in the lives of peasants. Each spring, farmers and laborers turned up burial urns and other prehistoric material as they tilled their fields. Rural legends portrayed these artifacts as part of the natural environment or even as magical objects. Hammer-heads and ax-heads were known as "thunderstones" that fell from the sky during violent storms, and local stories described burial urns as "magic crocks" that possessed the power to increase the fertility of seeds or to make milk fattier and perfect for butter making. Another legend claimed chickens that drank water from prehistoric vessels would never get sick.[22]

Myths also explained prominent markers in the landscape, like the former defense walls of the Roman Empire. Common people in southern German states called these ruins "the devil's wall" (die Teufelsmauer). According to legend, the devil had requested a piece of earthly territory, and God offered him as much land as he could enclose in a single day. The devil then went to work but at the end of the allotted time was still extending his wall instead of connecting it to surround his share of land. God mocked the devil's greed by crumbling all his walls, leaving them in the ruined state that locals witnessed.[23] As these stories demonstrate, local populations were certainly aware of urns, burial mounds, and ancient ruins. But the material remains from the distant past were part of a rural mythology that enchanted the landscape, not sources for the writing of history.

On the other end of the social ladder, royal houses showed some interest in archaeology. Rudolf II, the Holy Roman emperor from 1576 to 1612, excavated urns near the Silesian town of Greisitz (today Gryżyce, Poland) and marked the spot of his discovery with a wooden column. Duke Johann Friedrich of Württemberg (r. 1608–28) added coins, valuables, armor, and excavated antiquities that he acquired from a private collector to his royal cabinet. But the presence of archaeological objects in royal collections reflected the idiosyncratic tastes of their owners, not a commitment to the study of prehistory. Several burial urns in the Habsburg collection, for example, were decorated with jewels and precious metals. They were part of the court's *Wunderkammer*, a cabinet of curiosities that used an assortment of natural wonders and rarities to showcase the "unbelievable richness and variety" of the court's possessions.[24]

Archaeological objects did inspire scientific curiosity in several royal courts. Electress Anna and Elector August of Saxony were aware of the debate between those who claimed that urns were products of nature that grew in the ground and scholars who insisted that they were man-made. In the 1560s and 1570s, Anna and August called upon landowners and scholars to assemble a collection of urns and verify that they were indeed ancient.[25] In the Habsburg court, Rudolf II's physician compared the nature of prehistoric metals with contemporary samples and argued (wrongly) that ancient implements were iron tools that had turned to stone over time. Early measures for historical preservation accompanied this scientific interest. In 1605, Rudolf II decreed that all prehistoric material found in his realm be turned over to the royal collection, and Duke Eberhard III of Württemberg (r. 1628–74) passed a similar mandate later in the seventeenth century.[26]

Renaissance scholars turned to classical and early medieval historians for clues about ancient Europe, but they rarely combined their textual studies with the collection of artifacts. Royal cabinets of curiosities contained ancient objects, but they presented these items as intriguing possessions, not as historical sources. By the eighteenth century, however, a new relationship between texts and objects was emerging. A broader antiquarian movement connected the growing interest in the peoples of ancient Europe with the material culture they left behind. This methodological breakthrough transformed archaeology into a vital historical practice.

Enlightened Absolutism and the Historical Description

Following the Peace of Westphalia in 1648, energetic rulers in central Europe increased the productive capacity of their states. They collected taxes more efficiently, improved agricultural techniques, and ordered the construction of roads and canals. In addition to increasing the power and stability of the state, the projects of enlightened absolutism changed the way rulers viewed the countryside. Territory was no longer a static possession; it became an essential part of the realm with resources to be exploited for economic and political gain. Rulers hired topographers, agricultural experts, and economists to chart the contours of land, assess its

natural resources, and quantify its economic potential. These tasks produced a greater knowledge about the rural population and brought the
state into direct contact with archaeological material.

A prime example of this development occurred in Brandenburg-
Prussia. Friedrich Wilhelm, the Great Elector of Brandenburg (r. 1640–
88), drained large tracts of land in the Havel valley, connected the Oder
and Spree Rivers with a canal, and facilitated exports by improving roads.
Friedrich II (also known as Frederick the Great, r. 1740–86) continued
these practices. During his reign, the Plaue and Finow Canals improved
transportation between the Havel and Oder Rivers and provided a navi-
gable route from Berlin to the Baltic Sea. Land initiatives reclaimed an
area of three hundred thousand acres in the marshlands along the Oder
and Warthe.[27] These large-scale projects integrated once-isolated areas
into the economies of enlightened realms. They also brought laborers,
scholars, and even rulers into greater contact with antiquities.

The growing attention to artifacts was evident in a text by Gotthilf
Treuer, an antiquarian and archdeacon in Frankfurt an der Oder. In 1688,
Treuer wrote of the incredible number of "heathen urns" that were uncovered "in heaps" (*hauffen-weise*) during his day. He explained where
the urns were found and how contemporaries described them. Like some
of the earlier royal inquiries, Treuer hoped to dispel several false beliefs
that surrounded ancient objects. He disputed the idea that clay pots grew
in the soil, that they were used by dwarves who lived beneath the ground
(even though they were commonly called *Zwerg-Töpfe*, or dwarf pots),
and that they possessed the power to make milk creamier. Instead, he confidently stated that the vessels were man-made.[28] Although he could not
date the objects, he rejected the idea that the thickness of the moss that
had grown on the outer stones was a reliable measure of time. (He noted
that the famous Swedish scholar Olof Rudbeck believed that a layer of
moss the width of a finger equaled five hundred years.)[29]

Treuer also described the excavation practices of his day and provided
advice to others who wanted to engage in similar research. He noted
that locals set out each year on the Feast Day of Saint John the Baptist
(June 24) to look for fresh herbs and to explore burial grounds. The early
summer was ideal because the earth was dry enough for digging, but the
weather not yet so hot that it would dry out the clay vessels too quickly.
After locating a clay piece, Treuer insisted that a ring must be dug around
it so that air could get to the clay and harden it. When removing a clay
pot, he cautioned, one should reach under the pot and carefully lift it out

along with the sand and earth surrounding it. If these methods were not followed, the sides of the pot would break apart under the weight of the moist earth.[30]

Throughout his description, Treuer expressed the sense of wonder that would animate so many antiquarians in the eighteenth and nineteenth centuries. He was amazed, for instance, that some graves contained hair "that was as clean and well preserved as if it had been cut from a head today."[31] But the main goal of Treuer's text was not historical documentation. He did not use these sources to speculate about the settlement history of Brandenburg or the identity of the makers of the urns. Instead, he devoted his thoughts to human mortality and the practice of cremation. He compared the burial of ashes and bones in urns to Roman, Egyptian, Jewish, and Christian burial practices. He wondered if cremation, forbidden by the scriptures, had come about through the persuasion of the devil, or if, as others said, it originated with the ideas of the Greek philosopher Heraclitus, who said that everything began with fire and that all bodies must be cleansed with fire. Regardless of how this practice started, Treuer noted that "the monuments related to cremation continued to exist among the ancient Germans until they, along with other peoples, had made peace with the blessed belief in the remembrance of higher Christian powers."[32]

The interest in burial urns was also evident in Silesia. Since the middle of the sixteenth century, locals had traveled each spring to the hills outside the village of Massel to dig for urns and metal artifacts. In the early eighteenth century, Leonard David Hermann, a pastor from the area, excavated over thirteen thousand burial urns. In the description of these finds, Hermann's assistant poetically declared the historical potential of these objects:

Und jeder muß gestehen	And everyone must concede
daß Urnen und Gebein	that urns and bones
der Schlüssel aller Welt	are the key to understanding
und ihrer Sitten seyn.[33]	the world and its customs.

This new appreciation of archaeological material removed objects from the magical realm of peasants' tales and placed them in a new context that shed light on the distant past. Hermann also saw this as a matter of local pride. He wanted "to satisfy the learned and curious minds" interested in this Silesian village, "for it is a shame to be born and to live in a famous place without bringing it and the fatherland honor and fame."[34] Hermann's work earned him an invitation to join Prussia's Royal Academy of Sciences, and it established Silesia's reputation as a region rich in archaeological finds.

An even more remarkable attempt to generate interest in the distant past came from a country pastor outside Hamburg. Andreas Albert Rhode had participated in local digs with his father since he was eighteen. Following his father's death in 1717, Rhode began publishing a description of the burial customs and cultural practices of the ancient Cimbri, a people that migrated from Jutland (today's Denmark) as far south as Carinthia, Austria, in the first century CE. From January 1719 through February 1720, Rhode sold installments of his *Cimbrisch-Hollsteinische Antiquitaeten-Remarques* (Remarks on Cimbri-Holstein antiquities) to a Hamburg gazetteer. An edition appeared each Tuesday and sold for a schilling. Rhode clearly intended his description of the ancient Cimbri for a general audience. Each issue of the *Antiquitaeten-Remarques* was about eight pages in length and easily readable in the course of a week. Rhode did not bore his reader with heavy academic prose, and when he included quotations from other scholars that appeared in Latin or French, he almost always provided a German translation. He often strayed from the subject of archaeology to make jokes and local references. And each week he opened with a catchy two-line rhyme that invited his readers to contemplate the ancient past. The issue from January 31, 1719, devoted to treasures found in graves, for example, began:

Schaue, wo es Dir gefällt,	Come look what can be found
Was der Hügel in sich hält.[35]	inside this ancient mound.

The rhyme, stressing Rhode's desire to share the excitement of excavation with his readers, accompanied a cross-section drawing of a large urn hidden inside a stone chamber. In the summer of 1719, he received permission to excavate several promising mounds near Wandsbeck, and he called on "those gentlemen interested in antiquities" to experience firsthand the opening of several mounds on the fifth of July. The costs of the outing included the rental of a wagon and wages for two to four laborers for the day.[36]

Antiquarians like Hermann and Rhode left a record of their excavations, but they were not alone. Others were active throughout German-speaking Europe during the eighteenth century, performing excavations in Brandenburg, Schleswig, Silesia, the Rhineland, and Saxony and on the island of Rügen.[37] In England, John Aubrey (1626–97) sketched the impressive stone circles at Stonehenge and Avebury, and his work influenced William Stukeley (1687–1765), the doctor and clergyman whose descriptions gave Stonehenge the evocative (but false) reputation as a site of human sacrifices and Druidic cult practices.[38]

As Rhode's text makes clear, antiquarians were proud to showcase local archaeological material. They opened their readers' eyes to the historical treasures that surrounded them and invited them to explore the distant past. This link between territory and history would become an enduring feature of antiquarianism and domestic archaeology in the nineteenth century and continue to the present day.

Over time, rulers came to see their realms as entities held together by cultural forces, and they commissioned scholars to study the natural resources, cultural traditions, and history of their lands. The career of Johann Christoph Bekmann provides a striking example of this development. Born in 1641, Bekmann witnessed the vigorous changes that swept through Brandenburg-Prussia during the late seventeenth and early eighteenth centuries. After completing his university studies at the age of twenty, he received a travel stipend from Friedrich Wilhelm, the Great Elector, to study history in Holland and England. He then began a long career as a professor of theology and later university rector in Frankfurt an der Oder. In 1710, he published a history of the principality of Anhalt that served as a model for a new kind of scholarship. Previously, royal houses had treated archives as the secret holdings of political information, and they strictly limited access to them. Bekmann, however, received access to the Anhalt royal archive, and his work reflected an official desire to present the state's political development as part of an effort to define the principality as a coherent entity.[39] After completing this work for the Anhalt court, Bekmann received a commission from Friedrich I of Prussia (r. 1701–13) for a similar study of Brandenburg. In 1712, the king ordered that all historical material be turned over to Bekmann for evaluation, and the royal house provided ongoing financial support for this scholarship. Bekmann worked on this project until his death in 1717, and it was finally published in two volumes by his grandson as the *Historische Beschreibung der Chur und Mark Brandenburg* (Historical description of the Electorate and Mark of Brandenburg).

The *Historische Beschreibung* of Brandenburg demonstrates how the policies of enlightened absolutism created a new relationship between the state, its territory, and its history. Instead of a genealogy of rulers, Bekmann's work offered a textual panorama that included over one thousand pages of anecdotes, images, scientific reports, and essays that delved into Brandenburg's history. It began with the geography and demography of Brandenburg, including studies of the region's ancient borders and inhabitants, the development of its cities, and the customs of Brandenburg's Jews. It noted the area's natural resources and closed with a commentary on the region's most influential historians. This blend of economic and cultural information served as a link between the ruler and his territory.[40]

The section of the *Historische Beschreibung* dedicated to Brandenburg's antiquities demonstrated the new attention accorded to archaeological sites in the eighteenth century. Bekmann explained that

> we now want to turn to the things themselves and start with the very oldest events or, rather, the remains and *fragmentis* that make up the ancient histories of this land. And not via speculations about one passage or another from the old Greek or Roman historians, because there is little that is certain to take from them.... Instead, [we will] rely solely on those antiquities that have been preserved since ancient times and stand as irrefutable witnesses to the early inhabitants of these places, whoever they may have been.[41]

The antiquities that most interested Bekmann were megalithic chamber graves that were located throughout Brandenburg, Mecklenburg, and Pomerania. Bekmann fancifully called them the stone beds or graves of giants (*Steinbetten* or *Hünengräber*) because earlier interpreters had described them as the final resting places of ancient kings and princes, and legend maintained that the size of the stone slabs in the center of the graves (often over ten feet long) suggested that the royal leaders were of giant stature. Bekmann described fifteen of these sites and depicted twelve of them in drawings. He did not attempt to date these sites or their contents, and he ventured no guesses about the identity of the buried kings. Nevertheless, Bekmann argued that these graves and other antiquities were the most authentic sources for Brandenburg's ancient history.[42]

Over the course of the eighteenth century, scientific societies began to turn their attention to archaeological sources. Friedrich I of Prussia, following the French model, founded a royal academy of sciences in Berlin. Its historical division was charged with researching "German history as a whole and particularly the secular and church history of our lands [Brandenburg-Prussia]."[43] Archaeology became a natural part of this

FIGURE I. Johann Bekmann's *Historische Beschreibung der Chur und Mark Brandenburg* featured numerous illustrations, including depictions of the stone circles and chamber tombs he considered to be the oldest evidence from Brandenburg's ancient history. Bekmann identified this stone circle, located just east of the Oder River near Zehden (today Cedynia, Poland), as a *heidnischer Kirchhof*, a heathen burial site, and explained that the immense stone in the middle was "without a doubt the grave altar." (Illustration from Bekmann's *Historische Beschreibung*, 1:364. *Source*: Bayerische Staatsbibliothek, Res/2 Bor. 2–1.)

mission. In 1748, the Prussian academy sponsored a competition that asked scholars to consider "how far the ancient Romans had advanced into Germany." The winning response, written by a pastor from Hameln, used archaeological sources to argue that Domitian and his armies had passed through much of the Electorate of Brandenburg and almost reached Potsdam in the last quarter of the first century CE. The competition also inspired a two-volume study of Roman incursions into eastern Franconia by Christian Ernst Hansselmann, an archivist in Öhringen (Württemberg).[44] The 1759 bylaws of the Bavarian Academy of Sciences specifically instructed members of the historical division "to give precedence to the history of the [Bavarian] fatherland . . . and create a description of the land and prepare maps of ancient, as well as medieval and modern, times, and eventually compose a topographical dictionary as well."[45] The historical division published a study of the Roman presence on Bavarian territory in 1761, and in the following decades, academy members compiled lists of ancient sites, such as grave mounds and the ruins of Roman walls and fortifications.[46] As these examples demonstrate, history was no longer only a chronicle of the deeds of the royal house. The academies approached the landscape as a historical field that contained markers of the past and registered the settlement of different populations.

An engagement with this history would increase the glory of the state and inspire feelings of historical belonging.[47]

Historical descriptions signaled a new attention to antiquities. As Bekmann stated, it was time "to turn to the things themselves" in order to produce new knowledge about the distant past. This task embodied the English and German words for the Enlightenment. Authors shed light on the darkness of the past (*Dunkel der Vorzeit*) and clarified (*aufklären*) the blurry gray of early time. Gotthilf Treuer and Leonard Hermann sought to dispel stories of mounds "growing" in the ground and tried to teach people about the ancient nature of archaeological material. At the same time, antiquarians were intrigued by what they viewed as centuries-old stories about the supernatural powers of urns and the fantastic origins of grave mounds and walls. They believed, as the author in the Hannover gazette explained, that this "rural mythology" (*Bauernmythologie*) contained fragments of historical truth that could be useful in reconstructing the ancient history of central Europe and locating significant finds.[48] Antiquarians sought out peasants who could tell them about possible archaeological sites and place-names that possibly referred to ancient events. This communication took on added urgency once the objects became more valuable to antiquarians. They knew that most objects uncovered during planting and harvesting or during road construction went unnoticed or were discarded. The only way to save this material was to encourage people to view it as historically valuable.

Antiquarians and a Framework for Prehistory

Eighteenth-century authors were still a long way from using excavated material to shed light on Europe's prehistory. Treuer, Bekmann, and Hermann did not have command of a clear chronology that explained how old their objects were, and their comments about the peoples who produced the burial urns and gravesites that they described were very provisional. To identify their finds, they turned to their knowledge of classical sources. They applied general comments by Julius Caesar and Tacitus about ancient peoples, cultural practices, and religious beliefs to the sites they excavated. This combination of texts and objects produced a framework for prehistory that remained in place until the middle of the nineteenth century.

In the *Germania*, Tacitus separated prehistory into a mythic and a knowable past. Covering the mythic past, he reported that "[i]n ancient lays, their only type of historical tradition, they celebrate Tuisto, a god brought forth from the earth. They attribute to him a son, Mannus, the source and founder of their people, and to Mannus three sons" (77). Tacitus then noted that uncertainty surrounded the number of Tuisto's offspring. "Some people, inasmuch as antiquity gives free rein to speculation," thought there were more sons and "hence more tribal designations . . . and that those names are genuine and ancient" (77). Other myths connected Greek legends and Homeric epics to the Germans. Some claimed that "Hercules also lived among them," and "Ulysses as well, in his lengthy and storied wanderings, travelled into this part of the Ocean and visited the lands of Germania" (78). Tacitus closed this section on the mythic past by stating: "It is not my intention either to support these assertions with proofs or to refute them: each reader may withhold or bestow credence according to his own inclination" (78). It was not possible to gain certain knowledge about these archaic times.

Yet as Tacitus made the transition from the realm of myths to his description of the Germans of the first century CE, he made two infamous statements. Tacitus wrote: "The Germani themselves are indigenous, I believe, and have in no way been mixed by the arrivals and alliances of other peoples" (77). And in the chapter that followed the stories of Hercules and Ulysses, he explained: "For myself, I agree with the views of those who think that the inhabitants of Germania have not been tainted by any intermarriage with other tribes, but have existed as a distinct and pure people, resembling only themselves" (78). These statements are well known as the source for what later became the trope about German purity that extended from humanists in the sixteenth century, to Romantic patriots and radical nationalists in the nineteenth century, and to Nazi ideologues in the twentieth.[49] In addition to asserting German homogeneity, though, these passages performed an important narrative function for those who wanted to investigate the ancient past. In these lines, antiquarians found a starting point for a German history that was unencumbered by questions of geographic dislocation or by the cultural influences of other peoples. Tacitus did not speculate as to how long the Germans had been north of the Danube and east of the Rhine or how many generations had passed between the appearance of Tuisto and the living people he described. These questions were not problematic, though. No matter how long this had been, the Germans appeared to be autochthonous: they still lived in the same place where they originated, and they were culturally and genetically "pure." The *Germania* did not provide an exact chronology for the first Germans, but antiquarians assumed that these lines settled the question of origins.

As the quotations above make clear, the *Germania* is full of measured statements. By referring to speculation about the past and allowing readers to evaluate stories of origin for themselves, Tacitus left room for the possibility that information about the ancient Germans is debatable or even incorrect. Antiquarians interested in the history of the Germans, however, read the text as a clear guide. They turned Tacitus's personal and somewhat-measured claims (introduced by "I believe . . ." and "For myself, I agree . . .") into hard knowledge about the earliest Germans. This reading of the *Germania* imposed a solid structure on an archaic past, and it had two critical consequences that lasted into the middle of the nineteenth century. First, it removed the expanse between the beginning of time and Tacitus's day from historical consideration. This finite view of prehistory focused questions about origins on the relatively recent period when the texts by Tacitus and other Roman authors were produced (between the first century BCE and the first century CE). The narration of prehistory became, not a disorienting journey *backward* into an indefinite past, but rather a movement *forward* from this baseline to the "middle" of the historical narrative, which rested firmly on the well-documented events of the Middle Ages.

The creation of this kind of baseline followed a long-standing tradition in Western historical writing. Thucydides, for example, began his *History of the Peloponnesian War* with a brief section that dealt with a much earlier period than the conflict between Sparta and Athens that began in 431 BCE. This introduction, which Thucydides called an archaeology, quickly covered the tribes who had migrated across Greece in "ancient times" (i.e., ancient relative to the fifth century BCE) and provided a contrast between a chaotic primeval era and the wealthy city-states of Thucydides's day.[50] This beginning point helped to define the Peloponnesian War as a decisive moment and made these events truly worthy of historical narration. As historian Dan Smail has shown, a form of this convention was also present in eighteenth-century universal histories. Narrators used biblical events like the Creation or the Flood as starting points for their narratives. This act cut prehistory off from a notion of deep prehistoric time and allowed historians to trace clear narratives of progress forward from a defined starting point.[51]

Second, this framework for prehistory turned classical texts into ethnographic treasure maps that left little doubt about the identity of the finds uncovered in central Europe: if the ancient Germans were indigenous, as these sources claimed, then the grave mounds, bones, and metalwork that the underground revealed had to belong to them. This narrow scope of the distant past allowed antiquarians to identify the finite number of Germanic peoples mentioned in Tacitus's *Germania* and Strabo's *Geography* and to assign the nomenclature from these texts to the artifacts they uncovered.

Eighteenth-century literature on the ancient Cimbri nicely illustrates these interpretive shortcuts. The ancient world knew the Cimbri as the first northern people to cause considerable trouble for Roman armies. In 113 BCE, the Cimbri migrated from northern Germany southward along the Rhine. They broke into the Roman province of Noricum (roughly where Austria is today) and defeated a Roman consul at the battle of Noreia. By 109 BCE, these Cimbri, joined by the Teutoni, appeared in southern Gaul and were poised to enter the Italian peninsula. They turned toward Spain instead, met resistance, and eventually migrated back through Gaul toward what today is northern Germany and Denmark.[52] Julius Caesar made these conflicts famous in *The Gallic War*. In order to rally support for his own campaign against the Suebi king Ariovistus in Gaul in 61 BCE, he repeatedly referred to the earlier battles against the Cimbri and Teutoni as noble struggles carried out by a previous generation. Action in Gaul in his day could not be delayed, he argued, because earlier experiences showed the Germanic peoples to be violent and implacable. Caesar wrote that he "could see that the Germans were becoming gradually accustomed to cross over the Rhine, and that the arrival of a great host of them in Gaul was dangerous for the Roman people." Barbarians so fierce, he supposed, would not stop after seizing the whole of Gaul; rather, like the Cimbri and Teutoni before them, "they would break forth into the Province, and push on thence into Italy." It was therefore imperative that the Romans act with the same bravery that their forefathers had displayed in earlier battles.[53]

Tacitus, writing 150 years after Julius Caesar and 200 years after the defeat of the Cimbri at Vercellae in northern Italy, commented on the past glory of this people and their modest presence along the North Sea in the first century. He reported that the Cimbri were "a small state now but great in their glory; widespread traces of their ancient fame remain, and on each bank [of the Rhine] are extensive camps, by whose circumference one can even now measure the massive troops of the tribe and the evidence for so great a migration" (92).

When antiquarians from Copenhagen to Hamburg began to publish works on the ancient history of their area, they made great use of these ancient texts. Andreas Albert Rhode, for example, was confident that he was describing Cimbri sites in Holstein. Both Caesar and Tacitus had located this people in the north, and Rhode took these words as clear evidence for the identity of the people who had created the burial urns that he uncovered. While the long temporal expanse from the beginning of time until the appearance of Roman descriptions of the Cimbri certainly contained numerous demographic changes, Rhode's cultural study assumed that the material he excavated came from the original settlements of the Cimbri before they moved southward. Rhode did not provide concrete dates for the burial sites that he described. Instead, his remarks assumed a coherent Cimbri culture that lasted from well before the time of Caesar's writing until the date of Tacitus's description. His work turned these two ancient ethnographic snapshots into a set of timeless cultural characteristics.

Both Caesar and Tacitus portrayed the Cimbri as a serious threat to the Roman Empire, which made them worthy of historical attention. Yet for many antiquarians, it was not enough to hear the Cimbri praised as a formidable opponent. Antiquarians hoped to overturn the established contrast between a cultured South and a barbarian North through the interpretation of domestic sources. Rhode, for example, argued that the Cimbri practiced a complicated set of burial customs that paralleled those of the Romans, Greeks, and Egyptians. He noted that written sources described cremation among the first two peoples, but the world did not know about this practice among the Cimbri. Excavations around Hamburg, however, revealed ashes and burial urns, which, for Rhode, elevated the cultural level of the Cimbri to that of the Greeks and Romans.[54] Rhode also uncovered the hidden history of textile production, leatherworking, sewing, pottery making, and metalworking, using archaeology to show that the Cimbri "had all kinds of artists and artisans among them, just as other civilized peoples had."[55]

Rhode's text revealed a selective application of archaeology's potential as a historical tool. He used archaeology to verify practices that ancient sources did not describe among the Cimbri, like cremation and metalworking. Yet when it came to identifying the finds, he did not consider the possibility that the objects came from a prehistoric people that the ancient texts did not mention. Instead, he was certain that this material belonged to the people that Julius Caesar and Tacitus had made famous. This example shows how classical sources shaped the interpretation of archaeological finds. The use of Caesar and Tacitus as the beginning of prehistory ruled out the possibility that archaeological evidence was completely prehistoric—that is, from before the appearance of texts. This approach attributed all archaeological evidence to the peoples mentioned by classical authors.

Writing seventy years after Rhode, Ignaz Pickel also relied on ancient sources to identify his finds. Pickel was a former Jesuit priest who published an interpretation of three grave mounds found in the Weißenburg Forest (Franconia, northern Bavaria) at the end of the eighteenth century. In the summer of 1788, a curious forester had opened two grave mounds that revealed a startling amount of material, and these items entered into the scientific collection of the Bishopric of Eichstätt, which Pickel supervised. The bishop ordered further excavations, and Pickel witnessed the opening of a third grave mound. In many respects, Pickel's text is an impressive early archaeological report. His "true and unblemished account . . . of what was found" includes clear details about the location of the grave mounds and a complete list of their contents. When it came to interpreting the finds, he reservedly claimed only to connect them to "local circumstances" and hoped that others with better knowledge of antiquities would confirm or correct his assertions.[56]

In his interpretation, Pickel turned to the writings of Strabo, Caesar, Pliny the Elder, and Tacitus. These sources led him to ascribe great significance to this secluded spot in rural Franconia. Tacitus wrote that the Germanic peoples "consecrate woods and groves, and the mystery that they see only in their awe they call by the names of gods." When Pickel found mounds in the forest and recalled a local legend about spirits inhabiting the area, he concluded that this was the kind of holy site Tacitus had described.[57]

To date the grave mounds in the Weißenburg Forest, Pickel used information about Roman emperors, the construction of the *limes*, and early medieval kings. He reported that the Roman emperor Hadrian (r. 117–38) had ordered the construction of the *limes* in Bavaria and that it was later strengthened during Probus's reign (r. 276–82). He guessed that Germanic invaders or the armies of Attila the Hun had destroyed the walls around the year 450 CE. Pickel observed that the mounds stood on both sides of the crumbling defense barrier and reasoned that they must have been created after the wall's destruction because of their position. (The wall did not appear to disturb any of the mounds, which might have been the case if they were older than the wall.) This fact meant that the burial sites were created in the fifth century CE at the earliest. Pickel further deduced that the mounds could not be younger than the seventh century, because they contained evidence of pre-Christian burial practices. The English missionary Willibald had come to Eichstätt and established the bishopric in the mid-eighth century, and this development would have eradicated such customs.[58]

Here we see the interpretive steps so important for the eighteenth-century approach to prehistory. Antiquarians rarely spoke about the distant past in strict chronological terms. Instead, they treated prehistory as a single static period. Pickel used Tacitus's ethnic labels and cultural descriptions from the first century CE to identify finds that he dated as belonging to the fifth, sixth, or seventh century. Pickel thereby compressed the four to six hundred years in between and turned Tacitus's *Germania* into a general treatment of all of central Europe's pre-Christian history.

Eighteenth-century antiquarians divided prehistory into three broad periods based on religious practices. They frequently spoke of *die heidnische Vorzeit* ("heathen" prehistory), the era of the Romans (in southern and western areas of central Europe), and a Christian era that blended into the Middle Ages. Gotthilf Treuer framed his entire description as an investigation of heathen burial practices, and Christian Detlev Rhode, Andreas Albert Rhode, and Leonard David Hermann, all pastors, frequently commented on their understanding of ancient religious beliefs. These concerns were also very clear in Pickel's description of Willibald's missionary work as a turning point in the history of the area around Eichstätt. As religious men, antiquarians were deeply interested in identifying the moment when Christianity arrived in their areas. Additionally, rulers in the eighteenth century legitimated their rule by tracing their lineage back to the moment when the realm converted to Christianity. For the Bishopric of Eichstätt, Pickel could use Willibald's arrival, which was documented in church chronicles, as the moment when pagan history ended and the Christian history of the territory began. This determination, like the use of Julius Caesar or Tacitus, provided a starting point for the history of the area and removed the expanse of deeper prehistory from consideration.

The use of Tacitus's *Germania* also allowed antiquarians to relate eighteenth-century political circumstances to ancient times. The first half of Tacitus's ethnographic account describes common features of the ancient Germans, suggesting that the individual peoples shared an overarching identity. Eighteenth-century antiquarians, following humanist scholars who had accepted the cohesion of all non-Roman and non-Slavic ancient cultures, assumed that specific cultural aspects unified these peoples and that one could therefore apply theories about finds in Bavaria to material excavated in the Rhineland. They employed overarching labels that suggested this cohesion. Ignaz Pickel defined his historical actors as "ancient Germans" (alte Deutsche), and with this nebulous national label, Pickel related what he found in the Weißenburg Forest to finds from throughout central Europe. A review of Pickel's work treated this question in a similar way. It suggested that the "ancient Germans" had migrated south to Bavaria from a possible original home near the Baltic Sea.[59] This very basic interpretation of early history recognized that Germans were not indigenous to the southern states, but it did not account for multiple migrations through the area, nor did it comment on relationships between Germans and Romans or among the separate barbarian groups. The simple phrase "alte Deutsche" hid a long history of movement, settlement, discord, and cultural transfer. Andreas Rhode connected the Cimbri to other "Germans." Even without a detailed discussion of the relationship between the various peoples, he repeatedly used parallel constructions such as "Our ancient Germans and therefore without doubt also our Cimbri" to link his local finds to graves excavated outside Holstein. Additionally, many texts interchanged "German" and "Germanic" (deutsch and germanisch) with no commentary. This cultural shorthand was not very scientific, but it was extremely important. It simplified the ancient past and placed the ancient Germans within the same historical trajectory as the antiquarians themselves.[60] Andreas Rhode made this connection when he described the people he excavated as "our ancestors and compatriots," and he treated his search for information about them as a duty: "As their descendants, we want to honor their memory and by all means seek to rescue their dignity."[61] The seamless shift from Germanic to German produced an ethnic category that included all the peoples of ancient central Europe, but it also allowed antiquarians to view their local investigations as familiar and personally meaningful.

Yet at the same time, the Germania did describe important differences among the ancient Germans. The second half of Tacitus's text explains that the various peoples beyond the Rhine and Danube were related, but that they had specific customs, characteristics, and languages. Antiquarians viewed regional variation and the political divisions that separated German speakers in the eighteenth century as the outgrowth of this earlier diversity. Ancient peoples occupied specific territories and practiced different customs, just as the inhabitants of Holstein and Brandenburg spoke different dialects and enjoyed different cuisine. These differences did not upset the notion of a general category of "ancient Germans," just as contemporary regional differences did not disrupt the idea of German cultural unity.

Archaeology was not always related to the search for "ancient Germans." Workers tearing down a house near Saint George's Church in Neubrandenburg in the 1790s hit upon stone and then the remains of an archway about half a foot below the ground. J. C. P. Kortüm heard of the discovery and began investigating. After finding reddish-brown and green glass fragments, pieces of cast iron, pottery fragments, a scissors, fish scales, and the head of an ax, he was convinced that he had discovered a significant site. Further digging revealed that the archway formed the exterior of an oven. While others in the town supposed this to be the place where medieval craftsmen had cast the church bell, Kortüm eventually came to another conclusion. He wrote of his investigation, "So, I gave up this thought and soon found myself certain in declaring it [the oven] an original Wendish smelting oven."[62] With this conclusion, Kortüm claimed that the town of Neubrandenburg was much older than popularly believed and its history more complicated. According to Kortüm, the site did not represent the town's medieval past or the origins of its visible landmark (the church bell). Instead, he believed the oven revealed a part of the area's past that had become invisible by the eighteenth century: the occupation of the land by Wends, the Slavic people who had inhabited much of central Europe during the early medieval period. Kortüm raised the stakes of his argument by suggesting that this foundry was built on the spot of Rhetra, the place where ancient Slavs stored their military flags and where their leaders came to consult the gods about battles. Germanic warriors destroyed the temple there in the eleventh century.[63] As with the use of Tacitus's ethnography by other antiquarians, Kortüm linked what was known about the Slavic past to this anonymous site. The oven in Neubrandenburg became not just a Slavic site but one of the holiest sites in ancient Slavic culture.

Kortüm's "discovery" did not diminish the pride he felt toward his hometown. If he could make Neubrandenburg the possible site of ancient Rhetra, "then our discovery [of the foundry] would gain historical significance for all those who know and love Wendish history."[64] As Kortüm's interest in the Wends suggested, archaeology revealed the past of multiple peoples. His text served as an important reminder that while assumptions about ethnicity lurked behind archaeological interpretation, there was more to archaeology than finding German ancestors. Kortüm saw the history of the fatherland as the aggregate sum of the events that occurred on a specific territory. For him, the key to bringing renown to a place was the discovery of significant past events, whether they were related to ancient Germans or not. For Brandenburg, this included not only a period of German settlement but also a heyday of Slavic culture.

The Development of Archaeological Methods

In one of the most ambitious antiquarian projects of the eighteenth century, the French collector comte de Caylus called for an approach that departed from the idea behind royal cabinets of curiosities. In his seven-volume comparison of Egyptian, Etruscan, Greek, Roman, and French antiquities, he proclaimed: "I should like us to seek less to dazzle than to instruct, and to join the Ancients more frequently in their method of comparison which is to the antiquary what observation and experiment are to the physicist. The inspection of several monuments, carefully compared, may reveal their purpose, in the same way that the ordered consideration of several effects of nature may reveal their principle."[65] Caylus compared antiquarian work to that of the physicist, not the historian or storyteller. Collecting and comparing finds were analogous to the initial steps in the scientific method. Just as scientists could reach conclusions only after they formulated a problem and collected data from experiments, antiquarians would need better methods and more information before they could address questions of ethnography and chronology. Over the course of the eighteenth century, this information began to appear across Europe in two forms: new, larger collections that were devoted specifically to antiquities, and surveys of the literature that announced finds and described collections.

In German-speaking Europe, Friedrich Wilhelm I of Prussia (r. 1713–40) established a central office that received official reports about artifacts found on Prussian soil, and the court turned to members of the Academy of Sciences to evaluate these finds.[66] The historical division of the Bavarian Academy of Sciences sponsored excavations that delivered items to the Bavarian monarchy's collection of antiquities. At least one artifact uncovered near the town of Esting, a beautiful bronze wagon set from the Hallstatt period, can still be viewed today in Munich's Archäologische Staatssammlung (Bavarian state archaeological collection). In Württemberg, the royal house separated its wide-ranging cabinet of curiosities into more specific collections dedicated to natural specimens, armor, ancient coins, and excavated antiquities.[67]

Despite this new attention to collections, conditions for visitation and preservation were far from optimal. It was still a privilege for a scholar to gain access to royal cabinets, and their status was often provisional. The Württemberg collections, for example, changed locations five times during the eighteenth century, and information about specific objects was lost. There are almost no records for the first three-quarters of the eighteenth century for the Prussian collection.[68] Conditions in Berlin did improve near the close of the eighteenth century, though. After his ascension to the throne in 1797, King Friedrich Wilhelm III authorized the purchase of several well-known collections of antiquities from Posen, Saxony, Bavaria, and Brandenburg. These acquisitions improved the reputation of the Berlin collection, and individuals began to donate items because they saw "no better place to preserve fragile objects for the fatherland and for science than the royal collection."[69] The increasing breadth of the collection allowed antiquarians in the Prussian capital to compare finds from different regions for the first time.

Those who shared Bekmann's desire "to turn to the things themselves" were concerned with methods of excavation and observation, and they began to read one another's findings. They viewed their descriptions as written conversations with future excavators who would be inspired by their stories of discovery. Gotthilf Treuer included advice for safely handling objects. Andreas Rhode explained that he learned to excavate without damaging artifacts by observing his father's careful work, and he thoroughly described his procedure so that others might learn from him. His *Antiquitaeten-Remarques* also included drawings that noted the size and position of burial mounds. Other aspects of Rhode's work contained fundamental deficiencies when compared to later practices, though. He recorded items only after he had removed them from the earth, not analyzing their depth in the ground or their relationship to other objects. And Rhode and his father kept only single examples of object types. They discarded damaged pieces and often gave "doubles" away because they did not consider the value of comparing multiple objects or studying the frequency of an object.

Ignaz Pickel spent several pages of his description relating the shape, height, and diameter of the mounds he found in the Weißenburg Forest.

Once they were opened, he drew attention to the recognizable layers within the mounds. Pickel read the proximity of bronze triangles near corpses as evidence that they functioned as jewelry. Laborers set aside the items from the graves that were considered "valuable" (three knives, five spears, an iron horse bit, several fibulae, and rings and triangles made of copper), but Pickel also studied the pottery found at the sites and tried to organize it according to three basic shapes. His attention to a wider range of grave contents showed that antiquarianism was more than treasure hunting.

Pickel also performed experiments to see if he could figure out how excavated items were made. He examined shards with a microscope and compared them to contemporary ceramics. He then combined local ore, lime, and quartz in different proportions and baked these samples for varying lengths of time, hoping to replicate the coloration and consistency of the fragments he found at the site. Pickel also dissolved samples of the stone and soil mixture that covered one of the mounds in a solvent to see if they reacted similarly to soil samples from other locations in the forest. Although he was unable to replicate the ancient pottery exactly and remained unsure about the effects of time on the materials, he expressed his joy at the thought that he was the first after so many centuries to fire pottery from this clay as the ancient inhabitants of the forest had done.[70]

The discussions antiquarians had with each other about methods and finds are further evidence of archaeology's development. Rhode, for example, drew heavily on earlier work on Cimbri religion, and Kortüm strengthened his assertion that Neubrandenburg was a "genuinely Wendish" site by comparing his finds with those described in other works about the Wends.[71] Pickel was unsure about the classification of pottery types and hoped that "experts" would confirm or correct his assertions. One of the most impressive signs of a community of readers interested in archaeology was the lengthy review that publicized Pickel's excavation and critically assessed his description.[72]

By the last decade of the eighteenth century, at least one author took stock of antiquarian knowledge and evaluated the future of the field. Bernhard Friedrich Hummel, a rector at a Latin school in Altdorf (near Nuremberg), had collected archaeological material for over twenty-five years. His *Bibliothek der deutschen Alterthümer* (Library of German antiquities, 1787) filled four hundred pages with notes, references, and criticisms of literature from across the German states, and his *Beschreibung entdeckter Alterthümer in Deutschland* (A description of antiquities discovered in Germany, 1792) fed the young interest in archaeology by evaluating numerous antiquities collections. Hummel concluded that the study of local antiquities had made some progress in the preceding decades, but it still was not a fundamental part of historical research. He regretted that most scholarly correspondence was never published and that most objects were held in closed, private collections. For most people, these conditions made objects so inaccessible that they might as well have remained undiscovered beneath the soil. All this led to his sad conclusion that even though domestic antiquities should not be ignored because they were related to the history of the fatherland, they largely remained "outside the view of the reading public."[73] Hummel's bibliographies partially addressed this problem by listing the titles of many antiquarian pieces and the locations of numerous collections. As Hummel explained, "When possible, I have named these [objects], thus giving travelers direction as to where they can see them for themselves."[74]

Over the course of the eighteenth century, clergymen and educators took notice of archaeological objects and began to view them alongside classical sources that described conditions in central Europe during ancient times. States supported these efforts to investigate the domestic past. Yet this activity remained in a preliminary stage, as antiquarians were limited in the number of objects they could view, and most of their descriptions focused on a single archaeological site. In the early nineteenth century, a much wider public came to view artifacts as more than curiosities, and archaeology blossomed as antiquarians sought to realize its potential as a historical practice.

Preparing Artifacts for History

Archaeology after the Napoleonic Wars

In general, we should begin to turn our eyes to beneath the surface of the earth in order to gain a more correct and higher understanding of what happened aboveground during the time of the Germanic tribes. — Leopold von Ledebur, 1837[1]

During the early decades of the nineteenth century, archaeology in central Europe flourished. In increasing numbers, people interested in local history began to investigate the megaliths, grave mounds, Roman mile markers, and medieval ruins that had dotted the landscape for centuries. Pastors, teachers, administrators, and other enthusiasts founded over forty historical and antiquities associations between 1819 and 1850. These groups excavated in fields and forests, collected artifacts that turned up during construction projects, and publicized their activities in antiquarian journals and newsletters that circulated throughout German-speaking Europe. Friedrich Kruse, a historian teaching in Breslau and one of the most prolific writers about ancient German history in the 1820s, found this wave of activity exciting, even magical. It was "as if someone had cast a spell" (*wie durch einen Zauberschlag*), and "princes and subjects, mighty and lowly, learned and uneducated, individuals and entire societies set out to research the traces of our ancestors."[2]

This new interest in domestic archaeology was part of a broader engagement that contemporaries described as *vaterländische Altertumskunde*, the study of the fatherland's antiquities. Traditionally historians have linked *vaterländische Altertumskunde* to the rise of Romantic nationalism and treated *vaterländisch* as referring to a German fatherland. This important line of interpretation begins with the German philosopher Johann Gottfried von Herder (1744–1803) and his idea that each *Volk*, or people, possessed a national spirit. Each people's spirit was unique, Herder proposed, and it maintained itself through the ages through a common language and shared customs. Inspired by Herder's ideas, intellectuals throughout Europe embarked on cultural projects that celebrated their national communities. They created national museums, established canons of national literature, and standardized national languages. Architects and art historians examined medieval architecture as a symbol of national vitality, and scholars founded the *Monumenta Germaniae Historica* in 1819, a massive effort to assemble all the documents related to the legal, ecclesiastical, and political history of Germany. Jacob Grimm and Wilhelm Grimm pioneered the fields of linguistics and folklore studies with their *Deutsches Wörterbuch* (German dictionary), *Deutsche Sagen* (German legends), and *Kinder- und Hausmärchen* (Children's and household tales).[3] Theorists of nationalism have emphasized the creative nature of these activities. Nineteenth-century intellectuals did not just document or rediscover national stories—they defined and invented them.[4] These interpretations cast *vaterländische Altertumskunde* in a particular light, arguing that nationalism was the primary motivation for this cultural activity.

It is true that nationalist thinking flourished in the early nineteenth century. The experience of the Napoleonic invasions, French occupation, and Wars of Liberation turned Herder's cosmopolitanism into a conception of history that emphasized the distinctiveness of the German nation. Johann Fichte's *Reden an die deutsche Nation* (Addresses to the German nation) called for the preservation of German culture but also elevated the Germans above the French as a people who supposedly had not lost touch with their original genius. Ernst Moritz Arndt urged the German people to direct their national character against France, and Friedrich Jahn created a gymnastics movement that was to inspire national awareness and moral regeneration through physical activity. All these authors and activists valorized the ancient Germanic past and viewed their efforts as tapping into spiritual principles associated with it. In this context, Heinrich von Kleist wrote *Die Hermannsschlacht* (Hermann's battle, 1808), a drama that depicted the triumph of ancient German unity over Roman imperialism.[5]

But the moment of national enthusiasm was short-lived, and after 1815, *vaterländische Altertumskunde* was not exclusively, or even primarily, national. Restoration politics opposed the nationalism that had arisen during the Napoleonic occupation. The German states fostered a conservative vision of loyalty to regional monarchies, not an emancipating celebration of the *Volk*.[6] In response to the Wartburg Festival in 1817 and the murder of the dramatist August von Kotzebue by a nationalist student in 1819, the German Confederation banned Jahn's gymnastics movement, censored the press, and monitored university professors and other intellectuals who were suspected of propagating nationalist ideas. In the years after 1815, therefore, *vaterländische Altertumskunde* was mostly devoted to the history and antiquities of local areas and regional kingdoms. The fatherlands here were Prussia, Saxony, Hannover, Württemberg, Bavaria, and the rest of the states of the German Confederation. These identities displaced a broader notion of German nationalism in the public sphere, as regional monarchies used the teaching of history and geography, the building of monuments, and the sponsorship of festivals "to strengthen popular awareness of the characteristics of the particular Fatherland."[7] This regional focus was clear in the names of archaeological collections in Berlin and later Münster, Kiel, and Stuttgart. They were called *vaterländische Sammlungen* (regional collections), and they carried this name not because they were national collections but because *vaterländisch* expressed the connection between a particular territory and its history.

The *Altertum* (antiquity) researched by *vaterländische Altertumskunde* was similarly flexible. It was not pinned to a specific chronological period; rather, it covered the entire cultural heritage of a particular territory. *Altertumskunde* included the study of artifacts from the deep past, architecture from the Middle Ages, church history, and heraldry. It also investigated villages and rural life, as observers viewed dialects and folk customs as the manifestations of timeless traditions that grew out of the ancient past. By collecting excavated objects, architectural treasures, old documents, place-names, and folktales, antiquarians hoped to present a complete portrait of their fatherland's past. This meant that domestic archaeology arose not as a separate discipline but within a broad cultural movement that sought out a wide range of historical material.

Recent literature on Romanticism has explained this new attention to history as a response to the sweeping changes that occurred during the late eighteenth and early nineteenth centuries. The political upheaval, dislocation, and destruction caused by the French Revolution and Napoleonic Wars brought a sense of rupture, cutting individuals off from the past. Additionally, the demographic and economic changes of early industrialization made it seem that traditional ways of life were disappearing and that the predictability of history had broken down. In these unsettling times, poets, painters, novelists, and historians tried to reconnect the present to the past through artistic representation and historical interpretation. As art historian Stephen Bann has explained, their "desire for history" represented an attempt to recover what was lost and infuse the present day with historical meaning.[8]

For antiquarians, the "desire for history" was more than a national question. It also posed a methodological challenge. Historian Franz Schnabel noted long ago that *vaterländische Altertumskunde* was marked by a paradox. In the three decades after 1815, thousands of individuals devoted time and energy to the study of ancient sources, but these actors produced few narratives about the distant past. Antiquarians "loaded up on the past and all kinds of historical material without really connecting with the past or attaching the future to this past."[9] The great achievements of *vaterländische Altertumskunde* were catalogs, not histories. Administrators and art historians produced inventories of ancient ruins and architectural treasures, the Grimm brothers compiled linguistic evidence and folklore in dictionaries and edited collections, and the *Monumenta Germaniae Historica* was an assemblage of historical and legal documents, not a narrative of Germany's ancient past. In the same manner, antiquarians dug up thousands of prehistoric objects, but they did little with them beyond listing them in inventories and storing them in makeshift museums.

Scholars in more established fields mocked antiquarians as hobbyists who seemed unaware of broader questions.[10] But antiquarians were doing something new: they were preparing artifacts for historical interpretation. Their hesitancy was not a lack of serious scholarship but rather a sign of patience and caution. Antiquarians viewed their work as a first step that preceded the act of historical interpretation. They excavated in their local areas, assembled larger collections, and read about finds in other parts of Europe in order to gather preliminary information that would eventually add up to something meaningful. They believed that the process of accretion—the idea of bringing fragments together and turning them into a coherent whole—would one day transform anonymous and obscure artifacts into historical evidence and shed light on undocumented eras.

This forward-looking perspective was a central feature of *vaterländi-*

sche Altertumskunde. As historian Susan Crane has emphasized, grave mounds, ruins in the landscape, and medieval architecture had been "hidden in plain sight" for decades or even centuries, but the new attention that was accorded to these places and objects turned them into monuments (*Denkmäler*).[11] They needed preservation not only because they were fragile, old, and rare but also because they could be saved for later generations and filled with historical meanings in the future. The leaders of the *Monumenta Germaniae Historica* made this point explicitly when they described their efforts as the fertile ground from which future historical work would grow.[12]

The process of accretion also helps explain how the personal interest in the local past developed into a collective effort. Individual antiquarians formed historical associations that extended an awareness of historical objects to their local communities. These groups in turn saw the benefit of sharing information with other groups. By the middle of the nineteenth century, this communication led to the creation of a national corresponding society and larger museums that contained objects from throughout central Europe. Finally, antiquarians hoped to provide a more complete vision of the distant past by connecting archaeology to other cultural activities, such as the study of place-names, runes, legends, and Roman history.

By returning to Kruse's magic "spell" and the widespread excavations in the early nineteenth century, we uncover many sources of inspiration for antiquarians and gain a greater appreciation of the ways that personal curiosity intersected with local history, regional politics, and national thinking. We also see that antiquarians were so committed to the stewardship of the past because of their desire to prepare archaeological material for future interpretation.

Ancient History in the Evocative Landscape

Archaeology brought individuals into contact with authentic pieces of the past, and it encouraged people to see their surrounding areas as places that contained an abundance of historical material. In this way, archaeology fostered new relationships between people and the landscape and presented regional kingdoms, cities, and towns as places with an ancient past. This was certainly the case with Andreas Buchner, a history professor from Munich, who set off in 1818 to follow the Roman walls that stretched over three hundred miles through Württemberg and Bavaria. He hoped to shed light on the nature of the relations between the Germanic peoples and Romans during the first centuries of the Common Era, but his travel account also displays the participatory nature of historical work. Buchner expressed the exhilaration he felt while actually being in the same forests, breathing the same air, and seeing the same landscape as the earlier inhabitants of Bavaria had. This sense of history heightened his interest in nearby towns and the countryside as the possible sites where historical events had occurred. The fact that the once-strong walls were now ruins, however, created a feeling of loss and made Buchner's experience all the more valuable. He sadly wrote, "The day's last ray of sun illuminates the ruin that no one after me will capture!"[13] He wished that the crumbling watchtowers and fading sun would remain this way forever so that others might experience the same emotional response to ancient history.

Other investigators found archaeological sites to be the ideal setting for contemplation and the composition of poetry. Friedrich Wilhelm Huscher, a student of theology and active member of a historical association in southern Saxony, had read much of the literature about the domestic past, but he had become "sick of all this old German stuff."[14] In 1829, however, Huscher "discovered" a grave mound in a secluded forest outside Ansbach. A stone covered with runes marked the mound. (This was of particular interest because rune stones were common in Denmark and Schleswig but rare so far south.) Huscher spent hours at this spot and wrote of the "feelings and magic" aroused by his investigation. The theology student had "seen many beautiful things and revered the sublime," but "never have I viewed dignity and loveliness so intimately united, . . . never have such feelings shaken my breast, never have such thoughts carried my spirit above the clouds, as when the sight of this solitary, unadorned grave mound overtook my heart." Huscher's hands trembled as he tried to depict the site and record his feelings. He became lost in thought as folklore, fantasy, past, and present blended together in his head: "What dreams from the heights of Walhalla come to comfort the lonely hiker . . . who lies down upon the moss and eavesdrops on the voices and legends that whisper around him?"[15] These were feelings Huscher could not experience in the library or archive. "Oh how different the voice of the past sounds when it comes from these monuments," he wrote. "The dead characters of books cannot compare!"[16]

Landscapes enchanted by ancient monuments were a frequent motif for Caspar David Friedrich, Germany's most celebrated Romantic artist. Friedrich was born in the town of Greifswald in 1774, and he studied at the Academy of Art in Copenhagen. Several of his teachers were influenced by Gotthard Kosegarten, the Danish theologian who was largely responsible for the fame of the Neolithic burial chambers near the Baltic coastline and on the island of Rügen. Friedrich traveled to Rügen several times between 1801 and 1809, and its ancient monuments figured in many of his works. In *Dolmen in Autumn* (1820), a dark, forbidding sky sends down a wind that stirs the trees and grass, but the ancient stones stand as durable markers. The painting invites viewers to contemplate the long expanse of time that had elapsed since the monument was erected. In *Neubrandenburg* (1817) and *Two Men Contemplating the Moon* (1819), ancient stones, along with deep color contrasts and unstable skies, evoke emotional responses like the longing, nostalgia, and humility recorded by Buchner and Huscher. Friedrich also executed several paintings of the jagged sandstone formations in Saxony and Bohemia that contemporaries believed to have been Germanic holy places. These landscape scenes emphasized the sanctity of these sites and often depicted individuals lost in solitary contemplation.[17]

During the first half of the nineteenth century, states and historical associations sought to direct the emotional responses to these holy and historical encounters in several ways. Historical associations offered the possibility that these feelings could be shared and acted upon. With state support, they promoted a greater awareness about ancient monuments with the hope that more material would be protected from neglect and destruction. Antiquarians also wanted to transform the sensations of discovery and communion with bygone ages into the writing of history. They wanted to explain the battles and religious ceremonies that occurred at these places and connect these events to the early history of their towns and villages. States supported these efforts, in part because of the belief that historical awareness could act as a unifying force that strengthened cultural connections and monarchical traditions. Before any of these tasks could be achieved, however, historical associations would need more material and new methods for evaluating ancient objects.

Archaeology, Historical Associations, and the State

In the early nineteenth century, the states of the German Confederation sought ways to channel the enthusiasm of people like Buchner and

Huscher toward regional cultural identities. They employed preservationists to document cultural heritage, opened archives to historians, and created libraries and museums as public places that shared the state's treasures with a wider population. Domestic archaeology was a central component of these cultural policies, as several German states initiated efforts to register and preserve antiquities and dozens of historical associations participated in these projects.

The management of antiquities produced a flourish of communication between state capitals, provincial cities, and local governments. In Prussia, Konrad Levezow, a professor of history and curator of antiquities for the royal collections in Berlin, offered a commentary on the scientific value of antiquities in 1825. In 1835, Karl Friedrich Moritz Paul von Brühl, the first director of the Prussian museums, circulated a statement on excavation techniques that came directly from the Prussian Cultural Ministry. The interest in antiquarianism also extended out from Berlin to Prussia's provinces. In 1820, Wilhelm Dorow, an appointee of the Prussian Cultural Ministry sent to the Rhineland, published several works on excavation methods and the sites that he excavated. Friedrich Kruse and Johann Gustav Büsching led a network of antiquarians in Silesia. Outside Prussia, the Bavarian monarchy supported historical societies that excavated hundreds of sites in the 1820s and 1830s. A royal comptroller in Saxony named Karl Preusker provided a scientific society in Görlitz with a commentary on the purpose of antiquities research in 1829. And in Freiburg, Catholic priest and librarian Heinrich Schreiber composed a handbook for archaeological finds in southern German states. A common theme in all these texts was the belief that the collection of artifacts would lead to the writing of the ancient history of local areas and territories held by regional monarchies.

Texts about excavating noted that the appearance of archaeological material was nothing new. Konrad Levezow, for example, commented that floods, farming, and roadbuilding had brought objects to the light of day for centuries.[18] Many authors viewed their efforts as a continuation of the work of eighteenth-century antiquarians, and they often referred to earlier studies by Johann Christoph Bekmann, Leonard Hermann, and J. C. P. Kortüm. Yet preservation became more urgent after 1800, as the modern age directly threatened historical material. Policies of secularization in revolutionary France and the Napoleonic Empire stripped the Catholic Church of much of its wealth and left many churches, cloisters, and abbeys in a state of neglect. Wartime disruption forced the Bavarian Academy of Sciences to close in 1799, and it did not reopen until after the founding of the Kingdom of Bavaria in 1806. In the following decades, it appeared that a future of urbanization and factory production would increasingly replace the traditional world of village life and artisan work. These developments exposed the fragile nature of the past and brought forth a desire to preserve customs and practices that might be displaced or destroyed by modernization.

State administrators and members of historical associations repeatedly expressed concerns about the fragile and irreplaceable nature of artifacts. The Gesellschaft für pommersche Geschichte und Altertumskunde (Society for Pomeranian history and antiquities studies), for example, set out "to rescue the monuments of prehistory in Pomerania and Rügen" from the elements of nature and human acts, like the tilling of fields and roadbuilding.[19] Poorly executed excavations were also a threat. C. F. Hofmann, an antiquarian in the Rhineland, knew that laborers were tempted to work faster when objects appeared in the ground, and he therefore hired only people who could remain calm despite the excitement of discovery. As a precaution, he instructed his workers to leave the dirt that encased an object in place. He would then gently scrape away the dirt surrounding the object himself.[20] Konrad Levezow was even more alarmed about improper excavation techniques. Pottery incautiously handled fell into fragments, and metal implements thinned by rust and decay broke easily. "The ignorance and carelessness of the 'first finders' has damaged much, perhaps even most [archaeological material]," he wrote. Levezow also warned that some people hunted archaeological objects for their economic, not historical, value. He lamented that "the fiery force of smelting ovens has fully destroyed the original form of so many pieces."[21]

These authors understood that construction, farming, and carelessness threatened ancient material, but that modernization projects also revealed many artifacts. They therefore urged local administrators, engineers, foresters, and customs officials to be on the lookout for prehistoric material. A clear example of this communication occurred in the 1830s when the Kingdom of Prussia undertook a major road-building campaign to connect several regional towns to Prussia's isolated eastern provinces. In the midst of this construction, Karl von Brühl issued the Prussian Cultural Ministry's instructions "regarding the antiquities of heathen prehistory" (1835). These guidelines presented road construction as an opportunity to look for traces of the land's earliest inhabitants. Workers should scan the landscape for "isolated hills . . . that had been created by man," the ruins of old walls, and piles of stones that stand at sharp angles or appear to be laid out in discernible patterns. These were the places where "it is highly probable" that antiquities will be found. Foremen should watch for soil that contained clay fragments or was mixed with ashes, coals, or reddish dirt—these were signs that the area may contain buried urns. They should also ask local workers and farmers if urns were common in the area. If locals did not understand what an urn was, workers should remember that many people referred to urns as "pots" (*Töpfe*).[22] Brühl's instructions urged those engaged in road construction to pay close attention to the ground and the history it possibly contained. This call for care and preservation would be repeated throughout the nineteenth century as road construction, the creation of railroads and canals, and the expansion of cities brought more and more antiquities to light.

As the Prussian Cultural Ministry publicized Brühl's instructions, dozens of *Vereine* (historical associations) were extending the concern for antiquities to a wider public. Enlightenment societies had disseminated information about agricultural techniques, science, and culture in the eighteenth century, but the number of these groups exploded in the early nineteenth century.[23] They engaged members of the rising middle class, and history was one of their primary occupations.

Governments supported associations because they believed these groups could impart a sense of historical coherence to their states. In Bavaria, King Ludwig I (r. 1825–48) encouraged the use of history to bind new lands together with core holdings. The treaties and settlements that ended the Napoleonic Wars turned Bavaria into a kingdom and added the Palatinate and some Austrian borderlands to its territory. Ludwig I, after noticing the efforts of three historical associations in the 1820s, decreed that "[e]ach town shall receive its own historian," and the government founded an association in each of the kingdom's eight districts. Ludwig believed that "knowledge created affection," and he hoped that these acts would "give the [Bavarian] constitution a more solid historical grounding."[24] Embracing the state's program, the Historischer Verein von und für Oberbayern (Historical association of and for Upper Bavaria) pledged to organize historical material "in a practical and useful way, and thereby respect ancient and recent times, the history of Upper Bavaria as well as the Royal House."[25] The pursuit of local history was an expression of loyalty to the Bavarian kingdom and an opportunity to acknowledge the patronage of the Wittelsbach dynasty.

Connections to the state were fundamental to historical associations. In an era of censorship, states granted associations the right to convene and to publish their proceedings. They provided some financial support for printing and mailing costs and for the acquisition of books and artifacts. In the Kingdom of Bavaria, Ludwig I used his personal dispensation fund to subsidize the publication and excavation costs of Bavaria's historical associations, and calls for membership and excavation notices appeared in the official *Kreis-Intelligenzblätter*.[26] Associations included high-ranking state officials. The Bavarian interior minister Carl von Abel belonged to the Upper Bavarian association, for example. Many members were *Landräte* or *Regierungsräte*, titles that suggested the connection between the *Vereine*, regional governments, and the state bureaucracy. Judges, court clerks, customs officials, railroad planners, foresters, and other civil servants were represented in the membership lists of these groups as well. These state employees were often the first to hear about finds that turned up during the construction of roads and canals, and they facilitated greater communication between provincial governments, museums, academies of sciences, and historical associations.

Yet the relationship between the state and antiquarians was one of cooperation, not domination. States relied on historical associations because the task of collecting fragments of the domestic past was too large for a single administrator or for the limited resources that states provided for these efforts. Prussian chancellor Karl August von Hardenberg called for the collection of antiquities in the Rhineland in 1811, but this directive did not yield lasting results until Wilhelm Dorow and the leaders of the provincial government advanced this effort. Over the course of the 1820s and 1830s, the provinces of the Rhineland and Westphalia had three historical associations with over four hundred members.[27] In Bavaria, it was the regional historical associations that carried out King Ludwig's order to collect and preserve antiquities. And the vast majority of an association's resources came from membership dues, not royal subventions. The Württembergischer Alterthums-Verein (Württemberg antiquities association) in Stuttgart, for example, had an annual operating budget of 2,223 *Gulden* in 1845. Only 100 *Gulden* came from its patron, Crown Prince Carl of Württemberg, and over 2,000 derived from the dues of its 498 members. Smaller amounts came from subscriptions to its journal (70 *Gulden*) and interest earned (28 *Gulden*).[28]

The devotion to archaeology was overwhelmingly provincial in other European settings too. In France, efforts to preserve historical material appeared in the 1810s and 1820s but increased dramatically during the July Monarchy (1830–48). One commentator noted in 1842 that archaeology was in its infancy a generation before. In his day, however, "archaeological digs are undertaken everywhere. One plows almost as much for history as for wheat."[29] The geographical focus for most of this activity was the *pays*, which antiquarians viewed not only as an administrative unit like a county but also as a "natural region." It encompassed an area no farther than a day's journey (meaning home was always close by) and became associated with local customs that had evolved in relation to the natural environment. The French believed the *pays* had deep historical roots in the Celtic and Roman communities of premedieval times.[30] During the 1830s and 1840s, county archaeological societies proliferated in southern England, inspired by Sir Walter Scott's fascinating depictions of the Middle Ages and by a middle-class desire to recover British heritage. County associations took advantage of the expanding railway network and covered roughly the distance one could comfortably travel to and from in the course of a day. This local focus allowed English historical societies to produce total histories of their local areas. As historian Philippa Levine has described, "the ideal county history was one which embraced accounts of local superstitions and customs alongside discussions of medieval land holdings, of monasteries dissolved under Henry VIII and transcripts of epitaphs on old tombs."[31]

Associations in German-speaking Europe shared this interest in a wide range of local material. In addition to their archaeological work, they published medieval charters and church edicts, studied heraldry, and documented churches, cloisters, and city halls as architectural monuments. The historical association in Ansbach (Bavaria) archived city chronicles, documents related to the history of the nobility, and drawings of architectural monuments and maintained a collection of coins, vases, and statues. The Historischer Verein von und für Oberbayern possessed similar items, along with a collection of archaeological artifacts that grew over the course of the nineteenth century. Most associations documented their activity with an annual report and a journal. Acquisitions for the Ansbach collection regularly appeared in the *Jahresbericht des historischen Vereins im Rezat-Kreis*, and additions to the Upper Bavarian collection were listed each year in the *Jahresbericht des historischen Vereins von und für Oberbayern*. The size, circulation, and regularity of these publications varied considerably. Some were small, short-lived affairs that did not circulate beyond the region where they appeared. Others, like the journals of the associations in Upper Bavaria and Lower Saxony, were substantial and became important organs for regional history. A single volume of an association's journal could include treatments of different historical periods and activities, from genealogical investigations, to the publication of town charters, to the listing of the contents of an excavated gravesite. What unified this array of material was a focus on the local area.

The local orientation of *vaterländische Altertumskunde* was so important because it created a connection between people and their local surroundings. Members of historical associations searched for and found the distant past near their homes, the schools where they taught, and the churches where they preached. The places discussed at association meetings were common to all present. This lent familiarity and an element of local pride to the work of associations. Names like the Historischer Verein der Pfalz, the Gesellschaft für pommersche Geschichte und Altertumskunde, the Historischer Verein von und für Oberbayern, or the Historischer Verein für Niedersachsen indicated not only where the association met and where its membership lived but also the entities that the associations made historical. The collection, display, and interpretation of the past transformed the Palatinate, Pomerania, Upper Bavaria, Lower Saxony, and other regions, as well as localities, into historical landscapes with a shared sense of identity and unity. The ethnographer Wilhelm Heinrich Riehl (1823–97) would later formalize this relationship in *Land und Leute* (Land and people, 1854), arguing that natural feelings of belonging were rooted in both geography and history. His ideas, discussed in the next chapter, were fundamental for the creation of museums that exhibited local and regional artifacts.

Archaeological excavations were carried out throughout central Europe, but they were pursued most energetically in historic borderlands. Prussia's western provinces, which were greatly expanded with acquisitions after the Napoleonic Wars, had an early history that was different from the rest of Prussia: both Germanic and Roman finds were scattered throughout the Rhineland, and the area was largely Catholic while most of Prussia was Protestant. During the 1810s and early 1820s, the Prussian state sent cultural officials to the Rhine province and Westphalia hoping they could use *vaterländische Altertumskunde* to steer cultural activity away from these historical differences.

The same pattern occurred in the southeast of Prussia. The province of Silesia had a mixed population of Slavs and Germans and had remained largely Catholic. The Prussian Cultural Ministry supported the antiquarian efforts of Johann Gustav Büsching (1783–1829) and Friedrich Kruse (1790–1866) in Breslau, but it followed Büsching's editorial work for the

Schlesische Provinzialblätter closely and prohibited the publication of research about the area's Catholic history that could draw attention to present-day religious divisions.[32] Instead, as the following chapter will show, Kruse and Büsching emphasized a common German past in their studies of Silesian antiquities.

The Prussian state's dedication to domestic antiquities in the Rhineland and Silesia faded over the course of the 1820s. Dorow began to receive less support for archaeological work from the Prussian state by 1822;[33] Kruse accepted a position at the University of Dorpat for world history in 1828 and began studying the early history of Russia; and Büsching suffered the onset of kidney disease in 1825 and died in Breslau in 1829. Although the Prussian state turned away from the German past, it would be a mistake to view this period as the onset of stagnation. Even as German nationalism fell out of favor, efforts in domestic archaeology increased, as antiquarians turned their attention to the history of their locales and regions.

This historical activity was not exclusively a top-down process. Historical associations often focused on local concerns or traditions that had little to do with an overarching German, Prussian, or Bavarian identity. In the late 1830s and 1840s, for example, the historical association in Bamberg focused on the early modern history of a local margravate that had ended during the Napoleonic Wars. This activity recovered a history that was lost, more than it sought to integrate the area into the Kingdom of Bavaria. Similarly, the historical association in Bayreuth spent much of its energy researching the area's connections to the House of Hohenzollern. Associations in places with strong civic traditions, like Nuremberg, naturally devoted much of their work to the medieval heritage of their towns. Historical activity could even flourish in opposition to an overarching identity. Much of the vitality of the Historischer Verein der Pfalz arose in reaction to the area's integration into the Kingdom of Bavaria.[34]

The local nature of domestic archaeology created an emotional bond between historical associations and the sites they excavated, but it also limited archaeology's historical potential. Antiquarians in the Rhineland were wary of Dorow's intentions because they feared that artifacts would be removed from their towns and placed in a central museum. Thomas Wright, a British antiquarian and the author of *The Celt, the Roman, and the Saxon* (1852), claimed that artifacts lost their meaning when they were taken out of their local context, and he therefore advocated the local display of archaeological artifacts.[35] This strong sense of possession created a thorny issue for collectors and administrators as they tried to fill museums, but it also represented an interpretive problem. If artifacts were viewed only as local material and evidence of local traditions, could they truly speak to broader questions of prehistory and tell the story of wider geographical areas? Antiquarians grappled with this question explicitly, and they responded in a way that resembled the celebrated method that was being developed in the field of history. Like the pioneers in historical research, antiquarians treated archaeological material as fragments of the past that needed to be combined with other material in order to arrive at broader meanings. They hoped these efforts would deliver knowledge that extended beyond the results of individual local associations, making the whole of their research effort greater than the sum of its parts.

Archaeology and the Definition of History

The ultimate goal of domestic archaeology was to access previously unknown epochs and chart the early history of central Europe. Rhineland antiquarian C. F. Hofmann had noted several times that "the main objective of excavation is always to clarify the ancient history of our local area that remains completely shrouded by the deepest darkness."[36] Excavations produced the sources needed to write this history, and if these sources were lost, the potential to produce such a narrative would disappear. This desire to attain knowledge about the prehistoric period occurred against the backdrop of a wider confidence in the ability to reconstruct the past. During the same decades when states registered antiquities and historical associations excavated and publicized their finds, the German founders of the historical discipline emphasized the collection of historical fragments as the first step toward the narration of history.

The desire to investigate the ancient past would seem to unite historical and antiquarian practices, but cultural elites effectively separated these two fields of knowledge. As historian Suzanne Marchand has explained, classical archaeologists benefited from the established academic practices of philology and the status of classical art and architecture in the early nineteenth century. A series of dichotomies distinguished them from antiquarians interested in domestic sources: they were professional classicists, not amateur antiquarians; they focused on classical civilizations, not the barbarian past; they pursued humanist inquiry, not narrow provincialism.[37] This dedication to classical ideals was evident in Berlin's cultural institutions. The Altes Museum, opened in 1830, exhibited Greek and Roman sculptures and European paintings. The ancient past was represented by classical civilizations, while northern European culture began only after the Middle Ages. Barthold Georg Niebuhr's lectures on ancient history at the University of Berlin during the 1810s focused on the Roman Empire, not central Europe in ancient times. And Leopold von Ranke (1795–1886) ruled out European prehistory for methodological reasons. He wrote that "it is permissible to confess our ignorance" about prehistory because historical methods cannot shed light on epochs that had no written records.[38] Writing in the 1880s—after several decades of widespread archaeological activity and the creation of cultural history museums that included archaeological objects—Ranke still viewed the study of prehistory as fundamentally different from his craft. In the opening lines of his *Universal History*, he explained:

> History cannot discuss the origin of society, for the art of writing, which is the basis of historical knowledge, is a comparatively late invention. The earth had become habitable and was inhabited, nations had arisen and international connections had been formed, and the elements of civilization had appeared, while that art was still unknown. The province of History is limited by the means at her command, and the historian would be over-bold who should venture to unveil the mystery of the primeval world, the relation of mankind to God and nature. The solution of such problems must be entrusted to the joint efforts of Theology and Science.[39]

European history, for Ranke, began in the Middle Ages and had to rely on texts. His seminar at the university in Berlin investigated the history of the great powers and of the church; it did not include local history or the prehistory of Europe.

Despite the line drawn between historical writing and prehistory, it is productive to return to this founding moment of modern historical scholarship and consider the rise of domestic archaeology in light of Ranke's comments about historical methods. Historians distanced themselves from antiquarians by defining their domain as the realm of time covered by texts and by emphasizing their scientific methods. This made the study of prehistory a foil that helped define what became professional history. Yet domestic archaeology had much in common with the emerging practice of scientific history. The early stages of archival work and antiquarianism shared the evocative experience of discovery. Researchers described archival documents that needed to be rescued because of their age and inaccessibility. In the early nineteenth century, archives were dusty and chaotic, and "decay was everywhere: small animals gnawed at the documents, which were fouled with dead bugs, rodent excrement, worms, hairs, and nail clippings." Historians viewed themselves as valiant knights who entered these spaces for the first time in decades or perhaps centuries and rescued documents from oblivion, and they compared their work to the efforts of private investigators who decoded "illegible writing, strange languages, shorthand, and secret codes."[40] The emotions associated with these discoveries and rescue missions were similar to the excitement that accompanied the hands-on experience of removing objects from the ground.

Fundamental concepts of interpretation also linked the foundation of scientific history and the early phase of domestic archaeology. Both historians and antiquarians spoke of the interpretive work that transformed fragments of the past into narratives. To prepare his lectures on Roman history, Niebuhr sifted through the records of property transactions "to discover the built-over and hidden foundations of the old Roman people and their state."[41] This archaeological metaphor revealed the author's search for order among the documentary chaos, an order that would lead to a coherent story about landownership in the Roman Empire. Wilhelm von Humboldt (1767–1835), a language scholar, leading educational reformer, and important voice on issues of aesthetics, spoke of the active interpretation of fragments even more directly. The historian's task, he wrote in 1821, is made up of two processes that repeatedly check one another. First, the historian is the collector of facts and events. His research and observation lead to the presentation of what actually happened, and this is "the primary, indispensable condition of his work." Yet observation and presentation are not enough. Humboldt continued:

> One has, however, scarcely arrived at the skeleton of an event by a crude sorting out of what actually happened. What is so achieved is the necessary basis of history, its raw material, but not history itself. To stop here would be to sacrifice the actual inner truth, well-founded within the causal nexus, for an outward, literal, and seeming truth; it would mean choosing actual error in order to escape the potential danger of error. The truth of any event is predicated on the addition . . . of that invisible part of every fact, and it is this part, therefore, which the historian has to add. Regarded in this way, he does become active, even creative—not by bringing forth what does not have existence, but in giv-

ing shape by his own powers to that which by mere intuition he could not have perceived as it really was. Differently from the poet, but in a way similar to him, he must work the collected fragments into a whole.[42]

At the heart of historical writing, then, stands the process of accretion—an approach that finds a meaning that is greater than the sum of the historical fragments. To reveal larger truths about humanity, the historian, according to Humboldt, has to go beyond the presentation of the fragments of the past found in the archive. He has to stand apart from those fragments and craft a coherent narrative that reveals "the causal nexus" between events. In short, he must venture into the realm of narration. Anything less would be an error.

The idea of accretion also stands at the center of one of historiography's most famous phrases. In the introduction to *Geschichten der romanischen und germanischen Völker* (Histories of the Romanic and Germanic peoples, 1824), Leopold von Ranke asserted that he intended to write about the past as it actually was (*wie es eigentlich gewesen*), and he announced that his reliance on documents separated his work from that of the novelist, who fictionalized the past, or of the philosopher, who worked from abstractions. For decades, historians in Europe and the United States idealized this statement as an expression of the historian's dedication to objectivity.[43] According to this view, the historian diligently amasses archival documents and orders the events that make up an institution's or a nation's long history. Yet, as Georg Iggers has stressed, Ranke's famous phrase meant more than a commitment to objectivity. Iggers translates *eigentlich* not as "actually" but rather as "essentially," meaning that the historian captures the *essence* of the past, not the objective historical reality. This crucial distinction draws attention to the fact that Ranke viewed history as much more than the documents in an archive.[44] According to Ranke, the historian contemplated documents and, through intuition, found relationships between these fragments of the past that revealed the essence of history. Ranke explained that "the study of the particulars, even of a single detail, has its value, if it is done well.... But ... specialized study, too, will always be related to a larger context.... The final goal—not yet attained—always remains the conception of and composition of a history of mankind."[45] Like Humboldt, Ranke strove to describe a reality that existed beyond the documents. The particulars of history provided an outline of this reality, but the historian could arrive at the "larger context" only through intuition.[46]

Whereas Ranke viewed prehistory as something distinct from history, antiquarians in the early nineteenth century found many parallels to his process of turning fragments into history. Antiquarians unlocked the underground as an archive of historical material, and they hoped to bring order to the antiquities they collected. Like the historian, they believed that the accumulation of more material would lead to the accretion of higher meanings. Two major differences separated history and "the darkness of prehistory," however. First, the fragments from each era were very different. History relied on texts that appeared to have clear meanings. One merely needed to amass a critical number of laws, charters, and other documents before one could begin to interpret the past. Antiquarians, however, collected objects without obvious meanings. They had to develop methods to determine the age and purpose of their material before they could begin the task of interpretation. Second, history had the advantage of researching well-defined entities like kingdoms, nation-states, or the church. Here the outcome of historical processes was more evident, and one could therefore speculate about the "causal nexus" (Humboldt) and "larger contexts" (Ranke) of these institutions. The study of the premedieval past, on the other hand, lacked a framework that made sense of the fragments, and one could not speak of a "larger context" for prehistory until the basic contours of the distant past were established. Thus, *vaterländische Altertumskunde* did not initially lack narratives because antiquarians were not interested in larger questions. Rather, this absence of narratives existed because of the paucity of knowledge about conditions in ancient Europe. Antiquarians first had to find, assemble, and prepare fragments of the past. They composed inventories and descriptions of excavated objects as a preliminary step. Only after the accumulation of more material could they begin to interpret objects and weave them into historical narratives about the ancient history of central Europe.

The Organization of Artifacts

Pioneers in domestic archaeology recognized that their work represented the most preliminary stage of historical writing. They had not yet reached what Humboldt described as "the skeleton of an event" or "a crude sorting out of what actually happened." They still needed to assemble the bones of the skeleton and the fragments of distant events. Archaeology offered a solution to these problems. Leopold von Ledebur (1799–1877), who would become the director of the Prussian collection of antiquities (the largest of its kind in German-speaking Europe) in 1838, recognized the potential that archaeology held for shedding light on prehistory. Surveying the literature about ancient Germany from the 1820s and 1830s, Ledebur noted that archaeological excavations were becoming as important as the study of classical authors. He recommended that "in general, we should begin to turn our eyes to beneath the surface of the earth in order to gain a more correct and higher understanding of what happened aboveground during the time of the Germanic tribes."[47] Konrad Levezow explained that the fewer texts there were that documented an ancient epoch or the prehistory of a specific area, the higher the historical value of archaeological objects. Early conditions in Brandenburg and Prussia's eastern provinces, for example, were more obscure than encounters between barbarians and Romans along the Rhine River, which were documented in classical sources. The prehistory of the eastern provinces therefore depended much more on archaeology.[48]

The desire for history led to the improvement of archaeological methods and record keeping. Dorow repeated Hofmann's assertion that it was no longer sufficient to know that there was once a Roman settlement here or there; "rather, one must also provide exact information about where items were found, in what building, at what depth, under what circumstances, and among what kinds of other items. It is therefore necessary to keep a daily log during an excavation and to record every circumstance briefly, but exactly."[49] Levezow was impatient with antiquarians who did not document the provenance of their artifacts. "All specific scientific information and the entire historical value [of the objects] vanish" when owners could no longer match objects with excavation sites. Levezow went so far as to assert that the "naked artistic value [of most objects] is negligible," but when one can clearly and authoritatively prove where a single object or groups of objects were found, the objects possess "a higher meaning" that elevates them "to the rank of historical monuments."[50] Many historical associations repeated this conclusion and used their journals and yearbooks to document excavations and record the provenance of their material.

Antiquarians were aware of the tentative nature of their knowledge of early history. Yet they were confident that their fieldwork would one day yield historical results. Therefore, they were not just collecting objects for the sake of collecting. They were saving archaeological fragments for a near-term future when they would be able to shed light on questions about the ancient past.[51] Members of the Thüringisch-Sächsischer Verein für Erforschung des vaterländischen Alterthums und Erhaltung seiner Denkmale (Thuringian-Saxon association for research into the region's antiquity and preservation of its monuments) in Halle recognized that they stood at the threshold between collecting and interpretation. Their statutes explained that with *vaterländische Altertumskunde* "the breadth has been measured, but the depth has not." This referred to the enormous amount of collecting work that had happened throughout the German states in recent years. These efforts, however, generated no specific information about the peoples of prehistory. Future work would need a "historical sensibility and purpose" to provide more depth and to create a body of knowledge that would be more firmly grounded than fantasies about prehistory.[52] Georg Christian Friedrich Lisch, the keeper of the ducal collection of antiquities in Mecklenburg, explained that the sole purpose of publishing a catalog for the collection was "to prepare wider access to the larger project and to sharpen the general attention given to current and future excavations."[53] The Gesellschaft für pommersche Geschichte explained that its mission was to "seek out and gather Pomeranian antiquities and carefully preserve them for the present and future generations." The goal was to make excavated objects *"gemeinnützlich,"* useful and beneficial for the wider community. By this word the members expressed their hope not only that they could inspire a local awareness about prehistoric material but also that their preliminary studies of artifacts would "help future researchers."[54] An interesting archaeological practice further illustrates this forward-looking commitment. When excavators found a series of grave mounds, they sometimes opened only a few of them, leaving the rest untouched not only so that others might experience the joy of discovery but also so that future excavators could investigate the mounds with more advanced knowledge about ancient peoples.[55]

Antiquarians employed many techniques to move from collecting to clarifying the distant past. Eighteenth-century authors had already noted the importance of place-names and legends, but after 1815, Jacob Grimm began to treat linguistic information as a historical source. Grimm believed that a Germanic past lay beneath Germany's medieval and modern history and that folktales, mythology, and the German language itself still contained fragments from this earlier era. He therefore considered the language and traditions of rural people as living clues to the past and regarded them as historically rich and even sacred.[56] Brühl included this idea in his excavation instructions, advising antiquarians to know not only the location of antiquities but also what they meant to the rural population. He commented that locals were familiar with peculiar place-names that suggested connections to ancient peoples and religious practices. Examples included variations on terms such as "fortress," "bride's stone," "heathen's cellar," "devil's cellar," "giant's tomb," "sacrificial stone," and "Wendish (Slavic) graveyard."[57] Others, like Buchner and Huscher, tried to connect local descriptions to Nordic mythology, giving archaeological sites a religious allure. They wanted to confirm the mythic past with concrete evidence but still preserve the emotions evoked by the discovery of an ancient religious site.[58] These approaches displayed how *vaterländische Altertumskunde* created connections among philology, folklore studies, and domestic archaeology and how antiquarians hoped that the combination of different kinds of fragments would shed light on the darkness of prehistory.

What did the preliminary step of collecting and recording look like? The first volume of the *Oberbayerisches Archiv*, the journal of the Historischer Verein von und für Oberbayern, nicely illustrates the pervasiveness of archaeological activity and the limits of antiquarian studies. Over the course of 1839, the journal included three inventories that listed more than 150 archaeological sites. Each entry included the number, size, and location of the grave mounds that were excavated, as well as references to other antiquarian literature.[59] Shorter articles provided further details about individual sites and their local contexts. Joseph von Hefner, for example, provided this description of a site he investigated near Weilheim:

If one follows the county road that leads from Weilheim to Murnau, one arrives in one and a half hours at the village of Etting. From there, one travels another quarter hour to a valley that opens up on the left side of the road. The valley is somewhat swampy because the nearby spring does not have an adequate outlet. To the south, a hill gradually rises and is covered by a group of mighty oak trees and a *rampart* [*Umwallung*] that, differing from the Roman style, stretches out in a half square.

Grave mounds are spread out along the top of the hill and make up a so-called pagan cemetery [*Heidenkirchhof*]. Cultivation by farmers has leveled about half of the mounds; the rest, around forty, are on the property of the Zistelbauer and had not been touched before the excavations that I undertook in September 1838.[60]

It is hard to imagine anyone except for a fellow antiquarian or a local inhabitant veering off the rural road to seek out the hill containing these grave mounds. Yet readers could see that impressive evidence of the distant past was nearby.

Hefner continued with a detailed description of the mounds. Each was covered by a three-inch layer of moss and six to twelve inches of a mixture of shards and ash. Each mound covered a stone chamber that contained between six and twenty-two urns. None of the mounds contained fibulae or coins, which Hefner interpreted as a sure sign "that the Romans never came into contact with the people who rested here."[61] Nine mounds held a large number of human and animal bones. Hefner also noted the presence of a sword (eight parts copper, two parts zinc) that measured about fifteen inches in length. Drawing on a scene from Ossian's poems, which Herder had popularized but were later proved a forgery, Hefner claimed that the site was the final resting place of a brave warrior. This would explain the animal bones as the leftovers from a feast that marked a hero's burial and the proximity of the other mounds, which must have contained his comrades-in-arms.[62] Hefner did not discuss the possible identity of the warriors he uncovered or the era when they lived. These historical facts lay beyond his empirical observations and mythological fantasies.

Hefner's description exhibits both the potential and the limitations of early archaeology. His five-page entry reveals his excitement about the excavation, and the impressive level of detail about the construction and contents of the mounds shows that he saw the site as something more than a treasure trove. On its own, Hefner's work provided little concrete information about Weilheim's distant past. Together with the dozens of other entries in the Upper Bavarian journal, though, it had the potential to contribute to conclusions about the settlements that existed in the region before the Middle Ages. This was the first layer of accretion that allowed for the interpretation of the distant past.

While historical associations documented fragments of the past from their particular localities, state authorities soon began to create inventories that covered larger areas. The Kingdom of Prussia created the permanent position of "conservator of *Kunstdenkmäler*" in the 1820s, and by the end of the 1850s, Baden, Württemberg, Hannover, Saxony, and Bavaria had similar posts for the preservation and cataloging of historical monuments.[63] Many state projects, such as topographical surveys for mining and strategic maps for the military, recorded grave mounds and Roman walls. And state officials compiled inventories devoted specifically to archaeology. Leopold von Ledebur sent out an archaeology questionnaire to the members of the Verein für Märkische Geschichte (Association for the history of the Mark Brandenburg) in 1841 and eventually compiled information from over one hundred localities in that Prussian province.[64] The Historischer Verein für Niedersachsen (Historical association of Lower Saxony) commissioned Johann Wächter, the chief forester and conservator for the Kingdom of Hannover, to produce a catalog of "heathen antiquities." Wächter's inventory, published in 1841, related the discovery and excavation of hundreds of sites and detailed their contents. A model of *vaterländische Altertumskunde*, it inspired greater oversight of antiquities in the Kingdom of Hannover. An Interior Ministry circular from 1844 ordered that "care should be taken so that local prehistoric antiquities, graves, and other sites receive proper treatment." Officials involved in forestry, agriculture, roadbuilding, and railway construction, as well as religious leaders, were to watch for antiquities. If sites were excavated and no property owners made claims on the objects, the Interior Ministry would assume ownership of movable items such as urns, weapons, and tools. In 1851, these instructions were extended to the Royal Railway Office, and the Interior Ministry offered compensation to railway workers who turned antiquities over to the state.[65]

Inventories that covered larger areas, like the Prussian province of Brandenburg or the Kingdom of Hannover, assembled the knowledge generated by many individuals, and they created new opportunities for comparative study. This is certainly how Ledebur viewed his work. He saw his Brandenburg inventory as a piece in the larger *vaterländische Altertumskunde* puzzle and hoped that scholars would use it alongside works on other parts of German-speaking Europe. Wächter also looked toward a future when there would be enough information to compose an overview of the earliest history of central Europe.[66] The works by Ledebur and Wächter suggested that new knowledge arose through the accumulation of local publications and the efforts of multiple associations. They highlighted the kinds of finds that were characteristic for specific regions and allowed for comparisons of material from across a larger area. The benefits of comparative study were even more apparent when the actual objects from an entire province were displayed together, and the creation of larger collections was the logical complement to inventories that covered wider geographical areas.

Prehistory on Display

After being removed from the ground, discussed at an association meeting, and listed in an inventory, many objects had finished their time in the historical spotlight. The collections they entered during the first half of the nineteenth century left much to be desired. Historical associations often rented rooms in civic buildings to house their collections, and these spaces quickly became overcrowded. In addition, the collections were subject to the financial situation of the associations that owned them and were sometimes moved from location to location, which resulted in the loss of provenance information. Despite these problems, though, these collections did mark an important development. They made ancient artifacts somewhat public for the first time. The collections were open a few hours each week, and teachers, pastors, and civic officials could visit them and exchange interpretations with the men who had discovered the material on display.

Furthermore, associations protected many artifacts from loss or destruction. A private collection could be dispersed after an antiquarian's death, but an association's collection would be maintained longer than the lifetime of any one individual. With this in mind, the district president in Würzburg turned his antiquities over to his local association so that they would not be sold to a foreign collector and leave Franconia when he died.[67] The Historischer Verein von und für Oberbayern ensured that artifacts remained in the area by publicizing its desire to have a first right of refusal for local antiquities. It wanted to prevent the sale of artifacts to traveling collectors and justified its offer by explaining that artifacts lost their historical value when they were taken out of their immediate geographic context or when they were part of a small, private collection instead of a larger, public one.[68]

Just as larger regional inventories combined the efforts of individuals and historical associations, larger archaeology collections appeared as states acquired more artifacts through excavation and acquisition. When the Prussian Königliches Museum vaterländischer Alterthümer (Royal museum of the region's antiquities, KMVA) officially opened in 1838 in Berlin, it offered dramatically new possibilities for viewing and comparing excavated objects. Antiquities from the royal house's *Kunstkammer* (cabinet of art) made up the original core of this museum, and King Friedrich Wilhelm III (r. 1797–1840) entrusted Leopold von Ledebur with the task of transforming his family's private possessions into a public display. Ledebur had purchased the collections of well-known antiquarians such as military reformer and traveler Johann Heinrich von Minutoli in 1823 and *Gymnasium* teacher Johann Friedrich Danneil in 1834. These acquisitions, along with others made by the Prussian monarchy in the late eighteenth century, brought well over a thousand items to Berlin and covered a wide geographical area that included Prussian lands, the Grand Duchy of Posen, Saxony, and Bavaria. The *vaterländisch* in the name of the museum referred primarily to the Kingdom of Prussia and the role played by the royal family in acquiring the artifacts and patronizing the museum.[69]

Two factors led to the opening of the KMVA. First, German states were responding to a wider European discussion about history, politics, art, and museums. As historian James Sheehan has explained, the French Revolution's attack on royal power had transformed art and antiquities "from an old-regime luxury, traditionally associated with conspicuous consumption and social privilege, into national property, a source of patriotic pride and an instrument of popular enlightenment."[70] This manifested itself in Paris, where the Louvre Museum was converted from a royal collection into a public display. This example inspired German intellectuals, and in 1830, both the Kingdom of Prussia and the Kingdom of Bavaria turned royal cabinets of art into public collections. Karl Friedrich Schinkel designed the Altes Museum in Berlin, which stood across from the Hohenzollern Palace on Unter den Linden, and the Glyptothek opened west of the Wittelsbach Palace in Munich. Both museums were dedicated to aesthetics and public education and featured Greek and Roman sculpture. This same spirit led to the decision to separate domestic antiquities from the royal cabinet of art and create the KMVA in the 1830s.

Changes in Copenhagen offered a second significant context. Rulers in Scandinavia had supported antiquarian research since the seventeenth century. In the early nineteenth century, national pride and the desire to prevent the destruction of ancient material intensified these efforts in Denmark. A royal commission charged Christian Jürgensen Thomsen (1788–1865) with the task of organizing the state's prehistoric monuments, and in 1819, he transformed the Royal Collection into a public museum and personally led tours of the artifacts each week.[71] The significance of Denmark's National Museum of Antiquities came not only from the breadth of the collection but also from Thomsen's work on relative chronology. Over the course of the 1810s and 1820s, Thomsen had studied the types of artifacts found in northern Europe and their decorative styles. From his careful work, he established what we know today as the Three Age System, which explains the rough chronological order of remains of the prehistoric past in northern Europe on the basis of the use of different materials. Thomsen argued that stone items were the oldest; they were followed by things made of bronze; and the use of iron did not occur until the Common Era. Thomsen also created guidelines for recognizing stone items made in later periods and bronze items manufactured after the introduction of iron. Thomsen published his ideas in his *Guidebook to Nordic Antiquity* in 1836, and he used this chronological system to organize the artifacts in the Danish museum.[72]

German scholars followed Thomsen's work closely. Georg Christian Friedrich Lisch, the archivist for the Grand Duke of Mecklenburg, traveled to Copenhagen and corresponded with Thomsen. He applied these new ideas in his arrangement of the ducal collection of antiquities in Ludwigslust and in the catalog that accompanied this display. Ledebur was also in contact with Thomsen and was inspired by the scale and scientific standards of the Danish National Museum as he prepared the Prussian collection for public display.[73]

The KMVA was housed in Monbijou Palace, which was across the Spree River from the Altes Museum in the center of Berlin. (The palace was destroyed during World War II, and the site is now Monbijou Park.) The antiquities collection was already in Monbijou before 1838, but its condition left much to be desired. According to one contemporary description, visitors had to pass by the rubbish discarded by the palace to get to the collection.[74] In 1831, King Friedrich Wilhelm III, the minister of culture and education Karl von Altenstein, and the director of the Königliche Museen Karl von Brühl began discussing the possibility of improving the collection and opening it to the public. Brühl proposed the renovation of the palace's glass gallery for a display of domestic antiquities. After three years of debate and delay, the king authorized over four thousand thalers for the collection, which represented less than half of Brühl's initial request. Meanwhile, Ledebur composed a detailed catalog of the collection. Berliners began visiting the museum in 1837, and it officially opened in 1838, when Ledebur's guidebook was published.[75]

The KMVA housed the largest collection for domestic archaeology in all the German states and therefore was one of the first places, along with the National Museum in Copenhagen, where one could compare objects and begin to draw broader conclusions about the prehistoric past. The collection presented an overwhelming number and variety of objects. All 4,655 items were displayed in, on, and even under six cabinets in Monbijou Palace's rectangular glass gallery. Each item was labeled with an inventory number and the place where it was found. Three hundred "doubles"—items similar to others in the collection—were arranged on consoles along the walls.

The size and accessibility of the KMVA were impressive, but it faced the same problems as the collections supervised by historical associations. The inadequate space accorded to the collection in Monbijou overflowed with artifacts, and the gallery was not designed for the historical or scientific study of ancient objects. Although an exact reconstruction of the room is difficult, it appears that the glass gallery kept the rococo style of the palace. The placement of roughly hewn urns and ancient jewelry upon ornate consoles must have made a disjointed impression.[76] An 1842 museum guide reported that the exhibit, despite its breadth and value, unfortunately did not generate much public interest. The palace also lacked proper security measures, which became painfully evident three years after the collection opened. Brühl had proposed the installation of metal bars on the glass gallery's windows in 1831, but the king balked at the added expense. Robbers broke into Monbijou in September 1841 and stole thirty-three gold and silver items from the collection.[77]

Perhaps the textual presentation of the collection was as important as the museum itself. Ledebur's *Das Königliche Museum vaterländischer Alterthümer im Schlosse Monbijou zu Berlin*, published in 1838, was much more than a list of the artifacts found in the museum. It revealed the vast amount of archaeological activity that had begun in the eighteenth century and exploded in the early nineteenth century. Ledebur provided the excavation history of thousands of artifacts and reviewed the wide range of archaeological literature that had appeared in recent years. As the author of several major surveys of antiquarian work himself, Ledebur was well qualified to evaluate these writings and other collections.

A typical entry from Ledebur's catalog, about a site in East Prussia, showed many similarities with the inventories published by historical associations:

In 1835 in the old province of Nadrauen, one struck upon a collection of weathered and apparently burned stones, intermixed with clay containers, some well preserved, others in pieces, while moving earth for the new county road. . . . Some of the containers were filled with human bones and jewelry. Of these items, two urns—one is 6½ inches high and pear-shaped (I. 1395); the other is formed out of two flattened spheres pressed together (I. 1396)—, as well as several bronze fibulae, became part of the collection (II. 2022–2025).[78]

This entry, like Brühl's advice from 1835, alerted antiquarians to signs in the landscape that indicated a prehistoric site, and the sighting of the "collection of weathered and apparently burned stones" again showed the connection between road construction and archaeological discovery. Ledebur's attention to the shape and measurements of the urns echoed Hefner's empirical approach. There was an important difference between Ledebur's entry and Hefner's, however. Ledebur described objects from East Prussia that had been transported hundreds of miles from their local setting to Berlin, while Hefner described a nearby place to Upper Bavarian readers. The readers of Ledebur's catalog no longer exclusively associated the objects with their immediate surroundings. Instead, the items belonged to a vast set of objects that had been assembled in a single place. Readers could compare the urns and jewelry to finds from throughout central Europe. The Prussian collection was not only a repository of local history but also a research institution.

Ledebur knew visitors to the Berlin collection were most interested in the origins of these objects. In the preface to his catalog, he wrote: "To which people, to which land, do these objects belong, and what does each individual piece mean? These three questions, above all others, are the ones that visitors will wish to have answered and they are at the same time the main issues that a collection such as this one must address." Yet Ledebur, keeper of the largest collection of material sources from central Europe's early history, did not believe that he could venture confident answers. He wrote that "the study of antiquities is not mature enough at the present moment to sort the artifacts according to their Germanic, Slavic, Latin, [or] Oriental origins." He continued: "up to this point it has not even been possible to separate with certainty the items that belong to the Germanic peoples from those that are Slavic. A chronology of the antiquities is extremely tentative."[79]

Ledebur was familiar with Thomsen's use of the Three Age System, and he had corresponded with German scholars, like Friedrich Lisch in Mecklenburg and Johann Danneil in the Altmark, who had also developed ideas about relative chronology. Yet he did not attempt to arrange the objects in the KMVA according to their relative age, as Thomsen had done in the Danish National Museum. He considered a display based on geography, arranging objects according to where they were found.

Thomsen had used this method as he classified finds. Ledebur knew this "would allow more confident conclusions about the borders between the different peoples who had inhabited the fatherland at different times," but he felt this still could not address questions of chronology. Ledebur therefore chose to organize the museum objects according to their presumed functions. He created categories like religious articles, items associated with domestic tasks, and those used in warfare or for trade. This arrangement required less space than a geographic exhibit based on Prussia's provinces. It also allowed for the inclusion of the many objects that had entered the collection with few details about where they had been found.[80]

This choice for the display is interesting because it addressed Ledebur's third question ("what does each individual piece mean?"), but not the two most closely associated with national origins or the location of "ancient Germans" ("To which people, to which land, do these objects belong?"). Scholars needed to accumulate more information and compare more finds before tackling these ethnographic questions. Ledebur's cautious display respected the limits of what antiquarians knew in the 1830s and tempered Romantic desires to identify ancient peoples as the ancestors of nineteenth-century Germans.

There were signs, though, that Ledebur thought *vaterländische Altertumskunde* would soon be able to shed light on patterns of settlement in earlier epochs. Unlike the exhibit, Ledebur organized the information in the catalog according to geography, and he related the artifacts to the unique prehistory of the eight provinces of the Kingdom of Prussia. Decorative items and jewelry made of silver found in East Prussia were significant, Ledebur explained, because they provided evidence of trade routes that brought precious metals north from the interior of the European continent in exchange for the amber that was known to be a famous export from the Baltic coastline.[81] Ledebur hoped that study of objects from Posen, Brandenburg, Saxony, and Silesia—the Prussian provinces that lined the German-Polish border—would reveal the complex history of the border between Germanic and Slavic lands.

Ledebur also suggested connections between early history and the medieval past. The entry quoted above included a reference to "the old province of Nadrauen," which was a medieval district in East Prussia that was taken by the Teutonic Knights during the thirteenth century. Ledebur viewed this conquest as the moment when the area was "re-Germanized" and converted to Christianity after several centuries of Slavic inhabitation. An entry on Brandenburg explained that although the area had been Germanic, it had been "fully Slavicized" by the fifth century CE. By the thirteenth century, Germanic peoples dominated the area again. Ledebur therefore expected that most of the antiquities from this region would be Slavic, but he would need a better understanding of chronology before he could identify objects confidently.[82]

The province of Westphalia, according to Ledebur, was also of particular interest because it offered something different:

This land between the Weser and Rhine is the one part of Germany where the consultation of Greek and Roman sources, which are rich for this area in comparison to the scattered reports for other areas, with the assistance of medieval geography and with the longer preservation of many customs, could successfully and in a clear manner solve the questions that are to be answered about the locations and borders of the peoples who were settled here during the Germanic era, questions that are more difficult and only partially answerable for other areas [of Germany]. Because of the great significance of this purely Germanic land—completely unmixed with Slavic influences, only fleetingly touched by Roman culture—it is particularly regrettable that, in comparison to other areas, so little has been done here to access and understand the material remains of heathen prehistory that remain in the earth.[83]

Ledebur's hope for more research in Westphalia and other provinces was not a nationalist search for German purity. Rather, it was a call for the further study of neglected artifacts that offered a potential methodological breakthrough. If antiquarians better understood these finds, they could begin to sort out what Slavic and Roman influences looked like in other areas. This was a simplistic schema, and it certainly was not the extent of Ledebur's understanding of ancient conditions. Ledebur recognized that the history he tried to reconstruct contained several layers of conquest, colonization, and settlement, and he acknowledged the interaction between cultures as a vital element of central Europe's prehistory. Most notably, he was not carried away by Romantic dreams about German ancestors. His patriotism inspired him to study domestic artifacts, but he did this in a careful, even cautious way. The KMVA in Monbijou Palace served as a public display of artifacts that awaited further investigation. It allowed for the comparison of material from the Kingdom of Prussia and beyond, and this opportunity increased the value of each individual object in the collection.

The Desire for National Networks

During the 1840s, many German liberals began to call for political reform and national unification. Their demands were manifested most clearly in 1848 at the Frankfurt National Assembly, where representatives from across central Europe gathered to draft a constitution for a united German kingdom. In 1849, the German states ordered the delegates to disband and used military force to suppress the urban revolts that had sprung up across central Europe. With this reemergence of the national idea, some antiquarians perceived the process of accretion in national terms and worked toward the creation of national institutions and a national archaeological collection. Their efforts, like the project of German unification, were not successful in these years because they ran up against the fragmentation and conservatism of German politics and the regional orientation of domestic archaeology.

Participants at a conference for German literary studies in Frankfurt in 1846 discussed the need for an overarching association of historical societies. In the summer of 1852, several leading antiquarians, including Friedrich Lisch and Leopold von Ledebur, followed up on this idea and founded the Gesamtverein der deutschen Geschichts- und Altertumsvereine (Combined association of German historical and antiquities associations). This umbrella organization had three sections that were devoted to, respectively, domestic archaeology and "heathen prehistory," medieval art and architecture, and historical methods. Planners hoped that the Gesamtverein would bring local and regional historical associations together at national meetings and that it would produce a newsletter that would cover antiquities research throughout central Europe. The combination of local efforts would yield greater enthusiasm for antiquarian activity and contribute to a more comprehensive version of ancient German history. Here, the idea of accretion applied not just to the meanings that would arise from the consideration of more artifacts but also to the combination of the efforts of scattered local groups.

The Gesamtverein made only limited contributions to a more complete national history, however. The organization gained financial support for specific projects like the archaeological investigation of the Porta Nigra in Trier, the completion of Ulm's medieval cathedral, and preservation work at Heidelberg Castle.[84] And it did hold annual meetings and publish a newsletter. The Gesamtverein did not become a truly national body that could direct domestic archaeology, though. Many of its leaders were Prussian cultural officials who balanced their interest in a national organization with their commitment to the Kingdom of Prussia. They viewed the Gesamtverein as a way to extend the geographic scope of *vaterländische Altertumskunde*, not as a way to promote a national historical consciousness. Additionally, local and regional groups remained driven by their interest in nearby areas, and it proved very difficult to transfer their enthusiasm for local history to the Gesamtverein. Bavarian *Vereine*, for example, largely ignored the Gesamtverein and did not participate in its projects.[85]

The national idea was more evident in two museums that were founded in 1852. The Germanisches Nationalmuseum (Germanic national museum, GNM) in Nuremberg began as the dream of the collector, patriot, and nobleman Hans von Aufseß (1801–72).[86] Aufseß had carried out archaeological excavations on his property since the 1820s, and he believed that a national institution was needed to unify the scattered efforts of the various local historical associations. Aufseß participated in the Frankfurt conference in 1846 and shared his assessment with the German philologists and historians gathered there. He noted that, up to that point, *vaterländische Altertumskunde* had only a local or provincial significance, but the future looked brighter. "I have no doubt that it is in keeping with the times that a version of this significance that extends to the entire fatherland will come to the fore and demonstrate its validity," Aufseß declared.[87]

Aufseß created the GNM from his own collection of antiquities, but he hoped that it would grow immensely. The use of "Germanic" in the name of the museum was not tied specifically to the Germanic peoples but rather reflected his desire to conceive of German history in the broadest terms possible. He envisaged a museum that included items from not only the German states but also Switzerland, Denmark, and Alsace-Lorraine. Reflecting the political mood of the 1840s, he described the GNM as a "national institution" that would be "the common property of the German people," intended to remind Germans of their past greatness and to foster a sense of national cultural unity.[88]

Despite Aufseß's desire for a national collection, the archaeological display at the GNM remained small. It included Aufseß's personal collection, items from a few excavations carried out in the late 1850s and 1860s, and some casts of objects from other collections. The main emphasis of the GNM was the decorative arts of the Middle Ages.[89] In the middle decades of the nineteenth century, the GNM was much more of a monument to the dream of national unification than a successful attempt to found a national institution for domestic archaeology.

Concurrent with Aufseß's efforts, Ludwig Lindenschmit (the elder, 1809–93), the leading figure in the historical association in Mainz, developed the idea of the Römisch-Germanisches Zentralmuseum (Roman-Germanic central museum, RGZM). This collection, Lindenschmit hoped, would extend beyond his group's regional boundaries and advance the study of Germany's early history by comparing a wider range of finds. The RGZM, founded in Mainz in 1852, focused on central Europe's pre- and early history and the Roman past, distinguishing itself from the GNM and its original emphasis on the Middle Ages.[90]

Both the GNM and the RGZM struggled to establish themselves initially. Even though they grew out of a national enthusiasm and the desire to document a national prehistory, their archaeological collections were no larger than that of an active regional association. Both institutions relied on contributions from historical associations and the patronage of various German states. The most generous support for the RGZM before 1871 actually came from France. Emperor Napoleon III was personally interested in the archaeological investigation of Julius Caesar's campaigns in Gaul, and he paid for Lindenschmit's travels to Vienna and Budapest, where Lindenschmit made casts for both the Museum of National Antiquities in Paris and the RGZM in Mainz.[91] The GNM and the RGZM expanded dramatically in the later nineteenth century, but they could not play a leading role at midcentury.

In the wake of the Napoleonic Wars, domestic archaeology brought antiquities to the attention of a wider audience, and antiquarians hoped to use this material to narrate central Europe's distant past. To advance this new research agenda, antiquarians needed to compare finds from multiple sites and wider geographic areas. They therefore began to assemble larger inventories and collections. But even Leopold von Ledebur, who supervised the largest of these collections, in the KMVA, stressed that it was too early to read artifacts as the cultural products from a specific people. Those interested in domestic archaeology knew they were performing an initial phase of accumulation, and they looked forward to a future when they would be able to connect excavated material to peoples and cultures.

The interpretation of ancient artifacts was also framed by Romantic nationalist assumptions and the local nature of archaeological finds. A desire to view antiquities as generally German or Germanic was present

throughout the first half of the nineteenth century, but it was always combined with provincial concerns. Antiquarians viewed artifacts as part of
their local environment, and as the following chapter will show, their work
engendered feelings of belonging that were simultaneously grounded in
the local landscape, directed toward regional monarchies, and connected
to dreams about a national history.

CHAPTER THREE

Archaeology and the Creation of Historical Places

In 1824, Wolfgang Menzel, an influential literary critic during the Pre-March period, published his *Geschichte der Deutschen bis auf die neuesten Tage* (History of the Germans up to the most recent times). Writing during the surge of *vaterländische Altertumskunde*, the author noted that historians faced two difficulties when trying to explain Germany's ancient past. First, they needed to separate historical truth from fantasy (*das Wahre* from *das Wunderbare*). Menzel recognized that leading intellectuals, including Johann Gottfried von Herder, Friedrich Creuzer, Joseph Görres, and Friedrich Schlegel, had explored the ancient German past, but their primary concern was to preserve its poetic and mythic spirit. Historians, Menzel asserted, needed to place the ancient past on firmer historical ground. Second, Menzel explained that the use of "the Germans" as a collective noun was problematic. It was difficult to offer "a clear narrative about a people that is divided religiously and even more so politically, [a people that is] everywhere divided into small independent units that go in their own directions."[1]

Menzel used two conventions from *vaterländische Altertumskunde* to address these historiographic issues. First, he relied on Tacitus's *Germania* and other Roman sources as a baseline for his history. His book opened in the primeval forests of central Europe and noted the religious ceremonies that were supposedly held in these holy groves. He then reviewed Scandinavian sagas and medieval chronicles that addressed the origins of the Germans, as well as Germanic mythology and linguistics research suggesting that Europe's peoples had migrated from India and the Caucasus. Menzel covered this material relatively quickly—in about forty pages—because he thought everything that came before Roman sources belonged more to the realm of fantasy than truth.

When Menzel came to the encounters between barbarian peoples and the Roman Empire, he began to provide concrete dates. The first example of this came with his description of the Senones, who, under the leadership of Brennus, crossed the Alps, asked the Romans for peace, were betrayed by the Romans, and then proceeded to sack Rome and annihilate the city elders. Modern scholarship counts the Senones among the Celts of ancient Europe, but Menzel described them as Germans, and he placed the year 389 BCE in the margin next to his description of these events in his book. He then closed his discussion of this encounter by invoking the idea that a lack of unity among the Germans was the only thing that saved the Romans from a much greater defeat. "Eventually [the Romans] even gained the upper hand over the Germans because they combined their valor with unity, a clear plan for conquest, crafty deceit and betrayal. The Germans could have easily annihilated the Roman state already in its first blossoming age if they had been united and if they had thought at all about pursuing a clear plan [of attack]."[2] As Menzel covered the deeds of other tribes, he continued to include years in the margins. From this point forward, the narrative could rely on Roman sources, and in the eyes of the author, it left the realm of speculation and entered the realm of truth.

A second feature marked Menzel's coverage of late antiquity. The author's use of the collective noun for the Germans (*die Deutschen*) stopped abruptly around the beginning of the Common Era. He explained that the idea of "the Germans" came from Roman historical writing and that this collective term lost its meaning after the end of the Roman Empire. Therefore, Menzel argued, historians writing about the early Common Era should avoid the use of *die Deutschen* and *die Germanen* and refer to specific tribes (*Stämme*) instead. "One must think of the German people [in the early Common Era] as a constantly swarming beehive." New tribes rose in power while others fell. They fought and mixed together, making it impossible to identify a single people. The historian, Menzel explained, had to accept that the tribe was the primary actor in this period, not the nation. Only with the unity brought by Charlemagne's empire in the ninth century, Menzel claimed, could one begin to refer to the Germans again.[3]

This call for clarity clashed with Menzel's strong desire for a national history. Menzel was active in the nationalist fraternity and *Turner* (gymnastics) movements during the Wars of Liberation and wrote about a common German culture in several literary works. He was frustrated that historical writing about Germany "has portrayed the component parts and the small details but neglected the whole."[4] Menzel longed for an early history that focused on the tribes but still held broader meanings. Instead of transporting his 1820s desire for German unity into the distant past, he wished for a middle ground between the antiquarian who documented local customs and archaeological sites and authors like Leopold von Ranke who wrote about a transcendent spirit moving through the long lives of European states or institutions like the papacy.

Antiquarians had a lot to offer to someone like Menzel, who sought to close the gap between myth and history and to narrate the centuries between Tacitus's *Germania* and the Middle Ages. Amid the enthusiastic excavating, collecting, describing, and displaying discussed in the previous chapter, some antiquarians began to move beyond the inventory and sort through central Europe's complicated early history. Their work relied on the comparison of multiple excavation sites, the interpretation of rural place-names and legends, the study of classical sources, and communication with fellow antiquarians. These authors shared Menzel's idea that the tribe (*Stamm*) was the key actor in early history and his desire to connect the history of individual tribes to a national story. This way of investigating early history was so alluring that twenty years after publishing his *Geschichte der Deutschen*, Menzel took up the spade himself to excavate several grave mounds in Württemberg.[5]

This chapter presents three case studies that illustrate how antiquarians began to narrate the Age of Migrations. In the sixteenth century, Conrad Celtis had planned a one-to-one mapping of ancient tribes onto existing villages. Nineteenth-century antiquarians turned this dream into a concrete program when they connected the bones and items they excavated to individual tribes and specific events. Inventories and museum labels explained where artifacts were found (their *Fundorte*). Increasingly, though, antiquarians wanted to go beyond the recognition that nearby places contained artifacts toward an understanding of what ancient events had transpired in the fields and forests that surrounded them. This information transformed archaeological sites into historical places. The *Fundort* became not just the information associated with an object; it became a place where historical events had occurred. This knowledge deepened the connections people had to the places where they lived and encouraged them to contemplate the traces left by migrations, settlements, and battles from ancient times.

The authors who wrote about archaeological sites believed that this tribal history lived on in the language and customs of the rural population, and they sometimes employed loose language that described current inhabitants and themselves as the descendants of ancient tribes. In the cases that follow, for example, some Bavarians wrote about the Baiovari, and antiquarians in Württemberg kept finding traces of the Alemanni. An emphasis on tribes was also an ideal historical model because it accommodated the coexistence of regional loyalties and nationalism after 1815. The tribes had different histories and different customs, which explained the separate identity of peoples within central Europe, but they also could be imagined as part of a national body. This is how antiquarians read Tacitus's comments about the "the origin and customs of the Germani as a whole" and about "the extent to which the particular tribes differ in their practices and rituals" (88).

Historians of early medieval Europe rightfully criticize the nineteenth-century fixation on tribes. Antiquarians viewed the *Stämme* as branches of a Germanic family tree and populations with clear-cut identities who continuously inhabited the same territory. Today's understanding of late antiquity, however, treats ethnic identities among the barbarian peoples as the outcomes of complex historical processes. Instead of emphasizing genealogical connections between ancient and modern peoples, historians explain the migrations, settlements, trade, and conflict that produced a world of ever-changing political loyalties and identities.[6]

Antiquarians, though, developed their ideas about early history as a response to the political conditions in central Europe in the decades before German unification in 1871. The Napoleonic Wars and subsequent peace settlements brought a dizzying sequence of territorial transfers. Between 1807 and 1815, several states added smaller principalities, bishoprics, and other political units to their holdings. Larger territories, like the newly created Kingdoms of Bavaria and Württemberg, as well as the Kingdom of Prussia, faced the challenge of merging new lands with their core holdings. The study of the period of migrations suggested that borders were always shifting and that tribal bonds continued, despite these territorial reconfigurations. By drawing connections between specific Germanic tribes and contemporary regions, antiquarians could also suggest that certain political boundaries and royal houses had deep historical roots. In this way, their engagement contributed to the task of political integration.

Making Silesia Ancient

The Prussian province of Silesia was made up of the land on both sides of the Oder River between the mountains of Moravia and the Oder's junction with the Neisse. Today, the majority of Silesia lies in southern Poland, with small parts of it on the northern border of the Czech Republic and the eastern border of Germany. This central European borderland has been settled, conquered, and transferred many times. Various peoples migrated through Silesia in late antiquity. In the second half of the tenth century, Mieszko I established a Polish kingdom that included these lands. The Duchy of Silesia, created in 1138, was part of the area colonized by German speakers in the twelfth and thirteenth centuries. During the late Middle Ages, this territory was separated from other Polish lands and came under the rule of the Kingdom of Bohemia. The Bohemian crown went to the Habsburgs in 1438, and Silesia remained under Austrian control until the 1740s, when the territory was lost to the Kingdom of Prussia.

During the eighteenth century, the powers that ruled Silesia certainly did not stress the German identity of this area. The Habsburg policy was to de-escalate any cultural activity that might ignite religious divisions. The monarchy monitored intellectual life and refused to support a university that might prove politically and religiously subversive. It made some efforts to reconvert areas to Catholicism that had become Protestant during the Reformation, but Breslau, the regional metropolis, had only a small school for Catholic theology. In 1742, when Silesia became the southeastern arm of the Kingdom of Prussia, the land contained pockets of German speakers but was mostly either thinly populated or inhabited by Poles. Frederick the Great sent more than sixty thousand people to the province to work in agriculture, mining, and a developing cottage industry in textiles.[7] Yet even after the arrival of these colonists, there were few efforts at fostering a sense of Prussian or German identity. Carl Georg Heinrich von Hoym, the conservative provincial minister in Silesia from 1770 to 1806, did not trust voluntary associations and maintained close political ties with Silesia's nobility. The Old Regime—whether Austrian or Prussian—expected obedience based on absolutism and suppressed discussions that might awaken religious or national divisions.

After the Wars of Liberation, however, the groundswell of historical activity that occurred throughout central Europe came to Silesia. The Prussian administration commissioned Johann Büsching to create an inventory of Silesia's cultural treasures, and in 1817, the royal archivist and curator became a professor for antiquities studies at the recently created Friedrich Wilhelm University in Breslau. Büsching was a founder of the Verein für schlesische Geschichte, Kunst und Alterthümer (Association for Silesian history, art, and antiquities) and active in the Schlesische Gesellschaft für vaterländische Kultur (Silesian society for regional culture). As a promoter of *vaterländische Altertumskunde*, he carried out archaeological excavations; coedited several volumes of German poetry, folk songs, and legends; and organized public lectures and art exhibits.[8] As editor of the journal produced by the Schlesische Gesellschaft für vaterländische Kultur, he informed corresponding members who lived throughout German-speaking Europe, Scandinavia, the Netherlands, and Bohemia about his activities.

Historian Friedrich Kruse enthusiastically supported Büsching's cultural work. Born in Oldenburg in 1790, Kruse had studied law, theology, and history in Leipzig and received his doctorate in 1813. He then moved to Silesia and taught at finishing schools in Liegnitz and Breslau. After becoming friends with Büsching, Kruse headed a separate section in the Schlesische Gesellschaft für vaterländische Kultur that involved around thirty gentlemen and was primarily dedicated to archaeology. With this group, he set out to "investigate the ancient conditions in Silesia and Germany."[9]

In the early nineteenth century, Büsching and Kruse not only made Silesia an important center for antiquarian studies but also established the idea that Silesia's German cultural identity stretched back to the early Common Era. Before the nineteenth century, Silesia was ruled as a province in an empire or kingdom, and few connections were drawn between the rulers and the ruled. With the rise of *vaterländische Altertumskunde* after the Napoleonic Wars, however, Büsching and Kruse cast the land as an ancient territory. They combined texts and archaeological finds in order to reconstruct the region's distant past, relate its inhabitants to the Germanic tribes, and present contemporary towns as the outgrowth of ancient settlements. This was a significant turning point. Nationalist figures would exploit this "German" connection after World War I, when part of the province was transferred to Poland. It is therefore important to return to the early interpretations of the area.

Büsching began excavating and acquiring artifacts shortly after his arrival in Breslau, and within six years, he had amassed a collection of over 2,100 items from Silesia and 300 "foreign" (non-Silesian) artifacts. Using these items, he published *Abriß der deutschen Alterthums-Kunde* (Outline of German antiquities studies) in 1824, which became a key reference text for historical associations throughout central Europe during the nineteenth century. The idea of accretion was central to the method described in the *Abriß*. Büsching explained how he hoped to generate new knowledge about the distant past by comparing a wide range of finds to see if general patterns emerged. He wanted to determine whether similar items were concentrated in small geographical areas and whether all items from the same site were made from the same material. He wondered, for example, whether all the implements from a given area were made of stone, copper, or iron. Büsching was sure that as more individuals participated in similar projects of comparison and compilation, they would eventually gain "a clearer and more decisive view into the study of German antiquities."[10]

Geography also played a critical role in the *Abriß*. Büsching wanted to link archaeological sites to the locations that Tacitus had assigned to specific Germanic peoples. He planned to plot his finds on a map of the contemporary district borders within the province of Silesia. The use of modern place-names, Büsching reasoned, would increase the accuracy of reports he received from other antiquarians. But it would also transform modern sites into historical places.

A healthy measure of skepticism and uncertainty accompanied Büsching's excitement about antiquarian research. He understood the complexities created by multiple layers at a single archaeological site, and he explained that it was extremely difficult to distinguish remains left by "earlier Germans" from those of "later Slavs." He also admitted that it was possible that antiquarians would never be able to distinguish between the individual Germanic tribes. Finally, Büsching realized that most archaeological evidence was still undiscovered, and he knew that his maps and his conclusions would be revised by later researchers.[11] As the *Abriß* makes clear, Büsching viewed antiquarian studies as a speculative venture that would be improved with additional finds and interpretations.

Over the course of the late 1810s and early 1820s, Friedrich Kruse applied Büsching's method to specific places in Silesia. One result was *Budorgis*, a book about "ancient Silesia before the coming of Christianity, especially during Roman times." "Budorgis" was a place-name that Ptolemy, the second-century Alexandrian geographer and astronomer, had included on his map of northern Europe. Kruse hoped to identify the modern location of Budorgis and other ancient places. The key to this endeavor was the combination of Tacitus's descriptions, Ptolemy's maps, and the interpretation of Silesian antiquities. Ancient graves in Silesia, Kruse explained, "can indicate the places that were once occupied better than the most precise geographer or historian and disclose power relationships, levels of wealth and of education. Ancient Germany can arise from them again, if more attention is paid to them than has been the case up to now and if more accurate comparisons are drawn between them and the accounts of ancient authors."[12] The main section of *Budorgis* presented an inventory of 126 archaeological sites in thirteen Silesian districts. It included evidence of Roman travel routes, finds of ancient coins, and burial grounds that contained thousands of urns. A typical entry reported all the details of the excavation that were known, including the date of discovery, a description of the surrounding landscape, measurements of mounds, the placement of items, and the names and occupations of those involved with the investigation of the site.[13] The most interesting feature of the text, however, was Kruse's attempt to move beyond a standard inventory and connect artifacts and ancient events to present-day places.

Kruse believed that he had identified seven places named by Ptolemy, including the ancient city that gave his book its title. In a long section devoted to Budorgis, Kruse considered possible locations for the ancient city and the evidence that supported each claim. Several kinds of sources suggested that the modern town of Laskowitz (east of Breslau) stood on the ancient site of Budorgis. Antiquarians had uncovered a considerable number of Roman silver coins nearby, and earlier authors referred to a place in this area called Budorgis. A local legend told of an ancient city that had been destroyed during times of war and sunk below the surface of the earth, only to live on in the memories of local inhabitants. One man recalled that "our fathers and grandfathers have told us about this since we were young." Another enthusiast for Silesian history spoke of a stone path outside Laskowitz that led in the direction of Öls. Kruse followed up on these stories and found not only the stone path but also "more than a hundred small rises in the ground, each capped with large stones," and a wall of rubble in the forest that extended back toward Laskowitz. As Kruse continued his journey to Öls, he noticed no rocks in the nearby fields, which made the presence of so many stones at the fabled place even more remarkable. All this seemed to indicate a former settlement and ancient traffic throughout the area.[14]

While Kruse and natives agreed that the area around Laskowitz was historically significant, not everyone was convinced this was the location of Budorgis. Some believed the city of Breslau itself to be the correct site. Others argued for a sandy area half a mile southeast of Laskowitz where excavations had also taken place. Kruse used old church records and medieval documents from the Laskowitz archive to conclude that the latter place was not ancient. He became convinced that his initial hypothesis about the forested area between Laskowitz and Öls was correct: "The legend of the ancient city, the number of smaller and larger stones, which is about the only thing in our area that could be the trade routes between ancient cities, the presence of Roman coins, and the conformity with the Ptolemaic map that indicates distances to five other places have convinced me to place ancient Budorgis here."[15]

Beyond the cartographic question of Budorgis, Kruse wondered about the ancient peoples who had lived in this area. Using Ptolemy's map, Tacitus's commentary, and archaeological finds, he concluded that a Germanic tribe called the Lugii occupied much of Silesia, including the area around Breslau. This connection was very tentative: Kruse did not identify individual items as the remnants of Lugii cultural production, and he suggested that the first settlement around Breslau was a Slavic fishing village, not a Germanic community. The passages that made these ethnic identifications were brief and provisional, suggesting that Kruse was interested in these questions, but that he understood that he was not on firm ground.[16]

Essential to the creation of historical places were attempts to connect prehistory to later epochs and ancient sites to present-day places. Kruse did this as he wondered whether modern towns had grown out of ancient settlements. He organized his list of excavation sites according to cities and towns, like Breslau and Liegnitz, and to the contemporary boundaries of princely holdings. For Kruse, these were more than geographical locations. He hypothesized that finds were concentrated around modern towns because these places had been inhabited since the time when Ptolemy drew up his map. In this way, his interpretation connected the ancient past to living communities. He then tried to fill in this long line of continuous inhabitation by consulting church documents and medieval city charters. He declared he would not have satisfied the wishes of his readers until his work was so thorough that they would be able to follow "a continuous line that connected his Silesian antiquities to the history of the Middle Ages."[17]

The efforts by Büsching and Kruse promoted the idea that Silesia had a distant past. Their publications announced the wide range of finds throughout the province, and many individuals, excited about these discoveries, began to report more of them. A school rector in Löwenberg, for example, wrote to Büsching in the fall of 1818 about an urn washed up on the shore of the Oder River that was currently in the possession of a doctor in Glogau. In that same year, Kruse traveled extensively throughout Silesia to follow up on reports in local publications. One lead from the *Schlesische Provinzialblätter* concerned thirty mounds discovered during the construction of a road and located about three-quarters of a mile southeast of Grünberg. Once word spread about Kruse's investigation into Budorgis, locals wrote to him about artifacts that came to light after heavy rains, when a blacksmith dug for coal, when farmers cleared fields, and when workers moved sand for a dam on the Oder River.[18] Furthermore, over four hundred people in Silesia subscribed to the installments of *Budorgis*. On the basis of this communication and support, Kruse reported that *vaterländische Altertumskunde* was alive and well in Silesia.[19] Kruse's publications indicated that Silesia had an ancient past, and many people wanted to participate in its recovery.

Kruse left Silesia in 1821 to assume a professorship for ancient and medieval history at the University of Halle (Saxony). While there, he was very involved in the Thüringisch-Sächsischer Verein, and he edited several publications associated with that society's historical and archaeological activities. Büsching continued as a professor for antiquities in Breslau until his health deteriorated in the later 1820s and he passed away in 1829. In the following decades, the version of Silesia's ancient past promoted by Kruse and Büsching found its way into respectable reference works, as *Brockhaus* and *Meyers Konversations-Lexikon* affirmed for their middle-class readers that the region's history began with the occupation of the land by the Lugii in the early Common Era.[20]

Silesia's ancient past, however, was a politically loaded topic. The rise of nationalist movements throughout central and eastern Europe inspired Polish protests in Prussia and the Austrian Empire in 1848 and in the Russian Empire in 1863. Polish nationalists pointed to the reign of Mieszko I as the region's founding date and argued that this early possession of the land justified Polish demands for political autonomy. During the 1870s, archaeologists began to document specific types of Bronze Age pottery and gravesites across Prussia's eastern territories. The anthropologist Rudolf Virchow identified these sites as the Lusatian culture, but he refrained from connecting such ancient artifacts to a people named in classical sources. (And contemporary archaeologists uphold this name and the

anonymous nature of this old culture.) In the ensuing decades, however, archaeologists of different national orientations interpreted these
finds as the legacy of proto-Slavic, Germanic, or Illyrian people. These
debates occurred against the backdrop of intensified Germanization policies that advanced Germans over Poles in Silesia and other eastern provinces. Educational opportunities were geared toward German speakers,
and Polish-speaking Catholic priests were replaced with men loyal to the
German state. Germany also struck at the Polish landowning elite with a
program that transferred their estates to ethnic Germans.[21] In this context, questions about the settlement of the land became highly politicized,
and nationalist figures from Prussia's eastern provinces turned to pre- and
early history for their arguments about the possession and development
of the land.

By the 1890s, nationalist pressure groups like the Deutscher Ostmar-
kenverein (German eastern marches society) and the Alldeutscher Verband (Pan-German league) waged a propaganda campaign about the history of the eastern provinces. They emphasized the presence of Germanic
settlements before the arrival of the Slavs and described the "recoloniza-
tion" of the east in the Middle Ages by medieval knights as a return to
original ways of life (i.e., "German" conditions in the early Common Era).
Rhetoric became even more vociferous after World War I, when Germany
lost much of its eastern territory. Archaeologists tried to substantiate their
irredentist claims by returning to the interpretation of ancient finds. In
Berlin, Gustaf Kossinna claimed that Bronze Age settlements more than
eight hundred years older than the Lugii migrations were still Germanic.
And Józef Kostrzewski, one of Kossinna's doctoral students and a professor of prehistory at the University of Poznań (Posen), described Bronze
Age sites as proto-Slavic. Both interpretations claimed the autochthonous
nature of German or Slavic peoples.[22]

These disputes over Silesia's ancient past cast a shadow over interpretations from the 1820s. Büsching and Kruse offered an ethnic interpretation of this area's complicated past, and they viewed their work as part
of a larger national project after the Napoleonic Wars. During the Third
Reich, archaeologist Ernst Wahle would claim the long-dead Büsching
as a nationalist who fought for German prehistory at a time when others
ridiculed the field.[23] The tone in the works by Büsching and Kruse was
fundamentally different from that expressed in later ethnic views on Germany's eastern provinces, though. Büsching and Kruse paid particular attention to what they considered to be German antiquities because of the

national enthusiasm in the years after the Wars of Liberation, but they recognized the provisional nature of their interpretations, and they acknowledged Slavic settlements in their area. Later authors would leap from Büsching's and Kruse's measured speculations to unyielding conclusions about Silesia's status as an ancient German territory. The discovery of Silesia's ancient past in the early nineteenth century did not cause these later assertions, but in the changing political context, arguments about settlements and territorial possession became central to expansionist views in the late nineteenth century and to irredentist claims after World War I.

Bavaria's Earliest Borders

The "bone field" of Fridolfing offers a remarkable story of discovery, excavation, and interpretation that linked nineteenth-century Bavarians to ancient peoples. In 1818, workers digging a trench outside the Upper Bavarian town of Fridolfing (six miles north of Salzburg) turned up ancient human remains. Others had found bones and artifacts near this small village before, and the unearthing in 1818 generated little attention in the community. In 1832, however, a two-page notice in a publication for state statistics announced the "discovery" of a cemetery that contained "the final resting place of several thousand war dead." Joseph von Koch-Sternfeld, a Munich alderman, philologist, and member of the Bavarian Academy of Sciences, named the site Fridolfing's "bone field."[24] Two years later, an enthusiastic antiquarian named Christoph Sedlmaier arrived in nearby Tittmoning (spelled Titmaning in several of the nineteenth-century sources) as the head controller for the border between Bavaria and Austria. Sedlmaier took a particular interest in Fridolfing and eagerly uncovered swords, jewelry, and over two hundred skeletons. In September 1835, he offered these items to the Bavarian king and suggested that the Bavarian Cultural Ministry establish a local association that would investigate and preserve these historical artifacts.[25]

In the meantime, Koch-Sternfeld had published a longer academic study that presented Fridolfing's "bone field" as the result of a bloody territorial fight between "ancient Bavarians" and an alliance of Ostrogoths and Romans. According to Koch-Sternfeld, the struggle took place during the sixth century CE.[26] His study launched a public debate that lasted over fifteen years. Initial arguments appeared in regional newspapers and the *Oberbayerisches Archiv*. But the bone field of Fridolfing also figured in a wider discussion about the presence of Celts in central Europe that engaged well-known authors beyond Upper Bavaria, including Heinrich Schreiber in Freiburg and Ludwig Lindenschmit in Mainz.

The first rebuttal of Koch-Sternfeld's hypothesis in the *Oberbayerisches Archiv* came in 1845 from Matthias Koch, an official from Vienna. Koch agreed that the cemetery resulted from a battle, but he asserted that the finds were much older than the sixth century CE. Koch saw the objects as irrefutably Celtic, perhaps even from the climax of Celtic decorative art. In his eyes, the battle at Fridolfing involved a highly civilized and permanently settled Celtic community that had defended its salt mines from an invading Germanic tribe, probably in the early fourth century CE.[27]

In 1850, Georg Wiesend, royal clerk for the district court in Tittmoning, presented a third version of the area's history. His essay compared the Fridolfing items to those found at other sites in the region. Wiesend, like Koch-Sternfeld, believed in the "Bavarian" origin of the cemetery, but he found no evidence of a battle. He interpreted the weapons as burial gifts and the high number of corpses not as the fallen in a violent fight but rather as the result of the site's long use as a community cemetery. As far as Wiesend could tell, the site was part of the history of the great migrations that had occurred between the end of the sixth and beginning of the eighth century CE. If his hypothesis about the origins of this graveyard proved true, then "the bodies of a warring tribe of Baiovari peasants who respected the Christian religion rest in the described graves."[28]

In the field of Bavarian prehistory, it is unclear whether any of these amateur archaeologists correctly identified the skeletons as Germanic, Celtic, or Baiovari or as being from the fourth, sixth, or eighth century CE. Most of the remains from the cemetery were destroyed soon after excavation. Sedlmaier turned over several metal items (which at the time were the only pieces valued as archaeological objects) to the royal Bavarian Antiquarium. Wiesend owned several items that were dug up later, but his collection of antiquities was lost after his death.[29] Yet the public discussion generated by the Fridolfing site displayed the significance of domestic archaeology in the mid-nineteenth century. Many people wondered what this ancient cemetery could say about the early settlement of the Salzach River valley.

One reason for the liveliness of the debate over Fridolfing was that the story had a ready audience in the 1830s and 1840s. The Fridolfing site fell within the domain of the Historischer Verein von und für Oberbayern,

the largest of the Bavarian historical associations established during the
1830s. This group was based in Munich, and Sedlmaier, Koch-Sternfeld,
and Wiesend numbered among its membership of over two hundred men.
In addition to paying for the excavations (along with a subsidy from King
Ludwig I), the association heard several presentations by Christoph Sedlmaier and Matthias Koch about Fridolfing and other excavations in the
region.

Civil servants were vital to the publicity that the Fridolfing site received. Christoph Sedlmaier had arrived in the area in 1834 as a border
patrol official for the Kingdom of Bavaria. His administrative itinerary
included daily walks along the Bavarian-Austrian border and frequent
travel to the royal capital. Sedlmaier alerted King Ludwig to the Fridolfing finds, and the king was sympathetic toward this eager antiquarian
whose funding appeals emphasized his Bavarian patriotism. Their correspondence provides many colorful details, including Sedlmaier's complaints about his rheumatism and his apologies for his tardy archaeological reports.[30] Georg Wiesend was also a state employee, coming to
Tittmoning as a district court clerk in 1838 and moving to Reichenhall
as a district judge in 1849. Mobile civil servants like Sedlmaier and Wiesend walked the state's roads, visited various towns, and met with their
inhabitants. They were in an ideal position to hear about archaeological
finds and to communicate this information to the state. Their archaeological studies were not part of their official work, but they were related to
the state's desire to foster a sense of historical belonging. Without their
interest, the discovery of the bone field may have remained a minor local
event. Instead, it precipitated a learned debate about Bavaria's borders
and the area's early inhabitants.

The stakes in the Fridolfing debate were raised further by political
changes wrought by the Napoleonic Wars. Fridolfing had belonged to the
Holy Principality of Salzburg. Following secularization in 1803, an Austrian regiment occupied the area, and the Peace of Pressburg (1805) made
Salzburg a province of the Austrian Empire. Napoleon's victory over Austria at Wagram in 1809 inaugurated a seventeen-month period of French
rule in Salzburg, followed by the transfer of the territory to the Kingdom
of Bavaria. In 1816, Austria got Salzburg back, but without the Inn- and
Hausruckviertel, a piece of Tyrol, and four other administrative districts.
One of these districts was Tittmoning, where Fridolfing was located.[31] This
stretch of the Salzach valley, severed from its traditional neighbors, remained in the Kingdom of Bavaria.

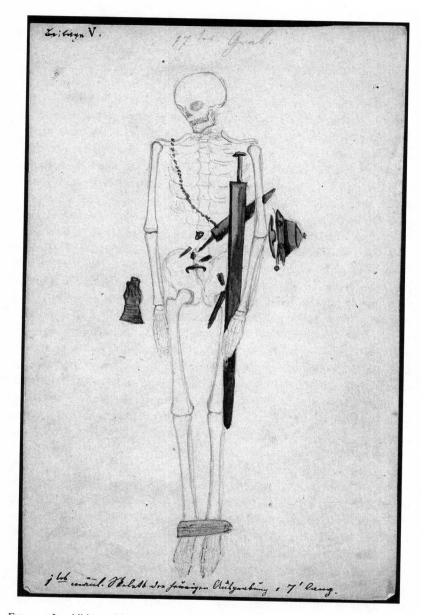

FIGURE 2. In addition to his interest in Fridolfing, Christoph Sedlmaier excavated a series of row graves near Nordendorf during 1854 and 1855 and helped establish these kinds of sites as belonging to the Migration Age. His detailed excavation diary included sketches of several graves and a map of the terrain. (Illustration from Sedlmaier, "Drittes Tagebuch," October 1, 1855, 9. Photo courtesy of the Archäologische Staatssammlung, Munich.)

After 1815, the Bavarian administration attempted to integrate new lands by redrawing its regional borders and naming each region after a local river. In this arrangement, newly acquired Fridolfing was combined with long-standing holdings around Laufen and placed in the Isar district (named for the Isar River). In 1837, however, the state attempted to foster a historical identity that suggested connections to the early medieval past. King Ludwig replaced the names of rivers with names that harked back to the Germanic tribes. The Isar district around Munich became Oberbayern (relating to the Baiovari tribe). Two districts to the north took Frankish names: the area around Bayreuth changed from Obermain (Main River) to Oberfranken, and the Untermain district around Würzburg became Unterfranken. The Oberdonaukreis, which included Augsburg (and was named after the Danube River), became Schwaben.[32] The Bavarian historical associations immediately adopted these names also. The excavations at Fridolfing occurred during and just after these changes, and they must have lent legitimacy to the administrative discussions about the tribal names that connected present-day inhabitants of the Bavarian monarchy to the Baiovari, Franks, and Swabians.

In addition to its significance for historical identity in Bavaria, Fridolfing became part of a passionate debate about the presence (or even existence) of the Celts in central Europe. During the 1830s and 1840s, an antiquarian named Heinrich Schreiber published a widely read handbook about archaeological finds in southern Germany. Like many in the nineteenth century, he believed that the Celts were the first inhabitants of central Europe, and he attributed all objects made of bronze to the Celts and those made of stone or iron to Germanic peoples.[33] On the one hand, Schreiber correctly understood that the Celts came before the Age of Migrations. On the other hand, his ideas thoroughly jumbled the chronology advanced by the Three Age System, which correctly identified Stone Age items as much older than those made of bronze or iron. Schreiber's ideas produced an oversimplified ethnic shorthand that used the presence of bronze to identify finds as Celtic. For him, material did not suggest the relative age of an object but rather the people who produced it. This idea sparked a wave of what critics called *Keltomanie* (Celtomania)—a rush to identify all sites with bronze items as Celtic and the assumption that Celts were the oldest ancestors of central European peoples.[34]

Schreiber's ideas clearly influenced Matthias Koch's interpretation of the Fridolfing finds. Writing in the *Oberbayerisches Archiv* in 1845, Koch described the graves as the result of a struggle over the rich salt deposits that stretched from Saxony to Austria. According to his interpretation, the bones and burial gifts were evidence of the Celts, whom he considered to be the first miners in the area. He wrote that

> when we look to the Dürrenberg near the town of Hallein, where great salt deposits lie on the border between Bavaria and Austria, we see that a grave with bronze items has been discovered there. Around twenty years ago, a similar grave with Celtic items was found at a site on a mountain near Hallstatt. Finally, not far from Weißenfels on the Prussian-Saxon border (and also near a salt mine) three more graves were found . . . and they *perfectly match* those of the Fridolfing battlefield. . . . All of these . . . are Celtic graves that contain buried Celts, and the bronze items are Celtic and in no way of Germanic origin.[35]

The finality of Koch's conclusion that "the bronze items are Celtic and in no way of Germanic origin" relied on Schreiber's guidebook, which Koch used not only to assert the Celtic nature of the finds but also to connect Fridolfing to many other sites between Saxony and Austria.

A vigorous rebuttal of the Celtic hypothesis came from Mainz. Wilhelm Lindenschmit, the older brother of the RGZM founder Ludwig, rejected Schreiber's general enthusiasm for the Celts in *Die Räthsel der Vorwelt, oder: Sind die Deutschen eingewandert?* (The riddles of prehistory; or, Did the Germans immigrate?) in 1846. He took issue with the way Schreiber labeled territories with ethnic terms. Lindenschmit fulminated against

> the unfortunate *circulus vitiosus*, or circular reasoning, whereby even an immovable object is to be nationalized [*nationalisirt*: i.e., be seen as Celtic] because it is on ground that is claimed to be Celtic; however, the ground only became Celtic because of the presence of the immovable object. These are entirely unallowable artifices. They bedazzle indeed, but they do not stand up to scrutiny. And yet one insistently proceeds in this manner with regard to hill forts and grave mounds. All hill forts should be baptized Celtic because the supposedly nomadic Germanic peoples did not need redoubts. [The notion] that the Germans of old lived by breeding livestock and were not fond of fortifications is believed by Herr S. [Schreiber] to be a special characteristic of their nomadism; and because they *were* nomads after all (!), so must all wall-like monuments be viewed as the fortifications of an earlier agricultural people and bear witness to the earlier Celtic nature of our land.[36]

The Lindenschmit brothers brought this critique to the interpretation of Fridolfing. In the early 1840s, they had begun excavating a graveyard in

Selzen at the behest of the Mainzer Altertumsverein (Mainz society for antiquities). The site revealed a rich collection of weapons, jewelry, and ceramics. On the basis of their knowledge of ancient coins and other items, the Lindenschmits concluded that the graves belonged to the Franks and the period of early medieval migrations. They then used this occasion to criticize Matthias Koch's interpretation of Fridolfing and his earlier articles in the *Allgemeine Zeitung* that described a similar site in Norendorf as Celtic. The Lindenschmits' prose became heated and sarcastic, especially as they described the discussion of Nordendorf. There, Koch had asserted the discovery of "the Celtic nationality"; Friedrich von Thiersch, a professor of literature and education reformer in Munich, claimed the finds to be Burgundian; and a third voice identified the site as the remnants of a Roman settlement. The Lindenschmits observed that "in order to reconcile these contradictions and satisfy all parties, the Swabian historical association in Neuburg decided that . . . a portion of the corpses there could be viewed as Roman, another portion as . . . indigenous (?) Celtic inhabitants, and a third portion . . . as Alemanni conquerors. It appears that if a Slavic scholar had taken part in the discussion, room would have been found for Slavic guests among these tolerant dead."[37] The authors hammered away at Koch's credibility, noting that Koch was confident that these graves were Celtic even though he had recently asserted that similar antiquities in Salzburg were Phoenician. Such a reversal was a worrying sign for research, the Lindenschmits wrote, for it was the product of "impatience and biased personal views, two things that cannot be incorporated into true science."[38]

The Lindenschmits raised a methodological issue—radical shifts in interpretation undermined the legitimacy of domestic archaeology—but this concern also had a political side. The Lindenschmits were ardent supporters of German cultural nationalism, and their position in the 1840s displayed this political conviction. They noted that current ideas about attributing graves to different nations on the basis of their external appearance led to conclusions about Celtic grave mounds and Slavic fortresses but said very little about the Germanic tribes. The tendency to claim "in the east of the fatherland" as Slavic what those in the west identified as German "disinherited our forefathers."[39] The Lindenschmits hoped to rectify this situation and proclaimed on the cover of their study that "graves with iron weapons are from the period of the tribal migrations" (and therefore Germanic).

Georg Wiesend's analysis of Fridolfing from 1850 fell in line with the Lindenschmits' ideas but expressed a particular excitement about the way the graves could be used to narrate history. Wiesend explained that "our graves stand on a transition line [*Übergangslinie*]," and he meant this in several ways.[40] First, he believed that the "warriors" had fallen in battle, proving that the region was a disputed geographical crossroads. Second, Wiesend was convinced that the Fridolfing objects originated from the time when the area was first settled by a Germanic tribe and that the graves contained evidence of this new occupation. He based this conclusion on Jacob Grimm's *Deutsche Mythologie* (German mythology, 1835) and Gustav Klemm's *Handbuch der germanischen Alterthumskunde* (Handbook for the study of Germanic antiquities, 1836), both of which discussed religious artifacts found in northern areas that were deemed undisputedly "Germanic." If Wiesend could show that the Fridolfing objects were similar to the items from these areas, he would prove the Germanic origin of the graveyard.

Finally, Wiesend saw Fridolfing as a site of religious conversion. He described the cemetery as "Germanic" even though it did not contain what most believed to be the telltale sign of Germanic graves: there were no traces of bones in buried urns, which existed throughout the northern and western German states. Wiesend nevertheless remained convinced of the Germanic nature of the cemetery, claiming that the ancient inhabitants of Fridolfing had experienced the spread of Christianity and that their cemetery therefore contained a combination of "heathen" and "Christian" burial practices. According to Wiesend, the community had avoided the use of urns because they had converted to the new religion, but they still built grave mounds and included weapons, jewelry, and household items as burial gifts. As other sites in the area showed, these practices were gradually replaced with the Christian practice of erecting gravestones. The Fridolfing bone field, Wiesend argued, provided material evidence of an aesthetic transition from the decoration of the internal (or underground) grave to the external Christian headstone.[41]

A connection between ancient events and the present day was implicit in Wiesend's discussion. He suggested that the contemporary borders of his study, based on the recently established Bavarian administrative districts, yielded evidence of the "Bajuwaren" (Baiovari). He then switched back and forth between the ancient Bajuwaren and contemporary Bavarians, revealing his assumption that the Bajuwaren were the ancestors of the current inhabitants of the core lands of the Kingdom of Bavaria. Similarly, he went back and forth between *die Germanen* and *die Deutschen* as he compared northern finds to his southern evidence, as if these two terms were interchangeable. This easy slide from ancient to modern peoples indicated that Wiesend was not only writing about transitions in prehistory. At Fridolfing, Wiesend felt he had found the area's beginning as a German, Bavarian, and Christian land, a transition that ushered in the conditions that still existed in the nineteenth century.

Matthias Koch blamed the lack of consensus on "national preference,"[42] but even his biography contains clues that may help explain his interest in the widespread Celts. Koch made several presentations to the Historischer Verein von und für Oberbayern. As his obituary explained, the group enjoyed his energetic spirit and German-national sentiments, but they also found his "hyper-Celticism" amusing. Perhaps his insistence on bringing together material from the triangle of salt mines that crossed the borders of Prussia, Saxony, Austria, and Bavaria was related to his political concerns in the 1840s. As the Italian national movement gained momentum in the 1840s, there were calls for the separation of South Tyrol from Austria. Koch, an official from Vienna, staunchly opposed this idea, and he even traveled to Frankfurt in 1848 to urge the representatives of the National Assembly to ensure the inclusion of South Tyrol in German affairs.[43] His interpretation of the Fridolfing site served as one way to suggest a common past for all German-speaking territories. Although the Lindenschmits and others viewed this Celtic thesis as very un-German, Koch saw it as a prehistoric version of a *großdeutsch* idea that revealed ancient connections among all the German states, including Austria.

By 1850, Fridolfing's bone field had become a historical place. Its investigators disputed what had happened there and who had lived there, but they hoped to clarify the early history of this area. They believed that their excavations, combined with the careful study of ancient written sources and material evidence from other archaeological sites, would allow them to address the ethnographic questions that Leopold von Ledebur had posed but left unanswered in the late 1830s: "To which people, to which land, do these objects belong?" The debate over Fridolfing also revealed how political concerns shaped the answers to these questions, as authors attempted to ground contemporary political, religious, and territorial realities in the distant past.

Alemannic Graveyards in Modern Württemberg

If excavations near Fridolfing and Nordendorf caused intense disagreements, archaeology in the Kingdom of Württemberg was producing a consensus during the middle decades of the nineteenth century. Since the late eighteenth century, scholars working with Roman and medieval sources had documented the expansion of the Alemanni between the third and sixth centuries throughout the area that today includes southwestern Germany, northern Switzerland, and the Alsace region of France. In the mid-nineteenth century, antiquarians added physical evidence to this historical literature. Not only did they interpret the graves they excavated as Alemannic, but they also related this ancient tribe to the customs and people of their day. This connection is still alive, as contemporary travel guides and visitor bureaus continue to describe the Alemannic origins of the Black Forest's unique culture and annual Carnival processions in Württemberg.

In 1846, the Württembergischer Alterthums-Verein commissioned Captain Ferdinand von Dürrich and Dr. Wolfgang Menzel (author of the *Geschichte der Deutschen*) to excavate a series of graves located in a field outside the village of Oberflacht, about seventy-five miles south of Stuttgart. The association, founded in 1843, was located in Stuttgart and received regular subsidies from the royal house of Württemberg. As with the Fridolfing site, there had been earlier notices of artifacts found in this area. A brick maker found items here in the 1810s when collecting clay, and a description of the area from 1823 mentioned some finds. But no one acted on these notices until Dürrich came to the area in 1844 to carry out a topographical survey for the kingdom. In the fall of 1846, he returned with Menzel, and they spent the next four weeks supervising the excavation of forty graves located on an elevation that "since time immemorial" locals had referred to as the Kreuzbühel. Their work received the support of local mayors and clergymen, and many spectators came each day to follow their progress. The association spent 343 florins on the excavations (most of this amount probably went toward wages for day laborers) and another 250 florins on tools and the preparation of the antiquities for display.[44]

Dürrich and Menzel called special attention to the wooden pieces found at Oberflacht. "It will not be lost on specialists in antiquities research that we are dealing with a completely new kind of find. The graves that we have opened distinguish themselves from all other heathen graves that have been found on German soil because of the magnificent wood found there."[45] The authors were most impressed by coffins made of carved oak that locals referred to as *Todtenbäume* (trees for the dead). The *Todtenbäume* in male graves were nine feet long; the ones for women

and children were shorter. Most of them contained a skeleton and several
wooden plates and bowls. Most of the skulls and longer arm and leg bones
had deteriorated and were now "soft as leather." Dürrich and Menzel
could salvage only four skeletons from the forty graves. The wooden items
were immediately taken to the Polytechnic School in Stuttgart, where
they were treated with an acid compound and dried so that they would
not crumble or shrink.[46]

  After describing the surrounding landscape and detailing the contents
of the graves, Dürrich and Menzel turned to their main questions: "To
which time period and to which people do these graves belong?"[47] They
noted that no medieval chronicle mentioned a churchyard and no Roman
sources referred to a cemetery here. The excavators reasoned then that
the graves "probably belong to the dark period soon after the migrations,
a time not covered by late Roman geography or by later Christian history."[48] Dürrich and Menzel would therefore have to rely on archaeological sources to shed light on this "dark period" in the area's history.
They confirmed that the site could not be Roman because it did not contain Roman coins, weapons, or clay vessels. It could not be Celtic because
Celtic sites present "roughly worked items of bronze and no iron." Nothing in these graves matched that kind of find except for a few brooches
and glass pieces that looked somewhat like the decorative items from "the
Celtic-Roman graves in places like Nordendorf."[49] Additionally, the authors argued that the site must be later than the Celtic presence in the
area because of the composition of the metal items. The Oberflacht items
were an alloy of copper, tin, lead, and zinc, whereas Celtic metal was a
blend of only copper and tin. And the buried certainly were not Chris-
tians, because snakes, a heathen symbol, were carved into the wooden coffins and the burial gifts included shoes, which Jacob Grimm had described
in his *Deutsche Mythologie* as a heathen practice. All this led the investigators to conclude that the graves contained heathen Alemanni from
between the fourth and eighth centuries CE. Dürrich and Menzel then
asserted a clear link between the graves and the present-day population
of the area. They highlighted cultural practices that they inferred from
the grave contents and that were still (*noch jetzt*) common in the area.
They noted that local inhabitants still referred to a coffin as a *Todten-
baum*; that candleholders (*Lichtstöcke*) still had the same form as the ones
found in the graves; that "still to this day" the people of the Black Forest use wooden plates and are known as skilled woodworkers; and that
the people of the Baar (the foothills region between the Black Forest

and the Swabian Jura) frequently bury their dead in their clothing. With these similarities in mind, which had supposedly been maintained for up to 1,500 years, "it appears that one may regard the buried dead at Kreuz-bühel as the direct ancestors of the current inhabitants. That is, they were Alemanni."[50]

Other investigations into Württemberg's past were carried out by the Verein für Kunst und Altertum in Ulm und Oberschwaben (Association for art and antiquities in Ulm and Upper Swabia), founded in 1841. The social backgrounds of this group's members reflected its bourgeois nature: Dr. P. L. Adam owned a bookstore and publishing house; Eduard Mauch was a drawing instructor at a public school; and Friedrich Eser served as a finance assessor for the Kingdom of Württemberg. Freiherr von Holzschuher, the district president in Ulm, became the association's first leader. By 1850, the group had over 250 members. Like other historical associations throughout central Europe, the association was involved in a broad array of cultural activities. During its first three decades, it published histories of several cultural sites, acquired an impressive art collection, and raised money for the restoration and completion of Ulm's most famous landmark, the steeple of the city's medieval cathedral.[51]

The Ulm association was also involved in several archaeological excavations. In the fall of 1848, it sponsored an outing in the forests near Ringingen. On Sunday, the eighth of October, members and their wives inspected an excavation that Friedrich Eser and others had begun a week earlier. Workers, supervised by the district forest ranger and hired with money contributed by association members, opened three grave mounds before an excited audience. In 1856, the association followed up on a request from Wilhelm, the Earl of Württemberg, to investigate a site near a new railroad line outside Ulm. This initial excavation was more like a treasure hunt than a scientific investigation. Only the "interesting pieces" (mostly precious metals) from over one hundred graves were saved for the earl's collection.[52] A year later a much more careful excavation was carried out by the association's leader, Konrad Dietrich Haßler.

Konrad Dietrich Haßler (1803–73) brought his love of history and desire for public outreach to the excavation site on the outskirts of Ulm. He had attended Ulm's *Gymnasium*, then studied philosophy and theology at the university in Tübingen and Middle Eastern languages in Leipzig and Paris. In the early 1830s, he returned to Ulm and became a leader in nationalist gymnastic and singing societies. He entered regional politics in 1844 as Ulm's delegate to the Upper House of the Württemberg Diet, and he was particularly engaged in the planning of the railroad connection between Stuttgart and Ulm. In the 1840s, he was gripped by national enthusiasm and strongly supported a German constitutional monarchy. He served as a delegate to the Frankfurt Parliament in 1848 and wrote the official stenographic report of the proceedings. When Prussian king Friedrich Wilhelm IV rejected the parliament's offer of a German crown, Haßler left national politics behind. Returning to Ulm, he taught history at a *Gymnasium* and joined the city's historical association. He served as its leader from 1851 to 1868.[53]

Haßler's historical interests and railroad sponsorship intersected in December 1857. During the construction of the line between Ulm and Blaubeuren, workers uncovered skeletons at the base of the Künlesberg, a hill just north of the Ulm train station. Haßler was called to the site, and he carried out a series of excavations over the course of 1858. His lengthy report, published in 1860, displays the increasingly professional nature of antiquarian excavations. It includes specific details about the number of objects found and their positions, size, and degree of preservation. Haßler assured his readers that his work was objective and accurate, and he invited them to assess his conclusions and compare his work to the results from other archaeological investigations. Haßler also worked hard to maintain the integrity of the excavation site, which was difficult because hundreds of men were working on the railway line and "sleazy antiquarian dealers" tried to buy artifacts from the workers. A few items escaped Haßler's watch in this way, and some ended up in archaeological collections in Munich and Berlin.[54] Most of the items from the excavation, however, remained in Ulm until Haßler's death in 1873. At that time, the items went to Württemberg's Königliche Staatssammlung vaterländischer Kunst- und Altertumsdenkmäler (Royal state collection for the region's art and antiquities) in Stuttgart.

The dominant theme in Haßler's report about the Künlesberg site was archaeology's growing importance for addressing historical questions. Haßler explained that much had changed in the previous twenty years. Antiquarians now had access to much more material; they compared finds from multiple sites; and they could consult significant works on premedieval cemeteries from Switzerland, France, and the German states. Perhaps most important for Haßler's interpretation of the site near Ulm were the debates about sites like Fridolfing, Selzen, and Nordendorf. Careful excavation of these places pointed to important differences between grave mounds and row graves and led to an acceptance of the relative chronology of these kinds of sites. By 1860, most antiquarians, following Ludwig Lindenschmit, associated grave mounds with earlier settlements and row graves with the period of migrations in the second to eighth centuries CE. Inventories, archaeological handbooks, and continued excavations were beginning to pay the historical dividends that the promoters of *vaterländische Altertumskunde* had hoped for earlier in the nineteenth century. This consensus, Haßler wrote, "made the evaluation of the contents [of these graves] in relation to historical questions substantially easier and much more certain."[55] He identified the row graves outside Ulm as the product of the period of migrations in the first centuries of the Common Era. Because most of the items were made of iron instead of bronze, he concluded that these were the remains, not of ancient Celts or Romans, but of the Alemanni.[56]

A new feature in Haßler's report was his interest in physical anthropology. Wilhelm Lindenschmit had included some comments about physical attributes (hair color, nose shape, etc.) in *Die Räthsel der Vorwelt*, but Haßler went further. He solicited the opinion of Dr. Volz, a military physician, who measured eighteen crania and jawbones. Although Volz's figures varied significantly, Haßler confidently determined that "we have nothing other than the skeletons of the purely white, that is, Germanic, race before us."[57] He did not pursue the connection between archaeology and anthropology further, although this would become much more important in the later nineteenth century, when skulls would be viewed as important evidence for ethnographic questions.

Haßler was a charismatic and enthusiastic teacher, and he treated his audience like students on a field trip. "Follow me, my readers," he called. "Descend with me into the opened graves of the cemetery. See and examine this for yourselves."[58] Haßler then provided meticulous descriptions of uncovered skeletons, weapons, knives, buckles, and objects of ornamentation. Throughout the tour, Haßler emphasized that this history was close by and that readers should note "the unique relationship" between the discovered graveyard and their locality. He was certain that this ossuary showed that Ulm had been a German settlement since before the year 600 CE and that it contained not merely the remains of earlier inhabitants of the region but the ancestors of the town's current residents.[59]

Beyond these genealogical issues, Haßler believed that the Künlesberg site contained several inspiring messages. In his opening lines, he described the Germanic past as a story of resistance to Rome's imperial ambitions, and he drew parallels to German-French animosity in the nineteenth century. He told readers that the graves represented German strength, even when Napoleon had subdued the area over fifty years earlier:

> It was on October 20, 1805, when the mighty uncle [Napoleon I] of the sly nephew [Napoleon III] stood at the farthest edge of the so-called Künlesberg.... As little as anyone else on that day did the proud emperor, nearing the pinnacle of his military fame and his humiliation of Germany, realize that in the ground ... in front of him rested entire generations of fully armed German warriors, before whose wild attack the rulers of the world, the Romans, ... had to yield and leave in the hands of the victors ... the province of Lower Raetia as their rightful hereditary possession—an event of world historical importance in that it meant that our area, which had been under Roman control for three centuries, did not become Latin. Instead, the German language and customs remained unadulterated, and God willing, they will remain preserved in the face of every new threat posed by Latin peoples.[60]

Haßler believed that this spirit of resistance could be resurrected, and the excavation of the row graves served as a reminder of the need to protect German culture from French (or Roman) influences.

Remaining in this theological mode, Haßler also described the archaeological site as a fulfilled prophecy. He claimed that the buried bones spoke to the current inhabitants of the area as the Old Testament prophet Samuel had spoken to the people of Israel. Quoting the book of Sirach, Haßler noted that "even when he [Samuel] lay buried, his guidance was sought," and the prophet explained the defeat of Israel at the hands of the Philistines. Haßler drew a parallel between Israel's situation and the fate of "the 25,000 German soldiers who were forced to lay down their arms shamefully" before Napoleon in 1805. The latter were the victims of "incompetent leadership, unfortunate circumstances, and a divided fatherland." But the image of the eventual victory of God's chosen people provided Germans with the hope for a triumphant future, which had come in 1813.[61] This version of German deliverance repeated the idea that was so common after the Wars of Liberation: that the success of the Germanic forefathers stemmed from their united stand against the Romans. In ancient times and in the nineteenth century, the mightiest army of the world could not subdue the spirit of the German people when they worked together. Haßler suggested that the success of contemporary Germans was inevitable if they would only learn the lesson of unity from the past.

Museums as Narratives

Throughout Germany's regions, antiquarians transformed archaeological sites into historical places, and dozens of historical associations followed the latest research on the early history of their local areas. Museum administrators wanted to tap into this enthusiasm, but they had trouble generating much public interest in domestic antiquities. According to an 1842 guidebook, visitors to the KMVA in Berlin were confused and bored, despite the collection's breadth and value. The rough nature of many items could not compare to the beautiful paintings and sculptures that one could see in other museums, and viewers had a hard time making sense of the objects.[62] What made these items interesting—the excitement of discovery, the evocative nature of the archaeological site, and the connections between the objects and the places where they were found—could not be transported to a distant museum.

In the 1850s, the Kingdoms of Bavaria and Prussia attempted to overcome this problem by exhibiting their large collections of domestic artifacts in new museums. In Munich, objects from the royal Antiquarium became part of the Bayerisches Nationalmuseum (Bavarian national museum, BNM). And in Berlin, the KMVA was integrated into the Neues Museum (New museum). These decisions were part of a major shift in the purpose of museums that occurred at midcentury. Whereas older institutions operated largely as places of safekeeping for rare and ancient objects, these new museums were accorded a much loftier task. Artifacts were to help visitors visualize historical scenes, and the museums themselves offered specific versions of cultural history.[63] A gallery of historical paintings in the BNM, for example, presented the items on display as evidence of the early history of the Bavarian monarchy. And murals and frescoes in the Neues Museum re-created the emotion of the archaeological site, connected the objects to Nordic mythology, and treated the domestic past as part of a broad narrative about world history.

During the Revolutions of 1848, demonstrations and calls for a constitution had caused the Bavarian king, Ludwig I, to abdicate in favor of his son, Maximilian II (r. 1848–64). In the wake of this turbulence, the monarchy redoubled its effort to foster an overarching dedication to Bavaria through the teaching of history. A Cultural Ministry circular from early 1851, glossing over the revolutionary activity, demonstrated how the state viewed deep-seated traditions as a bulwark against political unrest: "Recent history has again revealed the role Providence has given to Bavaria in Germany. Bavaria, as it has more than once in the past, has acted decisively and successfully against revolution and the desire to tear Germany apart. The Bavarian can be proud of his history. It stretches back to the most ancient past, and each of its pages shows how the people have remained true to their ancestral royal house and how this royal house has acted in the best interests of the people."[64] As this proclamation suggests, history was to offer a response to social change and political unrest. Several prominent intellectuals were motivated by this idea, and they hoped to resolve social tensions by emphasizing traditional forms of social cohesion. They turned to history and folklore, as well as economics, geography, and statistics, to explain the gradual and organic development of German society.[65] These authors differed from earlier antiquarians because they were more concerned with the applicability of their research than with the collection of information for its own sake.

The career of Wilhelm Heinrich Riehl (1823–97) provides a clear example of the desire to ground present-day social conditions in the past. During his student days, Riehl attended historical lectures by Ernst Moritz Arndt and Friedrich Dahlmann in Bonn. In the 1840s, he wrote for several regional newspapers and generally supported liberal demands for political reform. The Revolutions of 1848 exercised a decisive influence on his outlook, however. He was shocked by the street violence of 1848 and 1849 and felt that political radicalism threatened the possibility of constitutional reform.[66] In the wake of the revolutionary unrest, Riehl published his most famous studies—*Die bürgerliche Gesellschaft* (Bourgeois society) in 1851 and *Land und Leute* (Land and people) in 1853—which offered his views on the bases of social cohesion. These works caught the attention of King Maximilian II of Bavaria, who then provided the journalist–turned–social scientist with a series of influential positions related to the state's cultural and political affairs. Riehl became a professor of cultural history in Munich, the leader of the monarchy's statistical and topographical committee, and later director of the BNM and Bavaria's cultural minister.[67]

For Riehl, the key to warding off revolutionary activity was to emphasize how society had developed gradually and organically out of the past. In several works, Riehl spoke of the four fundamental links between the past and the present (known as the four Ss): *Stamm, Sprache, Sitte,* and *Siedlung* (tribe/genealogy, language, custom, and settlement).[68] These topics were grounded in pre- and early history: *Stamm* suggested regional identities as outgrowths from Germanic tribes, and *Siedlung* was directly related to the long history of German possession of the land. *Sitte* included present-day customs, but they were presented as long-standing practices that had developed out of ancient cultural traditions.

These ideas were clearly on display at the BNM, which was founded in 1855 and opened to the public in 1867. Karl Maria von Aretin, the original organizer of the museum, explained that the BNM was to present "everything that has a connection to the history of the life of the [Bavarian] people." King Maximilian II was the museum's strongest advocate, and the Bavarian state built a magnificent structure for it along the Maximilianstraße, the new royal avenue that would be lined with important cultural and governmental buildings over the course of the second half of the nineteenth century.[69]

The BNM was originally divided into three chronological sections. Visitors entered on the ground floor, and to the left they found a series of rooms dedicated to "Roman, then Celtic-Germanic artifacts, as well as antiquities from the Carolingian era, followed by works from the Romanesque period." The second section, also on the ground floor, was dedicated to Gothic art. Items from the Renaissance to modern times made up a third section, which was located on the museum's third floor. The second floor housed the gallery of 143 large-format paintings that illustrated the course of Bavarian history.[70]

The transfer of items from the royal Antiquarium to the BNM certainly increased the public's access to the monarchy's antiquities. The new museum had regular opening hours, its contents were detailed in a published guidebook, and numerous cultural journals reviewed its exhibits. Yet several factors limited the effectiveness of the BNM's presentation of pre- and early history. Items in the Antiquarium had been excavated before the nineteenth century, and many of them lacked even the most basic information about where they had been found. The state tried to improve the BNM collection by purchasing antiquities with better documentation and by sponsoring new excavations. The enormous amount of material excavated by the historical associations, however, remained in their local collections. Additionally, Aretin, the first director of the BNM, was less than enthusiastic about including barbarian antiquities in the BNM.[71] The 1868 guidebook, reflecting his lack of appreciation, stressed the inferiority of domestic antiquities: "Of course, only a smaller portion of the items found in graves were made domestically, notably all the urns and vessels whose rough form and underdeveloped technique offer no comparison to the achievement of the Roman ceramics. Most of the at-

tractive bronze items, in contrast, certainly came into the hands of our
ancestors through trade with Romans and Etruscans."[72] Indeed, the museum's chronological display began not with prehistoric items but with
a highlight from the Roman era. "Immediately upon entering [the mu-
seum], our eyes fall on a magnificent mosaic floor found in a Roman villa
outside the present-day town of Westerhofen." The discovery of this lav-
ish third-century mosaic had caused a sensation in 1856—over three
thousand people visited the site as it was being excavated. This treasure
was followed by a room full of Roman gravestones, sandstone sculptures,
and mile markers, and then by the "Celtic-Germanic" antiquities.[73]

The Celtic-Germanic objects were divided into two general categories
based on types of graves. Most of the material came from grave mounds,
and the guide explained rather imprecisely that "some of this material
reaches back to the Celtic *Vorzeit* and some of it belongs to the native
population who lived under Roman rule." The items were presented in
six cabinets in what must have been a crowded and jumbled presentation.
Some cabinets displayed items from a single excavation site, while others
mixed finds from multiple sites. One was devoted to weapons and tools
made of bronze and another to those made of iron.[74]

One special item, placed on a round cabinet in the middle of the room,
received a fuller description and even an illustration in the guidebook: "a
cone-shaped hat made of gold that was decorated with simple line and
ring patterns." A farmer in Schifferstadt (in the Bavarian Palatinate) uncovered this item in 1835 and reported it to local authorities. The hat was
transported to Munich so that the Academy of Sciences could study it,
and it then entered the royal Antiquarium and later became part of the
BNM display. The museum guide offered several explanations for this curious object: "It is either a shield boss or more probably a druidic head
covering. According to another interpretation, it was used as a cover for
a glass vessel."[75]

The "important and interesting finds from the row graves of Nordendorf, dated from the sixth to eighth century," made up the second category
of the Celtic-Germanic items. A display case was full of silver buckles,
brooches, a vase adorned with gemstones, keys, clasps, buttons, brass
pieces, combs, coins, and shells. Beneath these smaller items were spear-
heads and battle-axes, swords, a knife, and the clasp for a shield, plus urns
and other clay vessels. Additional clay pieces and a lantern were placed
on a nearby windowsill. The guide again pointed out the poor quality of
these ceramics, "which were dried in the sun, not baked in an oven," and
had primitive forms and decorations.[76]

The guidebook treated the golden hat of Schifferstadt as unique and described the Nordendorf finds as important and interesting, but overall the comments were quite muted. Other rooms held the most stimulating items. The thrills of discovery and excavation were lost in the larger museum display. The move to Munich had broken the bond between the object and the place where it was found.

Yet the BNM did try to present Bavaria as a historical place in a different way. The museum's second floor was devoted to a gallery that told the kingdom's story through 143 large-format paintings. The gallery's right-hand side depicted the history of the territories that had been annexed by the Kingdom of Bavaria in 1815 (the Palatinate, Franconia, and Swabia), as well as the period of time when a Wittelsbach monarch had sat on the Swedish throne. These scenes were followed by portraits of Maximilian I Joseph, Ludwig I, and Maximilian II, the modern rulers of Bavaria. On the left, the gallery presented the history of the kingdom's core territories, Upper and Lower Bavaria, with a chronological sequence rooted in the early Common Era. Five of the 143 paintings depicted scenes from before 800 CE.

The paintings were not merely to be viewed but to be studied. To this end, historian and geographer Karl Spruner provided an official annotation to the paintings that he hoped would inspire, educate, and entertain. Spruner conceded that the individual artist might dispute the aesthetic style of the paintings and that the historian might dispute the selection of one scene over another. Yet one must consider the gallery as a whole, Spruner explained. It was the king's wish that the paintings add a sense of vividness and life to the historically important objects. The cycle of paintings should leave "a strong and pleasing impression of the glorified deeds of the Bavarian people and of the praiseworthy accomplishments of the Bavarian royal house."[77] In the BNM, the non-Roman antiquities seemed static, but the paintings suggested the action that surrounded these objects and connected them to the monarchy.

The first image depicted interactions between Romans and barbarians. Spruner explained that the Romans had advanced to Lake Constance by 15 BCE and that this coincided with the destruction of the area's earliest settlements, which were dwellings along the shores of alpine lakes. In the area, the Romans won "a fight of destiny" that no historian had recorded. Spruner explained that the tribal peoples fought to the end, and women killed their children and threw their bodies into the faces of the Romans. The Roman imperial army then dominated the area, building military roads with outposts and guard towers. The Romans felt so secure that they even brought "the luxury of the Tiber" to these barbarian lands, as evidenced by the opulent mosaic from Westerhofen displayed in the museum.[78]

This narration reveals several interesting aspects of the museum's presentation of prehistory. First, the text established connections between the paintings and the objects on display. The reference to Roman luxury was tied directly to the Westerhofen mosaic that was on the floor below the gallery. Second, this scene employed the Tacitean baseline developed by antiquarians. The description of lake dwellings as the area's earliest settlements was an idea that prevailed after the discovery of these kinds of wetland sites in Switzerland in 1855. This interpretation, discussed at greater length in chapter 4, acknowledged a deeper prehistory, but the destruction of the lake dwellings provided a clear break in the history of the land. It did not matter how long the lake dwellings had existed before they were destroyed, supposedly at the hands of the Romans around 15 BCE. No historian captured their demise, but historians could narrate what came next. Bavarian history could begin with new migrations into the area and continue with conflicts between barbarian peoples and the Romans, the spread of Christianity, and the establishment of royal power.

The following scenes in the picture gallery followed this trajectory and documented moments of religious and political change in Bavaria. The second painting showed Saint Severin preaching the message of Christianity to the inhabitants of the region in the middle of the fifth century CE. This was followed by a depiction of the marriage of a Baiovari princess to a Lombard king in 589, which linked the ethnic history of the area to a royal one. Another painting displayed the monarchy's benevolence in ancient times. It showed Duke Tassilo III founding a school on the Herrenchiemsee, an island in the Chiemsee southeast of Munich. The anachronistic background to this painting showed lake dwellings and prehistoric people in canoes (who, according to the commentary on the first painting, had been "completely destroyed" by the Romans). The intention of the painting was not historical accuracy but rather to show that royal patronage of education and culture existed "already in the farthest reaches of antiquity."[79] The gallery of historical paintings did not offer a comprehensive narrative about the ancient past; rather, it sought to foster loyalty toward the Bavarian monarchy with these scenes of settlement, religious conversion, and royal munificence.

The Neues Museum in Berlin offered an even more elaborate interpretation of pre- and early history. During the reign of Friedrich Wilhelm IV from 1840 to 1861, leading personalities in the fields of architecture, law, philosophy, linguistics, music, and religion were inspired by new ideas about history and society. As historian John Toews has explained, they no longer viewed the national community as a closed ethnic fraternity, as many intellectuals after the Wars of Liberation had. Instead, they believed that political communities were the result of broad historical processes and that the monarch, as a benevolent patriarch, protected and fostered this historical development. The Prussian royal house enshrined this historical vision in a new "'sanctuary' for scholarship and the arts" in the heart of Berlin, and the Neues Museum was a central part of this project.[80] Completed in 1854, it unified several royal collections, including the Egyptian Collection and the KMVA from Monbijou Palace. In contrast to the BNM's focus on the Bavarian monarchy and the Bavarian people, the Neues Museum portrayed a more universal story of cultural development that included ancient Egypt, Greece, and Rome and medieval Europe, as well as the "primitive" settings of prehistoric Europe, the Americas, Africa, and the Pacific. The classical building, designed by Friedrich August Stüler and built over the course of the 1840s and 1850s, was conceived as an extension of Schinkel's Altes Museum. Visitors were to enter the Altes Museum to view original Greek sculpture, the coin cabinet, and the painting galleries. They then passed through a connecting gallery above the streets of Berlin to the exhibit of plaster casts of Greek and Roman sculpture on the second floor of the Neues Museum. The tour continued down an immense and breathtaking staircase that was decorated with historical murals by Wilhelm von Kaulbach and on to the Egyptian galleries, the hall devoted to central European prehistory, and the ethnographic collections from around the world.

The use of historical murals was particularly important in the Neues Museum. The building's central staircase was over 120 feet deep, 50 feet wide, and over 65 feet high, cutting across the entire width and through two stories of the museum. Its walls presented massive portraits of the "main phases of the cultural history of humankind." On one wall, the female figure Myth (*Sage*) proclaimed the oldest memories. She sat upon a *Hünengrab* (a megalithic chamber grave) and was surrounded by prehistoric ruins. History (*Geschichte*) decorated the other side of the wall. She recorded "the deeds of humankind" in her book. A cycle of giant paintings then depicted major cultural events, including the destruction of the Tower of Babel, Homer and the blossoming of Greece, Attila and the Huns at the Battle on the Catalaunian Fields, the Crusades, and the Reformation.[81] These monumental depictions extolled certain human deeds as turning points in world history.

The interaction between myth and history continued in the galleries, which were also decorated with elaborate murals and friezes. The walls in the room for Roman plaster casts, for example, portrayed imperial villas and cities. The Egyptian rooms exhibited artifacts from the graves of pharaohs, and the walls of each room were decorated with scenes from Egyptian mythology and death cults.

The antiquities from the KMVA in Monbijou Palace were moved into the Neues Museum and opened to the public in 1856. The gallery devoted to this collection occupied a large rectangular room on the ground floor that was divided into aisles by six columns. Smaller items were exhibited in oak cases that stood between the columns and the walls. Urns and larger ceramic objects stood on three shelves that ran along the walls between the windows.[82] The collection remained the largest of its kind in central Europe, and Leopold von Ledebur, now fifty-five years old, was still its director. Under his watch, the exhibit remained conservative in its approach. Despite his familiarity with the Three Age System and relative chronology, Ledebur felt that the variety of objects in the royal collection prevented a straightforward chronological display. The collection therefore remained organized by types of objects, as had been the case in Monbijou, but some of the display cases presented self-contained exhibits from a single location, like the island of Rügen or the lake dwellings at Lake Constance. The exhibit did not venture answers to questions about settlements, migrations, or daily life during prehistoric times, and it provided very little information about individual objects. The display also faced the overcrowding that plagued other institutions. The gallery was the only space allotted to a collection that increased from 3,540 items in 1838 to almost 11,000 by 1874.[83]

At the time of the move, the collection also changed its name from the Königliches Museum vaterländischer Alterthümer to the Sammlung der nordischen Alterthümer (Collection of Nordic antiquities). This designation reflected the significance of Danish research for German scholars, but it also found expression in the overall aesthetic design of the gallery. Stüler, the architect of the Neues Museum, explained that the paintings in the room were intended to enhance the appreciation of the antiquities. He wrote that "the walls have a simple gray tone that is not to detract from the color of the urns. The upper spaces that cannot be used for display are decorated with paintings of the most prominent gods and spirits of Nordic mythology and with landscape portraits and architectural drawings that depict graves as they have been delivered to us by the most distant past."[84]

The mythological paintings appeared as a double cycle along the longer walls of the room. On the right-hand side of the hall, the good gods and mythic figures of the Icelandic Edda saga were featured in light colors. The evil figures, painted in darker shades, stood on the left. This light/dark effect was increased by the presence of windows on only one side of the hall. Light shined on the good gods, and shadows covered the evil ones. A large painting of Odin, the eternal father of all the Nordic-Germanic gods, connected the two cycles. These mythological images provided a stark contrast to Ledebur's sparse labeling. They attached the objects to heroic stories and encouraged visitors to view the display as a confirmation of the Nordic sagas.

Other images that embellished the Sammlung der nordischen Alterthümer depicted actual archaeological sites, but with a Romantic flourish. Three paintings on the south end of the hall ventured beyond Ledebur's organization of the exhibit. They used three royal graves to illustrate the Three Age System and the customs and burial items associated with each period. The representations of the Stone and Iron Ages each showed a buried body, whereas an urn containing the remnants of a cremation ceremony stood for the Bronze Age. The two kings appeared at solemn rest after a lifetime of battle. They were barrel-chested and bearded, surrounded by the tools of war. The Stone Age king gripped a battle-ax, and his dagger and spear lay nearby. His legs were draped in a fur, an armband accented his bicep, and a necklace made of carved bone adorned his chest. The Iron Age king wore a cloak and gold headband. One hand was folded solemnly across his chest and the other rested on his sword. His jewelry was plentiful and ornate. The painting representing the Bronze Age depicted an archaeological bonanza, overflowing with pottery, shields, swords, and jewelry. The disparity between Ledebur's cautious presentation and these evocative paintings characterized the tension between the antiquarian's hesitation when dating and interpreting artifacts and his desire to transform prehistory into history.

The vestibule at the south end of the gallery featured two paintings of prehistoric sites on the island of Rügen. *Hünengrab auf Rügen* depicted a megalith and a grave mound. The other painting, *Opferstein bei Stubbenkammer*, presented a sacrificial stone in front of the famous chalk cliffs at Stubbenkammer on Rügen. These paintings, along with the mythological cycle and depictions of the Three Age System, brought the objects in the Sammlung der nordischen Alterthümer to life by reminding museum

FIGURE 3. *Metallzeit* by Heinrich Bögel was one of three lunette murals on the south wall of the Vaterländischer Saal in the Neues Museum that represented the Three Age System. The depictions were based on objects in the museum collection. This painting included an array of artifacts from the Bronze Age to the early Middle Ages. (Photo reproduced with permission from the Zentralarchiv der Staatlichen Museen zu Berlin-Preußischer Kulturbesitz. SMB-ZA, V/Fotosammlung 1.1.3./02814.)

viewers of the sites that had contained them and by connecting them to Nordic mythology.

The inclusion of the Sammlung der nordischen Alterthümer within the Neues Museum was also significant. As Stüler explained, the domestic antiquities and the ethnological collections were displayed together on the ground floor as "objects from a primitive artistic trajectory."[85] They were treated as exemplars in a cultural catalog that spanned many centuries and much of the globe. In the sequence of displays, domestic antiquities were not representatives of German ethnic nationalism or Prussia's territorial possessions. Instead, they were part of a panorama of cultural development that led to the world of the nineteenth century.

Neither the display in the BNM nor the one in the Neues Museum contributed very much to the scientific study of domestic antiquities. Shortly after the BNM opened, a new director removed almost all the prehistory section from view, and it remained this way until 1885. Guidebooks to the Neues Museum praised the architecture of the building and explained the historical paintings in detail, including the mythological cycle that accompanied the Sammlung der nordischen Alterthümer, but they said almost nothing about the objects themselves. They listed the kinds of artifacts that filled the nine cabinets in the room and offered a few comments about how they were organized. They made no statements about the

FIGURE 4. *Hünengrab auf Rügen* by Ferdinand Bellermann adorns the passageway just outside the Vaterländischer Saal in the Neues Museum. The placement of the megalithic tomb among ancient oak trees recalls Caspar David Friedrich's landscape scenes with dolmen. And the cattle seeking water and rest remind the viewer that present-day locations were surrounded by prehistoric sites. (Photo reproduced with permission from the Zentralarchiv der Staatlichen Museen zu Berlin-Preußischer Kulturbesitz. SMB-ZA, V/Fotosammlung 1.1.3./02824.)

age of the objects or how they might have been used in ancient times.[86] In 1880, when plans were under way to move these artifacts to the new Museum für Völkerkunde (Ethnology museum), administrators offered much more information about prehistory, but they had given up on the Sammlung der nordischen Alterthümer as an effective display. The greatly expanded guidebook from this year explained the Three Age System and how the artifacts reflected trade relations and technological innovation among ancient peoples. It noted, though, that this field of knowledge was changing rapidly and that artifacts would have a more appropriate home when combined with the ethnographic collections.[87] This evaluation suggested that an entirely new framework was needed to interpret domestic antiquities, a perspective that will be discussed in the following chapter.

During the first two-thirds of the nineteenth century, *vaterländische*

Altertumskunde inspired the collection of archaeological objects. As individuals and historical associations evaluated these sources, a new era of history appeared before their eyes. Domestic archaeology revealed evidence of the peoples that Tacitus and other ancient authors had described and suggested that the premedieval period of migrations could be reconstructed. It was difficult to fill the expanse of time in between the darkness of prehistory and periods documented in texts, however, because antiquarians viewed archaeological objects mainly in relation to their local areas. In contrast to the practice of archaeologists today, who discern multiple layers at a single site and can therefore suggest patterns of cultural change, it was hard for nineteenth-century antiquarians to extrapolate from a single site to a wider story. Other approaches to the ancient past, like museum collections that brought together objects from a wider geographic area, were also limited in their ability to explain the settlement and early history of central Europe.

The history produced by antiquarians, then, was largely a set of scenes that spoke about specific places. These interpretations could not fill in all of the time between the early Common Era and later epochs, but they did relate ancient events and peoples to present-day places and inhabitants. This was very clear in the connections suggested between the Lugii, Baiovari, and Alemanni and the nineteenth-century inhabitants of Silesia, Bavaria, and Swabia. These genealogies were simplistic and based on little evidence, and they were certainly more suggestive and evocative than dogmatic. These were not firm lines of racial purity or ethnic continuity but a way to explain regional differences and to establish a connection between the past and present.

The antiquarian engagement with central Europe's early history imagined only a relatively short chronology that focused on the early Common Era. By the 1860s, however, new ideas about the age of the earth and the origins of humankind challenged this framework. New discoveries revolutionized the conception of history and prehistory again, suggesting a timeline that was measured, not in hundreds of years, but in hundreds of thousands of years. The discovery of this even more distant past raised new questions for antiquarians, but it did not end their interest in local investigations or their focus on territorial settlement and ancient ancestors.

PART II

The New Empire and the Ancient Past

By the end of the nineteenth century, Germany had become a very ancient place. The state itself emerged from a recent process of unification in 1871. Yet almost immediately the new German Empire was connected to the distant past in a variety of ways. Theaters in Munich and Berlin revived Heinrich von Kleist's drama about Hermann (Arminius), the Cheruscan chief who had defeated the Roman legions in the Teutoburg Forest in 9 CE. *Die Hermannsschlacht* (Hermann's battle), written in 1808, captured the national spirit during the Wars of Liberation against Napoleon, but its message about German freedom resonated again after the victory over France in 1870. A monumental statue of Arminius was unveiled near Detmold in Rhineland-Westphalia in 1875. This public symbol placed prehistory at the geographic heart of Germany and reminded everyone of the event that had supposedly prevented the destruction of German culture at the hands of the Romans.[1]

Surveys of German history place great emphasis on expressions of national enthusiasm after 1871. Historian Thomas Nipperdey observed that, in the new empire, "the history of the Germans was given a national meaning and presented to the public in this way." Germanic pre- and early history gained greater prominence; historians and literary authors sought out historical events that could be presented as turning points in a story of national development; artists rendered ancient kings and battles for public display and for popular magazines; and even schoolbooks took on a national tone.[2] This evaluation of cultural life in the German Empire is certainly true, but it represents only one context for the development of pre- and early history and has exercised almost a complete monopoly on the history of the field. Two other frameworks were critical for the development of pre- and early history in the German Empire: the rise of anthropology as a new approach to questions about cultural development and the establishment of regional institutions for archaeology. Part 2 examines each of these approaches to the distant past and the interaction between them. The anthropological and regional approaches challenged the telling of a nationalist past in very direct ways. At the same time, however, national narratives drew on the other two approaches. The anthropological approach added deeper temporal and racial aspects to "German" prehistory, and regional institutions created a familiarity with the ancient past that could be steered toward nationalist narratives.

CHAPTER FOUR

Rudolf Virchow and the Anthropological Orientation of Prehistory

We must grow accustomed to the idea that the Germanic peoples, from their first appearance in history onward, were a mixed people [ein Mischvolk]. . . . They do not constitute a nationality in the sense of being related by blood; rather, they are a nationality because of political developments. —Rudolf Virchow, 1875[1]

During the middle decades of the nineteenth century, a series of discoveries exploded both the chronological sweep and the geographical scope of *vaterländische Altertumskunde.* Antiquarians in England and France uncovered human and protohuman bones alongside the remains of extinct animals. After reviewing these sites, the most prominent geologists of the day concluded that the earth and humankind were much older than biblical chronologies allowed. Initially, German scholars did not play a central role in these discussions, but questions about the age of humankind did arise in central Europe after another kind of discovery. In the 1850s, low water levels in several Swiss lakes revealed a series of ancient lake dwellings that suggested an epoch much earlier than the Roman Empire or early medieval migrations.

To engage the questions raised by these discoveries, scholars developed the field of anthropology as a broad science that would unite archaeology, ethnology, anatomy, and the study of culture under one banner. This new interdisciplinary endeavor had several consequences for domestic archaeology. First, archaeologists began to pay much more attention to sites that were fully prehistoric. Whereas earlier antiquarian work used Roman sources to relate row graves to the early history of towns and regions, archaeology of a more anthropological orientation focused on sites and cultures that were not known (or not well known) from ancient texts. This extension of archaeology's timeline disrupted the patterns of historical narration developed by *vaterländische Altertumskunde*. Reaching back before the migrations of late antiquity, archaeology came to study "prenational" periods when ethnic or national labels made little sense. Neanderthals or the inhabitants of lake dwelling sites could not be described as "our ancestors" in the same way as the skeletons of early medieval row graves.

Second, the discussion of lake dwellings introduced a new comparative dimension to archaeology, as antiquarians imagined that the ancient inhabitants of Switzerland (and of Mecklenburg, Bavaria, and the other regions where lake dwellings were sought) had lifeways that were similar to those of the current inhabitants of Pacific islands who also built lakeside huts. In England, Sir John Lubbock's highly influential *Pre-historic Times*, published in 1865, explained that archaeologists and anthropologists would come to understand prehistory only if they studied "ancient remains" at home and "the manners and customs of modern savages" abroad.[2] This comparison uprooted narratives that emphasized the growth of European towns and villages out of the distant past and suggested that one might learn as much about the earliest stages of European prehistory by observing present-day Polynesia as by excavating in central Europe.

Third, whereas *vaterländische Altertumskunde* brought archaeology, folklore, and the study of place-names together, the study of prehistoric sites required a new combination of scientific fields. Archaeologists increasingly worked with geologists, zoologists, and botanists to understand the human and animal bones, plant life, and geological layers at these prehistoric sites. Archaeologists were concerned not only with local finds but also with the connections one might draw between local finds and other European or even Asian sites. For the practitioners of *vaterländische Altertumskunde*, this broader approach moved the goalpost. If antiquarians thought they were getting closer to the narration of the distant past by the 1850s, anthropology's broader agenda made the field of research much larger, and the task of accumulating evidence would need to be practiced on a wider, perhaps global scale before the accretion of meaning could occur. Anthropological associations therefore returned to the work of collecting and creating inventories, eschewing narrative interpretations of central Europe's distant past.

Fourth, whether anthropologists were describing prehistoric sites or later periods, they began to emphasize a vision of the past that was more complex than the one-to-one mapping of tribes onto nearby places. Anthropology began to investigate patterns of migration, the mixing of populations, and the cultural transfer that came through trade relationships. Scholars also became more interested in questions about social dynamics and technological innovations in prehistory. This approach undercut simple conclusions that labeled wide stretches of territory as Germanic or Slavic. It also made archaeology a more professional discipline, as practitioners paid closer attention to archaeological sites. They began to study the relationship between different kinds of finds at a single site and came to value mundane objects as highly as ornamental items. In the anthropological reading, the archaeological site was more than a historical place. It was evidence of a much broader story of cultural development.

A Revolution in Time

The early modern European view of time held that the world was about six thousand years old, an estimate based on genealogies in the Old Testament that traced humankind back to the creation of Adam and Eve.[3] Over the course of the eighteenth century, two important shifts changed the way this time span was perceived. First, the authors of universal histories chose the story of the Great Flood recorded in Genesis as the beginning point for history. They reasoned that this catastrophe "set the civilizational clock back to zero," and therefore, the beginning of their narratives could be moved forward from the Creation to the Deluge.[4] Second, several French thinkers, inspired by the work of Georges Buffon (1707–88), began to detach questions about the age of humankind from the age of the earth. They argued that the six-thousand-year period applied to human history, but that the earth was much older, perhaps even eternal.[5] This change in the conception of early time was significant, but it did not disrupt the accepted age of humans: it confirmed an ancient earth and recent humans.

New archaeological evidence from the late eighteenth and early nineteenth centuries raised questions about the chronologies deduced from religious texts. Quarrying, canal dredging, and roadbuilding turned up bones and fossils that could not be matched with the animal life known in modern Europe. In 1820, for example, German geologist Ernst von Schlotheim wrote about human fossils that were mixed with hyena and rhinoceros bones in a quarry near Leipzig. And these curious sites proliferated. In the 1830s, Jacques Boucher de Perthes, a customs official based in Abbeville, France, began digging in the Somme River valley. Boucher de Perthes was particularly fond of man-made axes common in the area, but he began to think in much broader terms when he noticed that these items were often located in gravel layers with the bones of tropical animals, including elephants and hippopotamuses. He exhibited some of his finds in Abbeville and Paris and published a five-volume commentary on his theory of creation between 1838 and 1841. At the time, critics ridiculed Boucher de Perthes for the flamboyance of his language and his far-fetched claims.[6]

A major turning point in the reception of Boucher de Perthes's ideas came in 1859, after an important discovery in England. Quarrying for limestone near Torquay in southwestern England in January 1858, workers uncovered a set of fossilized bones in a cave. Word of this find reached the paleontologist Hugh Falconer, who secured financial support from the Geological Society of London for the investigation of Brixham Cave. The society created a scientific committee that included the most respected geologists of the day: Charles Lyell, Joseph Prestwich, R. C. Godwin-Austen, Richard Owen, William Pengelly, and Falconer. Under their watchful direction, excavations revealed the bones of cave bears, mammoths, and reindeer among man-made stone tools. Later that year, the committee announced that Brixham Cave contained human artifacts that had existed at the same time as extinct animal species. Several of these British scholars then traveled to France and confirmed much of Boucher de Perthes's work. Their findings led to a sea change in educated public opinion about the antiquity of humankind. As anthropologist Donald Grayson has explained:

> By 1860, then, a new resolution on the question of human antiquity had been reached in both Great Britain and in western Europe. It was now very generally agreed that people had coexisted with extinct mammals, that they had been on earth prior to the time that the earth had taken on its modern form, and that they had existed for a series of millennia that could not be encompassed within biblical chronology. Not everyone agreed, . . . but the situation was now reversed from what it had been just a few years before. Majority opinion now

held that people were both geologically ancient and ancient in terms of the number of years they had been in existence. Only a very small minority held otherwise.[7]

The discovery of hominid bones in a rock quarry in the Neander Valley near Düsseldorf in 1856 brought the discussion about the antiquity of humankind to central Europe. Hermann Schaaffhausen (1816–93), an anatomy professor in Bonn, compared the Neanderthal skull with Neolithic finds from northern German states and concluded that it belonged to a primitive people "that occupied northern Europe before the Germanic tribes. . . . [These people] were widespread . . . and related to the original inhabitants of the British Isles, Ireland, and Scandinavia."[8]

Questions about the primeval past spread with the news of the ancient Swiss lake dwellings. During the winter of 1853–54, the water in lakes around Zurich sank to extremely low levels. The people of Obermeilen, hoping to use this situation to their advantage, set about reclaiming land by building walls around the water. After digging just a foot into the soggy ground, workers uncovered an impressive antler rack, several large animal bones, and clay fragments. Further excavation revealed rotted posts in the ground that appeared to form the foundation of several buildings. A local teacher named Johannes Aeppli reported these finds to Ferdinand Keller, the president of Zurich's antiquities association. Keller first wrote about the lake dwellings in the *Züricher Freitagszeitung* on March 17, 1854, and he later published a series of reports in his historical association's journal. Keller assumed that the rotted posts were timbers that had been driven into the floor of the lake to support the foundations of buildings, and he therefore described the sites as *Pfahlbauten* (pile dwellings). He speculated that "ancient Celts" had built entire villages "on the water" and canoed from home to shore. He surmised that this arrangement made fishing easier and offered protection from wild animals.[9] The lake dwellings caused a sensation, and archaeologists throughout central Europe caught "*Pfahlbaufieber*" (pile-dwelling fever). They searched for similar structures in bodies of water from the Baltic and North Seas to Lake Constance and the Mediterranean.[10] (Twentieth-century research revised this image of huts on the water and demonstrated that the vast majority of these sites were along the shores of lakes and that changes in water levels gave the appearance that the sites were farther out in the water.)[11]

Brixham Cave, the finds in the Somme Valley, and the Swiss lake dwellings presented new questions that *vaterländische Altertumskunde* could not address. These discoveries provided evidence of peoples who lived well before the Germanic peoples and had ways of life that looked nothing like the descriptions offered by Tacitus or Julius Caesar. The writings of Schaaffhausen on the Neanderthal skull and Keller on the *Pfahlbauten* concluded that Europe's prehistory stretched back much farther in time.

These sweeping changes inspired greater scientific communication across national boundaries. British scholars, for example, traveled not only to Abbeville, France, but also to Hallstatt, Austria, where Eduard von Sacken excavated a series of rich Iron Age gravesites. Jacques Boucher de Perthes presented his findings in London. Key texts were translated into multiple languages. (Rudolf Virchow provided the preface to the German edition of John Lubbock's *Pre-historic Times*, which appeared in 1874.) Scholars also developed a series of national and international networks. The Société d'anthropologie was founded in Paris in 1858, and similar institutions appeared in London and Lisbon shortly thereafter. The first International Congress for Anthropology and Prehistoric Archaeology, organized by Gabriel de Mortillet, was held in Paris in 1866.

In German-speaking Europe, a scientific journal called the *Archiv für Anthropologie* (Archive for anthropology) appeared in 1866. Alexander Ecker, an anatomist in Freiburg and coeditor of the *Archiv*, captured the interdisciplinary nature of the new research agenda in the journal's first volume: "Driven by the desire for knowledge and carried by the free spirit of science, the connections between otherwise-distant disciplines gradually became stronger, and in this way the field of knowledge developed that today we call the *natural history* and *primeval history* of man or, more concisely, *anthropology*."[12] To advance this new science, German scholars founded the Deutsche anthropologische Gesellschaft (German anthropological society, DAG) in 1869, as well as branch societies in five cities throughout Germany. The full title of the DAG emphasized prehistory's connections to other fields: Deutsche Gesellschaft für Anthropologie, Ethnologie und Urgeschichte (German society for anthropology, ethnology, and prehistory).

Within anthropological circles, domestic archaeology became closely connected to anthropology as an auxiliary science that provided material evidence from the distant past. This orientation shifted from the questions of *vaterländische Altertumskunde* to new thoughts about daily life in prehistoric societies. Antiquarians wondered what ancient houses looked like, what these people ate, and how they procured their food. Beginning in the 1860s, domestic archaeology became much more than a search for "ancestors." It now had the potential to offer insights into the social dynamics of prehistoric settlements.

Rudolf Virchow and the Expansion of Archaeology's Horizons

Rudolf Virchow (1821–1902) was the most authoritative advocate for the anthropological orientation of domestic archaeology in the German Empire. He is well known as a pathbreaking pathologist, as a founding figure in the discipline of anthropology, and as a political antagonist to Otto von Bismarck. A central but overlooked feature of his remarkable career was his archaeological work. From the 1860s until the end of his life in 1902, Virchow excavated in central Europe and abroad and produced over five hundred publications devoted to archaeology and prehistory.[13] His intellectual biography clearly demonstrates archaeology's transition from *vaterländische Altertumskunde* toward a practice that responded to the scientific questions raised during the last third of the nineteenth century.

Virchow was born in Schivelbein (Pomerania) on October 13, 1821. His father, Carl Virchow, subscribed to *Baltische Studien*, the journal edited by the Gesellschaft für pommersche Geschichte und Altertumskunde, and he and his son read it together and often discussed their region's past. In the summer after his first year of medical training, Rudolf Virchow undertook an extended tour throughout Pomerania. He later recalled how he was captivated by the natural beauty and historical treasures of his boyhood homeland: "I had been to Prillwitz before and could not resist the temptation held by my memory of the beautiful landscape around Lake Tollense. . . . I can recommend . . . this place to all those who would like to unite the enjoyment of natural beauty with the small 'shiver' [one might get] from experiencing the past." His itinerary included a visit to a count in Neustrelitz who owned a collection of Slavic antiquities, and he recalled reading about an eighteenth-century antiquarian named Sponholz who had famously manufactured fake "prehistoric" items.[14]

For Virchow, antiquarian research was very personal. He observed that many place-names around Schivelbein sounded Slavic. And he noted that his surname reflected Slavic origins, but that his family was German speaking and held German cultural loyalties. As a medical student in Berlin, he wrote to his father about his forays into genealogical research. In a letter from February 22, 1842, he explained that he had traced the history of the settlement of their region back to only around 1000 CE, but he speculated that their Pomeranian ancestors descended from a branch of the Poles or the Lechen who had originally inhabited the region bounded by the Oder and Vistula Rivers and the Baltic Sea and that these Pomeranian ancestors had adopted the German language and customs only after contact with German knights and monks.[15] An American colleague recalled that this question was still on Virchow's mind fifty years later:

> I once discussed with him the question that his fathers were probably Slav in blood though they had been two hundred years or more on German soil. He said he must therefore be considered at least as much German as Americans who have been two hundred years in our country think themselves Americans. I reminded him that the older countries never quite gave up their claim on their children ... [and that] the best we can hope for from European writers with regard to them, was some such compound term as German-Americans.... He smilingly assented and said that under those circumstances the best he could say of himself was that he was a Slav-German.[16]

Perhaps Virchow's early interest in archaeology reflected the hope that he could shed light on the settlement patterns and ethnic combinations that had produced his family.

In 1848, Virchow took his place at the barricades in Berlin and demonstrated for revolution in Prussia. He and other democrats believed that a new political order would create better educational opportunities and raise the standard of living for Prussia's poorer residents. After the revolution failed, Virchow was exiled from Berlin. He turned his energies toward medical research at the University of Würzburg, and in 1856, the Prussian government called him back to Berlin to lead an institute for pathology research.

The 1850s witnessed a series of extraordinary achievements that created a strong faith in the power of scientific observation. In 1852, medical researcher Robert Remak advanced the theory of cell division in Berlin. The development of spectrum analysis by Gustav Kirchhoff and Robert Bunsen in 1859 allowed the identification of chemical elements in vapor form, and in that same year Darwin published *On the Origin of Species*. Virchow's contributions in the field of cellular pathology were pathbreaking for the study of disease, and he remained on the cutting edge of contemporary medical research.[17]

Virchow reentered Prussian politics as a cofounder of the German Progressive Party. During the Prussian constitutional crisis in the 1860s, he opposed the War Ministry's request for increased military spending. Otto von Bismarck saw Virchow as a personal adversary and challenged him to a duel in 1865, which Virchow declined. As a liberal critic of the state, Virchow did not share the patriotic enthusiasm that many felt after German unification in 1871.

These political and scientific commitments shaped Virchow's vision of domestic archaeology in several ways. First, he brought his faith in science to anthropology. He demanded a focus on research and refused to draw conclusions until adequate data had been collected. For Virchow, new knowledge about prehistory had to be rooted in science, not Romantic notions about German ancestors.

Second, despite his place at the forefront of new knowledge, Virchow viewed his study of science and medicine as a humanitarian duty. He researched the link between epidemics and poor water quality, unsanitary conditions, and food contamination, and he urgently called for public works that would improve living conditions in cities and stop the spread of typhus.[18] Virchow also believed that science had the power to reduce prejudice and enhance the lives of his fellow citizens. He therefore worked to make scientific knowledge more accessible.

Third, Virchow rejected the use of archaeology for the production of historical narratives that would legitimize the new German Empire. In his writings, he kept the concepts of race and nation separate and advanced prehistory as the science of humankind (not Germany) in ancient times. He reminded his contemporaries of the complicated history contained in the soil of the Second Empire. This approach to archaeology cannot be separated from Virchow's antipathy toward overtly nationalistic politics and his fear that nationalist sentiment would interfere with objective research. The end result was that Virchow preferred to pose new questions and initiate new lines of research. He was content with uncertainty and incomplete answers if the evidence did not allow for clear conclusions. This ran counter to the strong desire during his day to include prehistory in the writing of a national past.

These values were clear in one of Virchow's first presentations about prehistory. In December 1865, he delivered two lectures about megalithic tombs and prehistoric lake dwellings to an educational society for Berlin's artisans and workers. Virchow began with the quotation from Johann Christoph Bekmann's historical description of Brandenburg (1751) that encouraged those interested in the past "to turn to the things themselves." One had to turn to the objects, Virchow explained, "because historiography has certain limits; it is silent when we ask questions about time periods when there were no history books, periods that are not even captured in sagas." These words echoed the antiquarian plan to use excavated objects to shed light on the darkness of prehistory, but Virchow also stressed the need for scientific methods distinct from history and *Altertumskunde*. "At this point," Virchow continued, "the historian must defer to the investigator of nature, or if he does not want to do that, he must become such an investigator himself and learn to read from the book of nature."[19]

Virchow admired Danish archaeologists who, in the 1850s, had called upon geologists, zoologists, and botanists to investigate remains preserved in peat bogs. They treated these sites like "the strata of the earth's crust" and worked through layers of rubbish and the bones of mammals, birds, and fish. From this evidence, they suggested that these early people belonged to a sedentary society that hunted but did not farm.[20] Virchow called for this same approach to be applied to lake dwellings in central Europe. As settlement sites, they offered a look at daily life, and they allowed researchers the opportunity to study the social organization of prehistoric peoples. This approach dramatically expanded the scope of what needed to be collected. For archaeologists, supposedly mundane objects like pottery shards, food waste, and other traces of settlements became as significant as aesthetically exciting objects like shiny buckles, long swords, and ax-heads.

Virchow cast the genealogical questions of *vaterländische Altertumskunde* in a different light. Referring to the magnificent Brandenburg chamber graves, he stated that "these dead are our ancestors, and the questions we direct toward these graves concern our very own origins." But he framed this issue with a set of questions about human progress: "Where did we come from? What has been, from the earliest beginnings, the path to our contemporary level of cultural achievement [*der Weg unserer heutigen Bildung*]? Where will this lead us and our descendants?" Virchow's use of the German word *Bildung* in this context is significant. In the nineteenth century, the term most often referred to the fulfillment that came through education and self-cultivation, but in his lectures on prehistory, Virchow used *Bildung* to signify the accumulated knowledge of a culture. He described the Three Age System as three levels of cultural development (*Bildungsstufen*). The first step in humankind's mastery of nature was the ability to form antlers, bones, and stones into tools during the Stone Age. The shift from stone to metal represented great progress because it required skills like mining and smelting and knowledge of how to work bronze into decorative forms. The use of metal also revealed more complex social interactions, Virchow argued, because it depended on "the cooperation of many, the division of labor, and the development of trade."[21] This presentation demonstrated the difference between *vaterländische Altertumskunde* and the new anthropological approach. For Virchow, domestic archaeology not only revealed local events but also explored social interactions and cultural production in order to show the progress that human beings had made.

Virchow used this discussion to explain new ideas about chronology to his audience. He noted that individuals in Homer's time had access to metal (through mining or trade) and possessed the ability to produce ornamental shields and swords of bronze. In the nineteenth century, humankind worked with metal on a vastly larger scale, but this development had required a tremendous amount of time. "The experience of millennia was required to bring mining and metallurgy to such a high level of development that today no metal-bearing rock needs to be left unused. If we count back one or even two millennia in this experience, we encounter humans who had only a very incomplete knowledge of rocks and were able to work with them only with the greatest difficulty. Very slowly one metal after another came into use, only after each metal had been discovered and the technical means to process it had been developed."[22] If the development from Achaean bronze-working to Germany's industrialization took two and a half millennia, Virchow argued, then certainly the transition from a period with no knowledge of metals to one with the skills to manufacture bronze suggested a much deeper time than Homer's day.

This view of prehistory challenged the short chronology of *vaterländische Altertumskunde*. Virchow explained that earlier scholars had developed two alternatives to account for the earliest inhabitants of central Europe. One group maintained that the Celts were the first people to inhabit central Europe and that the Germanic peoples arrived later and conquered them. This view recognized Celtic people on the borders of Europe in Wales, Ireland, and Brittany as the descendants of a much earlier population who had lived throughout Europe. A second theory identified a Finnish-Lappish people as the original inhabitants of central Europe. They were displaced by a powerful wave of Germanic tribes but still survived on the fringes of Europe as Arctic nomads. Virchow was skeptical of both of these explanations and their common theme of strong tribes subjugating or annihilating a weaker people. He asked whether progress truly came about because, "as many assume in the rest of nature," a stronger and more intelligent people had conquered or destroyed another and supplanted that weaker culture with a more vigorous one. He spoke instead of scholarship that "sought progress not in the replacement of one race by another but rather in the actual progressive development of the native population."[23] This view acknowledged the tremendous changes brought by migrations and invasions but presented them as processes of interaction and synthesis rather than as a series of abrupt shifts or the wholesale replacement of one culture with another.

Virchow's alternative narrative about progress in prehistory—substituting cultural change for ethnic overthrow—unsettled the general story told by *vaterländische Altertumskunde*. Now the debate was no longer about the early history of a village or the connection between ancient royal graves and later princes; it went much further back in time. Virchow's ideas blurred the lines between a prehistoric and a Germanic epoch, pushing archaeology beyond the stories of battles and settlements to a broader study of primeval society. It was this version of prehistory that Rudolf Virchow elaborated through the DAG.

Anthropology and Science in the Deutsche anthropologische Gesellschaft

Anthropology, ethnology, and domestic archaeology were not established academic disciplines in the 1860s. A few German universities offered anthropology, but this training occurred in institutes dedicated to medicine or anatomy and it focused on physical anthropology, not the study of culture. Domestic archaeology was still the domain of local historical associations and largely isolated from interdisciplinary questions about cultural development. The founding of the DAG in 1869 fundamentally changed this situation. Its national organization and branch societies created a network of geologists, archaeologists, medical doctors, geographers, zoologists, and botanists who addressed the new research questions of the 1850s and 1860s.

The DAG was robust, with 2,350 members by 1884 and over 2,500 by 1900. Its Berlin branch, the Berliner Gesellschaft für Anthropologie, Ethnologie und Urgeschichte (BGfAEU), was the largest and most active with over 430 members in 1880.[24] The DAG published research by its members and reviewed work by other scholars in the *Archiv für Anthro-*

pologie. Its *Correspondenz-Blatt,* a supplement to the *Archiv,* reported on the DAG's annual meetings and the activities of the branch societies. The BGfAEU had its own publications. The *Zeitschrift für Ethnologie* published new research, and the *Verhandlungen der Berliner Gesellschaft für Anthropologie, Ethnologie und Urgeschichte* covered the society's meetings and other business.

Members of the DAG conceived of anthropology as a very broad field that studied the development of humankind and culture. This field combined physical anthropology, ethnology, and prehistory in a unified triad, not a hierarchy of fields, for anthropology could not achieve its desired breadth unless it had the international dimension supplied by ethnology and the chronological depth provided by prehistory.[25] The DAG collected data about the physical attributes of both Europeans and non-Europeans to develop the field of physical anthropology. Their surveys focused on skull measurements and the distribution of hair color, eye color, and skin tone. The field of ethnology relied on the DAG's acquisition of material culture from overseas. Most of the research in this area took the form of descriptions and comparisons of cultural practices among "primitive" societies. To advance the field of prehistory, the branch associations carried out excavations throughout Germany and published detailed local studies. Specific research questions strengthened the connections between the three fields. Anthropologists wondered, for example, if the study of skulls and other physical attributes could help identify prehistoric peoples and races. They also tried to understand the development of human culture through comparisons of present-day non-European peoples and prehistoric Europeans.

The DAG steered the methods of antiquarianism in new, more scientific directions by directing national projects, reporting important advances in the field, and organizing annual meetings that brought representatives from regional associations and museums together. Virchow, who served as head of the DAG for most of the period between 1870 and 1902, adopted a didactic stance toward local historical associations. He hoped to integrate their work into DAG projects, and time and again, he stressed empiricism and uncertainty in archaeological questions. Nowhere was the DAG's guiding influence clearer than at the annual congresses that brought scientists and historical associations together. Two of these meetings in particular, the 1875 congress in Munich and the 1880 congress in Berlin, highlighted the way the DAG advanced domestic archaeology as an anthropological practice.

The annual meetings of the DAG rotated throughout Germany, and leaders of the society always included a discussion of archaeological questions related to the local venue. The 1871 congress in Schwerin, for example, included a presentation by Virchow about the *Pfahlbauten* along the Baltic coastline. Similar to the approach in his public lectures, Virchow urged the participants to see that the study of these sites demanded attention to scientific and methodological questions.[26]

The proceedings of the DAG's 1875 congress, held in Munich, were particularly significant because Bavaria had a strong tradition of archaeological study, and the DAG planned a large exhibit of objects from the collections of Bavarian historical associations to complement the meeting. The settlement of Bavaria in ancient times served as a main topic, and the discussion revealed the degree to which Virchow's vision had reached the wider audience of the historical associations. Virchow acknowledged that antiquarians often posed genealogical questions, but he called upon his audience to avoid rash conclusions and to allow room for uncertainty. He repeated his calls for the combination of archaeological work and other disciplines, including comparative anatomy and chemical analysis, and he encouraged his listeners to consider discussions about periodization that were going on in other countries. In this context, he drew special attention to the difference between the physical anthropology of prehistory (*Urgeschichte*) and the question of lineage in early history. Virchow acknowledged that the questions of lineage were important, but he stressed that these two realms of time required different research methods. He did not want the questions of early history to direct the deeper questions about humankind posed by *Urgeschichte* (6–13).

Virchow's emphasis on *Urgeschichte* was part of an important vocabulary change in the last third of the nineteenth century that captured the new understanding of deep time. Antiquarians had used many terms to describe the period "before the evidence of history," including *Vorgeschichte*, *Vorzeit*, and *Urgeschichte*. Those who embraced the anthropological orientation in prehistory began to divide prehistory into two periods. The first (called either *Vorgeschichte* or *Urgeschichte*) referred to the expanse of time that was truly without textual records. It dealt with the origins of humankind, the Stone Age, and the Bronze Age. *Frühgeschichte*, or "early history," emerged as a new term for the epoch that followed. It took in the short chronology of domestic archaeology, covering the interaction between the Roman Empire and barbarian peoples and the establishment of early medieval kingdoms.[27] The binary designation of pre- and early history (*Vor- und Frühgeschichte* or *Ur- und Frühgeschichte*) began to appear in museum names and scholarly publications to mark this distinction.

The archaeological display at the 1875 congress offered further evidence that the DAG viewed and interpreted objects in ways that were different from *vaterländische Altertumskunde*. Joseph Würdinger, conservator of the Historischer Verein von Oberbayern's collections, noted that local antiquarianism was "groß im Kleinen und klein im Großen," meaning that it had devoted impressive energy to small details but had made little headway with broader questions.[28] He praised the display because it allowed members of associations to contemplate the kinds of anthropological questions Virchow had raised about the *Pfahlbauten*: he hoped the comparison of finds from throughout Bavaria would lead to insights about cultural development, and he highlighted questions about trade with cultures in southern Europe as particularly interesting. In his comments (22–26), Würdinger avoided language about "ancient Bavarians" or "ancient Germans." Instead, he asked questions about "the ancient peoples who once lived in our homeland." Later in the congress, Würdinger spoke again about the earlier population of Bavaria, emphasizing migration and trade in Lower Bavaria, Upper Bavaria, Swabia, and along the Rhine River, not the simple identification of ancient peoples (34–35). This coverage contrasted with the debate that members of his historical association had over the graves at Fridolfing twenty-five years earlier. Würdinger considered the life and times of Bavaria's earlier inhabitants without making genealogical connections between ancient tribes and present-day people.

On the second day of the Munich meeting, Friedrich Ohlenschlager, a *Gymnasium* teacher and also a member of the Upper Bavarian association, provided an example of the way local efforts could be linked to a larger and more meaningful project. He introduced the progress made on a comprehensive map of prehistoric sites in Bavaria. Even though the project had already registered over eleven thousand grave mounds (one thousand of which had been excavated), Ohlenschlager commented that this kind of work was still in its infancy and required the "dedicated and energetic collecting and diligent comparing" of local groups. He also offered cautious comments about the identification of ancient sites and peoples. The line between *Urgeschichte* and *Frühgeschichte* was very clear, he explained, and one could speak of ethnic identities like the Franks or other tribes only for the latest, post-Roman periods of early history, but these connections were lost when one went back into prehistory (36–39).

The success of the 1875 meeting suggested the possibility that what Munich had done for one region, Berlin could do for a united Germany. The DAG planned for a grand event in the capital city in 1880 that would include an "all-German exhibit" of archaeological objects. The DAG leadership asked hundreds of historical associations to loan artifacts from their small collections. The resulting exhibit brought together a great variety of objects from over two hundred private and public collections. Over 2,100 delegates attended the meeting during the second and third weeks of August, more than twice the number that attended the successful 1875 meeting in Munich. The dramatic increase reflected not only the growing participation in the DAG but also the importance of the German capital as a center for the study of prehistory. The Berlin branch was the largest and most active in the DAG, and Virchow, its head, was an internationally recognized scholar and the undisputed leader of prehistoric archaeology in Germany. Berlin was also the location of three very important archaeology collections: the BGfAEU's collection, the archaeological collection of the Märkisches Provinzial-Museum (Provincial museum for the Mark Brandenburg), and the collection displayed in the Neues Museum.

The significance of the 1880 congress was reflected in the press. Local newspapers covered the DAG's meetings each year, but the Berlin event received exceptional coverage. The city's dailies celebrated the dignitaries of science who were present, including Virchow; Heinrich Schliemann, who discovered Troy in Asia Minor; and Baron Nils Adolf Nordenskiöld, the Finnish-born explorer who had just returned from an expedition to the Bering Sea. Press reports followed the involvement of the Prussian crown prince and princess with special delight. The royal couple served as patrons of the display of archaeological artifacts, attended several of the conference sessions, and joined an archaeological outing to a site near Potsdam. The involvement of dignitaries gave the DAG proceedings a certain flair and affirmed the pride in the national capital as a place where science, high society, and the state converged.

The 1880 meeting was also unique because it brought the relationship between politics and domestic archaeology to the fore. Berlin was symbolically important as the capital of the unified German Empire. In its planning, the DAG called upon the state to make its eleventh congress a truly national event. Johannes Ranke, professor of physical anthropology in Munich and general secretary of the DAG, appealed to Robert von Puttkamer, the Prussian minister of the interior, by linking the work of the DAG to the history of the nation-state. Ranke described the society's main goal as "the scientific study of Germany's monuments from both the pre- and the post-Roman periods, especially where such monuments tell of a history that is not confirmed by written sources. From these prehistoric remains, [the DAG] reconstructs the migrations and settlements of different peoples on German soil. This is the story of the creation of our nation."[29] The interior minister's December reply was positive, and in February 1880, the exhibit commission—made up of Rudolf Virchow, chairman of the DAG; Dr. Albert Voß (1837–1906), the assistant director for the prehistory section in the Neues Museum; and Ernst Friedel (1837–1918), a Berlin city councilor and founder of the Märkisches Provinzial-Museum—sent its plans to the Cultural Ministry. The correspondence described the kinds of objects that would be exhibited and requested 9,500 marks to cover the costs for display cases, shipping and packaging of the objects, and printing expenses. A catalog was particularly important "so that the exhibit would achieve a lasting value."[30] To justify this sum to Karl Hermann Bitter, the finance minister, Puttkamer underscored the problem of fragmentation in German anthropology. He supported the plans because "the exhibit would be of great value for this field of study, which suffers from the division . . . of its material into countless private and public collections."[31] The Cultural and Finance Ministries approved the request and eventually allocated over 11,000 German imperial marks for the event. This kind of support for the DAG was not new. The society had received subsidies from the Cultural Ministry since its founding (beginning with 300 thalers in 1871 and increasing to 1,500 German imperial marks annually in 1873), but the generous backing in 1880 showed the state's commitment to the Berlin event.

State support for the 1880 congress went beyond subsidies. Puttkamer encouraged government authorities in Prussia to support the exhibit. (The commission estimated that over 150 collections existed in Prussia—around 60 of these were private—and another 120–30 in the rest of the German Empire.) The crown prince served as the patron of the congress, and the "all-German" exhibit was displayed in the Prussian House of Representatives. These political connections gave the congress and exhibit energy and prestige, but they did not mean that the German Empire or the Prussian state was fostering a national prehistory. Despite Ranke's statement about "the creation of our nation," the proceedings had a very mild national tone. There was no talk of a national narrative that stretched back into the prehistoric past. When the DAG did express national pride, it was most often related to the state's support for science. This was clear in Virchow's opening address to the 1880 meeting. With the cultural minister at his side, he cheerfully announced plans for a new Museum für Völkerkunde that would include the prehistory collection.[32]

The DAG's appeal to historical associations and the owners of private collections enthusiastically announced that the planned exhibit would, in contrast to the one in Munich in 1875, cover the entire German Empire. And it would not bring together just the most beautiful objects or oddities but, instead, would "offer an instructive overview of the most important objects that exhibit what is distinctive in specific regions and in the overall cultural development of the entire land. Even with the limited space, the exhibit will provide a complete picture of the prehistoric stages of development and the many relationships between the different parts of our fatherland that were so important for its cultural history."[33] The display created a temporary national collection of archaeological objects, but the request for submissions showed how much the agenda of domestic archaeology had changed since the middle of the nineteenth century. The emphasis was clearly on prehistory, not early history. The DAG wanted to address general questions about cultural development, not document who lived where in ancient times. It therefore divided objects into categories that reflected the discovery of deep time before the Common Era: first were the items from "the oldest of times, (Paleolithic) when the mammoth and reindeer roamed through central Europe." Items of interest included findings from caves in Switzerland and the Harz Mountains. Bird, fish, and animal bones, as well as mussel shells, were particularly important, for such food leftovers allowed conclusions about "past living conditions." The commission noted that much of this material might be found in natural history collections, but they wanted it as part of the archaeological exhibit.[34]

Items produced during the Stone Age (Neolithic) formed a second category and included flint tools, stones that had been bored through or polished, and decorative stones that perhaps had served as jewelry. The commission emphasized the importance of items from graves and settlements, including *Pfahlbauten*, and implored senders, "Don't hold back the skulls!" (Schädel nicht zurückhalten!) Only the third category dealt with the period traditionally covered by *vaterländische Altertumskunde*. Objects manufactured "after the discovery of metals" included bronze and iron items from before the Common Era up through the early medieval period. Even here the generic heading of "*Metallzeit*" (metal age) stressed that the exhibit would discuss broad questions of cultural development, not the formation of nations or regional identities out of premedieval tribes. And the commission encouraged historical associations and private collectors to look beyond the local significance of their items and consider how they might relate to Scandinavian, Slavic, and Baltic finds. For guidance, it listed several reference works that placed archaeological objects in a broader comparative framework.[35]

After the formalities of welcoming conference participants and thanking the DAG's hosts and sponsors, Virchow began the scientific sessions. He announced that the discussion would start with the most recent stages of prehistory. The congress would begin with antiquities that belonged to the epoch that was "half historic and half prehistoric, maybe even ¾ historic and ¼ prehistoric," and events that were familiar but required more precise historical study. The anthropologists would proceed back in time until they arrived at "the earliest beginnings, where then our friends, the geologists, can continue the discussion" (40–41). Two themes marked this survey of prehistory: first, the presentations confirmed an understanding of prehistory that went much further back in time than what was typically found in the publications of historical associations; and second, the speakers presented and debated lots of evidence, but they tended to leave major questions unresolved. The congress stressed the need to collect more data, not the creation of a consensus on specific issues.

Ernst Friedel gave the first presentation, which was on Slavic settlements in Berlin during the early medieval period. He based his comments on archaeological evidence, chronicles, and coins. Friedel concluded that the year 1141 marked the moment when Christian Germans began ruling the area. "At this point, the heathen prehistory of Berlin and the surrounding area along the Spree and Havel Rivers ended" (44). Presentations on the hill forts of Schleswig-Holstein, Merovingian row graves, and Carolingian architecture in the Rhineland followed, all of them stressing the combination of historical and archaeological methods. In his comments on Schleswig-Holstein, Heinrich Handelmann noted that "prehistory has no better partner in this area than historical criticism. This alone can clear up assumptions that are false but deeply rooted and vindicate the prehistoric character of these antiquities" (49–50). Virchow wrapped up the discussion of the most recent era of prehistory with comments about the Schlossberg in the Spreewald, the archaeological site that congress participants had viewed on the previous day. He explained that antiquarians had traditionally viewed sites like this as having a Slavic, then a Germanic layer. In the popular mind, the Schlossberg was a Slavic settlement that Germanic peoples had conquered and occupied ever since. Virchow argued, however, that his excavations had revealed a single cultural layer and that early medieval chronicles documented only a Slavic presence in the area. This suggested a long Slavic presence and only a very recent German occupation (75).

As the proceedings moved into epochs with fewer written sources, the discussion of methodology continued. Otto Tischler presented evidence from East Prussia that went back to the second century CE (80–85), and this was followed by Oscar Fraas's plea for a universal system of notations on prehistory maps (86–90). Count Wurmbrand commented on the beautiful finds from Hallstatt and speculated about the possibility of determining whether prehistoric peoples created new artistic styles or if they merely imitated the work of classical civilizations (95). Virchow noted that anthropology needed to study trade between peoples and viewed the Hallstatt culture as a conduit that transported cultural knowledge from southern to northern Europe (100).

The accessibility of the all-German exhibit provided some particularly colorful moments that emphasized the DAG's methods of observation. Johannes Ranke was curious about the age of artifacts found in caves in Upper Franconia and asked if reindeer had lived in Germany during the time of Julius Caesar. Virchow guided the proceedings not by determining the age of these particular caves but by calling for someone to bring in the antler racks. It was crucial for him to have the object under discussion in front of the audience, and he noted that the exhibit afforded a wonderful opportunity to compare reindeer antlers from across central Europe (125). The session did not answer Ranke's question. Instead, it seemed to emphasize that the observation of artifacts was a worthy end in itself.

The proceedings in Berlin must have been somewhat frustrating if participants wanted a clear story about Germany's ancient past. The experts spoke more about methods and evidence than conclusions. Their topics followed up on the long-standing interest in the chronological and territorial borders between Slavic and Germanic peoples, but they did not provide a coherent narrative that connected specific sites and ancient events to present-day peoples, and the rhetorical flourishes that had animated Konrad Haßler's description of the row graves in Ulm or Georg Wiesend's identification of "our Germans" at the Fridolfing cemetery were absent. Andrew Zimmerman has cited the inability to reach conclu-

sions as a general feature of the DAG's proceedings, and he views the re-
ferral of unsettled questions to an expert committee as an authoritarian
feature of a society that claimed to operate in democratic ways.[36] It is true
that DAG leaders expected that experts would have the final word, but at
least on questions related to pre- and early history, the interventions by
leading authorities kept the discussions open and expressed the need for
more evidence. The main effect of these interventions was to prevent the
discussion from falling back onto the conventions of *vaterländische Altertumskunde* and simple statements about German or Slavic artifacts. Despite the desire for a comprehensive or definitive version of European or
German prehistory, the DAG claimed that they were not ready to produce this knowledge.

The emphasis on methodology and empiricism was clearly evident in
the all-German exhibit. The DAG praised Albert Voß's organizational
efforts with a string of superlatives, noting that the exhibit presented "for
the first time an exact division of the cultural periods of the premedieval
prehistory of Germany." This scientific achievement "was the result delivered by the fortuitous cooperation between archaeological and anthropological research in Germany over the past ten years."[37] The catalog that
accompanied the exhibit was a further example of the DAG's preference
for observation over historical narration. Organized according to geography (in part to reflect the holdings of the regional collections that had
sent their objects to Berlin), it filled over seven hundred pages and provided detailed descriptions of individual artifacts. There was no attempt
to provide an overview of the settlement or cultural development of Germany out of the premedieval past. Additionally, Voß prepared a photo-
graphic album of 168 of the most important artifacts in the exhibit. This
volume, which the DAG described as essential for the study of premedieval antiquities, contained no text at all, inviting the close observation of
the individual photographs.[38]

Berlin newspapers accepted the scientific character of the congress and
all-German exhibit. The conservative *Neue preußische Zeitung* noted that
the rotating nature of the DAG's annual meetings sought to attract new
members and create a wider resonance for archaeology and anthropology,
but that the meetings were serious. "Attention-grabbing lectures are ruled
out. The discussions treat the objects with strict objectivity, not with sensa-
tionalism."[39] The *Berliner Tageblatt* was more animated in its description
of the balance between serious and popular understandings of science.
The paper advised its readers that the archaeological exhibit

engages less the eye than the intellect of the visitor. After all, this is not an exhibit of the most beautiful or most peculiar objects; rather, it offers an *instructive overview of finds from particular regions in Germany that reveal the course
of their cultural development.* . . . The visitor therefore will not get much out of
a casual stroll through the exhibit. He cannot nonchalantly pass by spear tips,
for they are witnesses to early forms of human existence. He cannot just give
a quick glance at fabric remains from the area around Lake Constance, for
these are the last remnants of life at the lake dwellings. And when he enters
the collection of skulls, he cannot allow himself some cheap reminiscence of
the gravediggers scene from *Hamlet.* Those are the finds from the era of mammoths and reindeer and the Paleolithic period. They represent the first traces
of humankind and the period when humans learned to shape stones into tools.

Yes, this rich collection will increase our knowledge about the prehistoric
era. It will reach its purpose and intent, though, only if it awakens a greater appreciation for the importance of prehistory among a wider audience. . . . People
must become more familiar with the concept of *Urgeschichte* so that the farmer
who finds an ancient urn or weapon while plowing does not just throw the
object away but rather preserves it and sees it as worthy of study and so that
misplaced piety no longer forces people to quickly rebury skulls and skeletons
that are discovered during construction projects. Only then will anthropology
have come closer to achieving its goal. May this congress be an effective start
to this process![40]

Finally, the *National-Zeitung* from August 5, 1880, noted the contrast
between the hot debates of the final days of the Prussian Diet's summer
session and the rational calm of the anthropology congress. The paper
observed that while politics created division and conflict, science delivered the beginnings of answers to questions that interested a wide Berlin audience: "It must have been particularly impressive to experience the
voice of science in a hall that so recently echoed with the loud disputes of
political parties. Science speaks with a confidence and verity that politics
cannot match. It is refreshing that anthropological research replaces the
Kulturkampf in this building. While religions have tried to solve the riddle
of humankind [for centuries], it is actually science, namely anthropology,
that now speaks to these questions."[41] All the newspapers described the
DAG's all-German exhibit as a scientific event and not as a presentation
of the nation's earliest history. The anthropological orientation of prehistory focused on objects, not narratives, and on human questions, not
German ones. This approach steered domestic archaeology away from

the focus on tribes and ancestors that was so central to *vaterländische Altertumskunde*.

Prehistory and Race

Because of the prominence of anti-Semitism and radical nationalism in modern German history, scholars have examined the discussion of race and science within the DAG very closely. Early biographers presented Rudolf Virchow's contributions to medical research and public health in a very positive light and viewed them, along with his liberal politics, as parts of a greater humanitarianism. More recent studies that focus on the history of anthropology contrast Virchow's emphasis on objective research and measured interpretations with the strident claims made by nationalist authors. Historian Benoit Massin, for example, has explained that the DAG "must be regarded as anti-racist" when compared with the rising tide of prehistory and anthropological works in the late nineteenth century that advanced claims of Nordic origins and Germanic purity. Similarly, Andrew Evans has argued for the persistence of a "liberal paradigm" within German anthropology until World War I that emphasized empiricism, the separation of science from politics, and the unity of humankind.[42]

An important challenge to this positive evaluation has come from historian Andrew Zimmerman. For Zimmerman, it is not enough to show that anthropologists publicly renounced nationalist conclusions and anti-Semitism. Instead, he frames these issues in terms of epistemology and probes the way anthropology sought to categorize human beings. At the center of Zimmerman's analysis is the challenge that anthropology presented to historical writing. "For most of the nineteenth century," he writes, "historicism had dominated the human sciences in Germany, setting the goals and methods of the study of humankind and offering a cultural and political identity based on the interpretation of textual sources of the European and classical past." In this context, "[t]he project of history involved the self explicating the self," meaning that the author or reader of history viewed himself and his present day as the culmination of a historical process.[43] The historicist approach became particularly problematic as international communication increased and Europeans were exposed to knowledge about the rest of the world during the imperial age. Europeans broadly believed in the unity of humankind, but the historicist paradigm could not include the histories and cultures of non-Europeans because they stood outside the trajectory that connected the classical past to the civilized, European present. Adolf Bastian (1826–1905), who studied cultures throughout the world as one of the leading scholars in the Berlin branch of the DAG, became the most ardent critic of historicism's limits. He argued that traditional history studied only the "high points" of the human story. It focused on the "quality" of history but dismissed the "quantity" of history. As Zimmerman explains, "Bastian thus proposed that anthropology rectify the narrow range of historical enquiry to create a discipline that would truly fulfill the goal of the humanities." Anthropology would use material culture, not texts, to investigate humanity, allowing scholars to study "peoples without history."[44]

Zimmerman describes Bastian's approach and the practice of anthropology in the German Empire as "antihumanist." This term refers specifically to the replacement of texts by objects in the study of humanity, but also more broadly to an outlook that believed in the ability of science (instead of the humanities) to explain the human condition. Antihumanism relied on scientific methods (the collection of objects as empirical data, measurements of bodies and bones, maps that displayed the distribution of finds) to capture the variety of humanity. Although the DAG may have set out to overcome Eurocentric historicism, Zimmerman argues that its treatment of human beings as scientific objects ultimately taught Europeans to see cultural differences as natural and permanent, making antihumanist anthropology a crucial precondition for the rise of scientific racism. Furthermore, Zimmerman shows that liberal anthropology was directly tied to an imperialist system of collecting and displaying that treated non-Europeans in inhumane and even murderous ways.[45] Was, then, the DAG a bulwark against nationalist interpretations in anthropology, ethnology, and prehistory, or was it responsible for the rise of scientific racism in Germany? Physical anthropology has been at the center of this debate because it was most clearly related to the rise of racial anti-Semitism, scientific racism, and eugenicist thought. Less attention has been paid to the place of prehistory even though the Germanic past and ideas about a supposed prehistoric Nordic race were critical to radical nationalist groups in the late German Empire, Weimar Republic, and Third Reich. By considering the DAG's contributions to the field of prehistory, questions about race and identity in German anthropology appear in a new light.

In Zimmerman's analysis, the study of objects (and humans as objects) and the study of non-Europeans occurred simultaneously, making non-European bodies the primary objects to be studied in antihumanist ways. Domestic archaeology, however, came out of a long tradition of antiquarians using objects to talk about themselves. They viewed the bones in row graves as "their ancestors." Yet their work was also outside the canon of historicism. Barbarian peoples were described in ancient Greek and Roman sources, but they did not belong to this cultural tradition. Domestic archaeology relied on material culture, but it was still a form of self-study. In this context, the work of the DAG was not primarily a precursor to scientific racism. It was first a challenge to *vaterländische Altertumskunde* and the way historical associations used archaeology to shed light on the prehistory of their towns and regions. The anthropological approach drew attention to a much deeper time, and it rejected simple conclusions that treated artifacts as ancestral remains.

Zimmerman is correct to point to the centrality of race within the DAG and the organization's belief that race was a legitimate scientific topic. Yet leaders of the DAG were very skeptical of attempts to use anthropology to write the prehistory of the nation. After decades of research, dozens of annual conferences that debated findings and methods, and countless detailed studies, the leaders of the DAG continued to assert that questions related to identity and race either could not be solved or required still more evidence before they could be solved. Virchow and other leaders of the DAG therefore rejected what they perceived as the hasty and biased conclusions of nationalist authors. They argued that anthropology and prehistory could describe but not narrate. This position did not lead to racism but rather to a dead end. By the 1890s, no way out of the dead end had been found, and many scholars came to see the DAG's research as unsatisfying. Racist ideologues refused to go down the same road and invented new lines of interpretation that drew clear conclusions about race and identity.

The scientific study of race goes back to the attempts by Carl Linnaeus (1707–78) and Johann Friedrich Blumenbach (1752–1840) to classify humanity according to a distinct number of races.[46] Against the backdrop of Darwinian thought and imperialism, this line of research took on even greater significance in the second half of the nineteenth century. Anthropologists across Europe set out to identify human races by recording and comparing the physical traits of peoples in Europe and from around the world. Many leaders in the DAG, including Alexander Ecker, Hermann Schaaffhausen, Johannes Ranke, Julius Kollmann, and Rudolf Virchow, were medical doctors and anatomists, and they played a critical role in developing physical anthropology in Germany. The DAG sponsored fieldwork and large surveys that amassed an extraordinary amount of data about body types, skull shapes, and facial features, as well as variations in hair color, eye color, and skin tone.

As with other pursuits of the DAG, physical anthropology cut across several fields of study. Rudolf Virchow, for example, amassed a collection of skulls during the 1850s for both medical and anthropological research. And from the beginning, physical anthropology was closely tied to prehistory. Several scholars, for example, immediately recognized the ancient remains found in the Neander Valley in 1865 as a transitional figure in evolutionary development on the basis of the formation of the excavated skull.[47] Even more important was the possibility that physical anthropology might address questions of identity in pre- and early history. If races were immutable, as many scholars believed, then the comparison of contemporary peoples and ancient bones might shed light on patterns of migration and settlement. Race would provide one more perspective, alongside the fields of linguistics, folklore, classical philology, and archaeology, to triangulate the question of origins and produce more certain conclusions about human development and identity. This line of thinking was already present in Konrad Haßler's reference in 1866 to "a white race" in his conclusions about the graveyard near Ulm.

Craniology, the study of skulls, was especially important for the connection between prehistory and racial thought. It was impossible to classify attributes like skin color and eye color at the archaeological site. The skull, however, endured the ravages of time and offered material evidence that might allow physical anthropology to trace racial categories back into prehistory. Scandinavian scholars had made scattered comments about prehistoric skulls since the 1830s, and in 1843, the Swedish doctor Anders Retzius published a craniological study that classified skulls as dolichocephalic (long) or brachycephalic (round) and that associated premedieval and modern peoples with four variations of these two types.[48] European anthropologists adopted this general language but eventually developed alternative ways to measure skulls. In the German Empire, the DAG led the way in the collection of data and the formulation of hypotheses related to race and prehistory.[49] Some scholars wondered about the relationship between the shape and size of the skull and the size of the brain as a factor that determined intelligence and human creativity, but the most pressing question was whether this data could be used to study the settlement of Europe. It was their hope that the measurement of excavated skulls and comparisons to the biometrics of contemporary Europeans would both firm up the definitions of racial types and establish connections between ancient and present-day peoples.

It is important to note that anthropologists were defining skull types and trying to apply them at the same time. Anthropologists in the DAG recognized the circular nature of this logic, but they also found the significance of these questions to be so great that they had to pursue them. They therefore proceeded with caution, allowing two ideas to frame their investigations. First, they refrained from connecting the discussion of race to contemporary politics and peoples. They stressed the need for more data before they could speculate about questions of development and descent. And second, they emphasized the unity of humankind. One could speak of differences within humanity but not hierarchies.

In other settings, however, the discussion of race, anthropology, and prehistory became connected to politics and the identity of modern peoples. In the mid-1850s, Joseph Arthur de Gobineau (1816–82), a well-regarded intellectual in France, published *Essai sur l'inégalité des races humaines* (Essay on the inequality of human races). This book proposed that the rise and fall of great civilizations, including the Persian, Greek, and Roman Empires, as well as European powers of the Middle Ages, had little to do with political developments, geographical advantages, or morality. Rather, their decline was tied to the "degeneration" of their racial stock. According to Gobineau, "a civilization will certainly die when the primordial race-unit is so broken up and swamped by the influx of foreign elements that its effective qualities are destroyed."[50] *Essai sur l'inégalité des races humaines* did not find a wide acceptance in France or the rest of Europe at the time of its publication, but the book suggested the idea that racial formations shaped historical outcomes.

Pseudoscientific variations of Gobineau's idea linked racial groups to modern nation-states. A famous example appeared in *La race prussienne* (The Prussian race) by Armand de Quatrefages, a prominent ethnographer who held a chair for anthropology associated with the Muséum d'histoire naturelle in Paris. In this text from 1871, Quatrefages accused Prussian soldiers of war crimes at the end of the Franco-Prussian War and claimed that this brutal behavior was related to the fact that the Prussians were descended from a barbaric Finnish tribe.[51]

Rudolf Virchow quickly rejected this text as a piece of political propaganda, but he did not stop with this dismissal. Instead, he engaged the idea of Prussian (and, more generally, German) descent from earlier tribes as a legitimate scientific question. Under Virchow's leadership between 1871 and 1886, the DAG conducted a survey of the hair, skin, and eye color of almost seven million schoolchildren in order to determine whether or not one could speak of Prussians or Germans as a racial category. In the presentation of his results, Virchow rejected the idea of a "Prussian race" and advanced the idea that all the peoples of central Europe were of "mixed blood."[52] This conclusion undercut Quatrefages's attack on German civility, but it also stood in line with Virchow's rejection of a simpleminded anthropological interpretation that would support ideas about eternal national communities.

Despite its nonnational character, Andrew Zimmerman has pointed to the school survey as a moment when Germans became aware of racial differences. He explains how teachers lined students up according to their eye and hair color and their skin tone. Teachers then tabulated the results on a form provided by the DAG. Zimmerman draws particular attention to Virchow's request that teachers separate Jewish and German pupils in the survey and argues that Virchow and the other organizers of the school survey saw Jewishness as a separate racial category. Or, in more provocative terms, they felt that Jews "posed a threat to the purity of the data." Although the survey contained neither overt ideological statements nor notions of racial hierarchies, Zimmerman argues that it encouraged teachers and millions of students "to *experience* themselves in terms of whiteness and brownness and to recognize racial distinctions between 'Jews' and 'Germans'" and that the study stands as "one of the major events in the history of German anti-Semitism."[53]

This evaluation of the survey runs counter to the actual discussion of race and identity that took place within the DAG. Virchow rejected the idea of either a German or a Jewish race. He did treat Germans and Jews as distinct groups of people, but these were labels that arose from historical circumstances, not biological markers. Virchow viewed every group as the product of racial mixing, so it was invalid to speak of national races. It is certainly conceivable that teachers and students did not draw the same conclusion as Virchow and other leading anthropologists. They could have seen the study as providing scientific authority to the idea that contemporary peoples (themselves included) could be categorized racially. If they did, they were guided, not by Virchow's refusal to use physical anthropology to explain national identity, but by other intellectuals who believed that nations and races were overlapping categories.

A lengthy exchange during the DAG's 1875 annual meeting in Munich illustrates the uncertainty and caution that permeated discussions of anthropology, prehistory, race, and nationalism. The event featured the impressive display of archaeological finds from across southern Germany, and in the meeting's final session on August 11, Alexander Ecker raised several questions about the title of the display, "Artifacts from Celtic and Germanic Prehistory," and labels that referred to "Celtic and Germanic skulls." Ecker was an anatomist from Freiburg and a cofounder of the DAG's *Archiv für Anthropologie*. In the 1870s, he was the closest thing to an expert in the developing fields of craniology and racial studies. He hoped that the meeting would answer some fundamental questions, which he formulated as follows: "When the Germanic tribes came to southern Germany, did they find that another people was already there, a people with a higher level of cultural development? Did they partially displace this people? Did all the older gravesites belong to this people or are they to be seen as original Germanic property [*ureigenes germanisches Eigenthum*]?" (73). These questions were based on the basic outline of central European prehistory that scholars had developed between the 1850s and 1870s. It recognized that Neanderthal and related remains were from the truly deep past. Lake dwelling sites were younger than those remains but still very ancient. After that came grave mounds that were often associated with the Celts and then row graves that were associated with Germanic peoples. Ecker's questions revolved around the peoples who created the grave mounds and row graves.

Scholars wanted to turn to ancient skulls to sort this out, but Ecker saw many problems with this endeavor. First, he noted that these questions had to be approached from multiple perspectives, and he was glad that the DAG included scholars from many different disciplines. But he noted that, especially when discussing the Celts, one has to ask, "Are you a philologist? an archaeologist? a craniologist?" because each discipline worked with a different definition of the Celts (73). Ecker explained that the Celts of linguistics were alive among the speakers of Irish and Gallic languages; the Celts of archaeology, as propounded by Heinrich Schreiber, were the producers of everything made of bronze; and the Celts appeared in the works of Greek and Roman historians as a people who occupied central and northern Europe in ancient times. Most recently, craniologists had joined the discussion about the Celts, but they could not agree among themselves (73). (Most scholars categorized Celts as dolichocephalic, but some English scholars described them as brachycephalic.)

Ecker maintained that the Celts of linguistics, archaeology, history, and craniology did not overlap, and scholars therefore found themselves "amid a great confusion" when they sought Celtic skulls in the gravesites of southern Germany. He therefore urged his audience to abandon the application of ethnographic labels to skulls and first develop a classificatory system for skull types. This, for Ecker, was the most sensible approach "because the concept of '*Volk*' [people] is in no way stable and unchanging over long periods of time; rather, it is, so to say, in constant flux, and we do not know if the conglomerations that present themselves as a people at a specific moment in time are of a more primary or a secondary nature. Language certainly changes quite easily, and even the skeleton, which, relatively speaking, is the most stable attribute, cannot remain unaffected by repeated mixings [of peoples] and migrations" (73–74). Certainly, the definition of a people from ancient times, like the Greeks, now refers to a different people today. "To name a single skull as Celtic, Greek, or Germanic only on the basis of its form is certainly misguided, and one would do well to keep *craniological* and *ethnographic* classifications fully separate" (74; emphasis in the original).

Ecker then applied this maxim to the row graves of southern Germany, where he and other scholars had documented a consistent type of dolichocephalic skull. He explained that one could say that these graves are both Germanic and dolichocephalic, but not that dolichocephalic skulls would always indicate a Germanic grave. Grave mounds, he said, were more complicated. They contained mostly brachycephalic skulls, but these specimens varied quite dramatically within the range of shorter skulls. Furthermore, Ecker was unable to comment on how far back in time this skull type might go because there were no skulls from lake dwelling sites.

This anthropological evidence contradicted the clarity in narrative accounts of southern Germany's early history. Ecker explained that some historians who wrote about the early settlement of the Black Forest region described a tall, blond-haired people of Germanic origin who lived there as a free people and Celts who appeared as a smaller people with dark complexions and lived in a state of serfdom (74). All this just could not be so clear because, as Ecker explained, the Celts of archaeology were wealthy and had impressive metalworking skills that were not compatible with a state of servitude. Also, "these supposed Celts [of history] were brachycephalic when, of course, the people of Ireland and Wales today are overwhelmingly dolichocephalic and at the same time dark-haired" (74). So many contradictions arose from these kinds of comparisons that

Ecker spoke of the great confusion and awkwardness that marked this scholarly enterprise. "We can only say that in this area everything is still completely in the dark," he commented at one point (74). And after circulating a set of images of skull types, he continued: "Gentlemen, you see therefore that I can tell you absolutely nothing about Celtic skulls, and I was right when I said at the beginning of my remarks that I come asking questions. I had thought that there would be a Celtic skull in the exhibit since it deals with Germanic and Celtic skulls. I have not learned, though, of any skull in the exhibit that can be described with certainty as the skull of a Celt" (75).

In the ensuing discussion, Virchow affirmed Ecker's hesitancy and skepticism, remarking that the many names that classical authors used for ancient peoples could not be viewed as referring to identifiable races but, rather, should be understood as names for groups of people who came together because of various historical and social conditions. "In my opinion, in general, it is a hopeless endeavor to investigate whether these collective names are based on a belief in racial identity." The names that ancient authors used, Virchow presumed, could not lead to conclusions about racial categories (75).

Ludwig Lindenschmit interrupted, claiming that Greek and Roman authors were very consistent in their descriptions of the white skin, blond hair, and tall stature of the Celts. Furthermore, he argued that the characteristics found in earlier classical authors were the same ones used by Tacitus three hundred years later to describe the Germanic peoples. He therefore claimed that both descriptions were really of one northern European group that had maintained these physical attributes over the centuries (75–76). This argument, which effectively subsumed the Celts under the Germanic peoples (or saw them as a case of mistaken identity), was in line with Lindenschmit's positions since the 1840s that the Celts did not exist as a separate people and that all prehistoric bronze was imported, not evidence of an ancient people before the Germanic peoples.

Virchow countered that this was not a correct reading of ancient sources and that one had to speak of Celts in different ways for different epochs. He then expanded on Ecker's thinking by questioning the existence not only of a Celtic skull but also of a Germanic one. He agreed that one could document a range of longer skulls among the remains found in the row graves of central Europe, but he bristled at *Germanomanen* (Germano-maniacs) who spoke of a Germanic skull type. This kind of conclusion was obviously problematic, Virchow argued, because the skull forms associated with these row graves were found all over Europe. If they were all Germanic, the *Germanomanen* would have to argue that graves found deep in the interior of Russia contained the remains from a Frankish tribe (76).

Virchow then hammered away at the point that must have been most troublesome for those who thought physical anthropology and prehistory could explain national identity. He agreed that row graves contained doli-chocephalic skulls, but he reminded the assembly of the accepted fact that this skull form largely disappeared in the archaeological record after the seventh century. He joked that *"the big bad brachycephalics gradually ate them up"* (76; emphasis in the original). "How do you want to reconcile this?" he asked.

> I can well imagine that a dolichocephalic tribe came to dominate in Germany in the fourth or fifth century and that this tribe subdued other tribes.... But this tribe found that a brachycephalic population was already living there when it arrived, and the brachycephalic population was still there after it disappeared, and it continues to expand. Look at how few dolichocephalics are here among us in this assembly. All the gentlemen who are enthusiasts for Germanic doli-chocephalics are themselves exquisite examples of short skulls. Where did they come from? You believe that the long skulls of the Frankish period became this way through their fight for existence. It is possible. But I must say: it is incom-prehensible that one wants to establish an original Germanic skull type when one cannot follow it further back in time than the Franks. (76–77)

Lindenschmit objected again, but Virchow continued:

> Let us not miss the significance of this. If those are Germanic people, then we must give up on the unity, the identity, of the Germanic ur-race; then we must grow accustomed to the idea that the Germanic peoples, from their first ap-pearance in history onward, were a mixed people [*ein Mischvolk*], that they were already a product of the mixing of earlier tribes [before they arrived in central Europe], that they *do not constitute a nationality in the sense of being related by blood; rather, they are a nationality because of political developments.* (77; emphasis in the original)

Virchow then insisted that his fellow scholars narrow their use of the term "Germanic." It should be reserved for the description of artifacts and sites documented through archaeology and never used in discussions of crani-ology.

Most of the anthropological discussion of prehistory proceeded along these lines, as scholars in the DAG described the arrival and mixing of various peoples. In some instances, they also referred to the "progressive development" of peoples, by which they meant some variation of Darwin's idea of adaptation. Halfway through the exchange about craniology and ancient peoples, Virchow commented that perhaps physical attributes like skull shape, hair color, and eye color were related not only to the migration of peoples but also to environmental conditions. He noted that the statistics from the school survey revealed a higher proportion of dark-haired children in German cities than in rural areas. At this point, there were calls from the audience of "The Jews!" Virchow responded: "'The Jews!' is being called out to me. Gentlemen, you will learn many wonderful things about the Jews from the school statistics survey. You will learn, for example, about the not-so-insignificant number of blond Jews—a perhaps surprising fact, but one that is verified, so much so that for certain areas there is a higher frequency of blond Jews than in others. . . . In any case, the increase in brunets is not to be explained by the presence of Jews, who in our statistics tables are treated separately" (78–79). This brief exchange shows that at least some in the audience at the anthropology congress were quick to connect the discussion of physical attributes to modern people. Virchow did not discuss German Jews further at this point. Instead, he offered an alternative explanation, wondering aloud if the higher proportion of people with a dark complexion in cities could be due to environmental factors or nutrition. He viewed this as a valid research question and hoped that the continued assessment of the school survey might lead to new ideas.

The discussion then returned to peoples and bones. Lindenschmit again advocated for the recognition of a clear and consistent anthropological definition of Germanic peoples. Julius Kollmann, an anatomist and anthropologist from Munich, agreed that there was an anthropological consistency among the bones found in row graves, and he acknowledged a similarity between these skull shapes and skulls found in Scandinavian graves and the measurements of present-day Scandinavians. But he then posed a question that emphasized Virchow's point more than the connection between northern Europeans and modern Germans. He asked, "Who are we then? If the only Germanic skull form since the time of the row graves has virtually disappeared among the Germans of today and can still be found only in Scandinavia, then to what type do we belong, ethnologically speaking?" (80). He then suggested that the disappearance of what was known as a Germanic type (based on Frankish and

Alemannic finds) meant that there must be another, more recent skull form that modern Germans possess and is brachycephalic.

Hermann Schaaffhausen, the early interpreter of the Neanderthal finds, did not think that the situation was as hopeless as Ecker had described and proposed a position that reconciled several strands of the discussion. He believed that craniology research showed that a dolichocephalic people had migrated from the north into Germany and that linguistic research showed that the German language had arrived from Asia. The territory of Germany was not empty before these Indo-Europeans came, and the mixture of these people with earlier inhabitants explained the different shapes and variation in skull measurements (80).

Later discussions of physical anthropology and prehistory unfolded along similar lines. At the 1880 congress, the "all-German exhibit" again included both ancient and modern skulls, and there was a lengthy discussion of "general and German anthropology" during the meeting. Delegates discussed material that was much older than the first centuries of the Common Era. Johannes Ranke described several skulls found in southern German caves (128), and Alexander Ecker spoke about the measurements of skeletons found at several Ice Age sites (139). When the delegates came to later time periods, they again stressed that modern nations were not connected to the peoples of late antiquity, and Virchow's presentation about the ongoing school survey and the collection of similar data among army recruits stressed the variety of physical characteristics that precluded the definition of clear races or types. As this coverage makes clear, the DAG was not teaching people to view national groups as permanent biological communities. The society was certainly interested in questions of national origins and the concept of race, but its members were encouraged to view these questions as open and speculative. Virchow especially worked to downgrade the "truths" asserted by others to the status of hypothesis. Even Ludwig Lindenschmit, who was the most insistent about correlations between historical, archaeological, and craniological information, conceded that his claims applied only to central European row graves from the first four centuries of the Common Era, not to all of prehistory.

A focus on the relationship between archaeology and historical narration places the work of Rudolf Virchow and the DAG in a new light. By the 1850s, antiquarians were crafting narratives that connected "their ancestors" to present-day populations and archaeological sites to local towns and regions. The anthropological orientation of prehistory disrupted these links. It called for the investigation of a much longer sweep of time and of new questions about human prehistory that would largely remain anonymous. This orientation required a European, not a local, perspective, and it demanded a greater degree of patience as scholars developed comparative methods and paid closer attention to the typology of artifacts before venturing new interpretations. If *vaterländische Altertumskunde* desired clear stories about origins, the anthropological orientation demanded the collection of more data.

The DAG's challenge to simplistic ethnic readings of the archaeological record anticipated the criticisms of present-day scholars who reject the nationalist ideas that grew in importance between the 1890s and 1920s and became dominant in the Third Reich. Like the DAG in its first two decades, scholars today stress the need to combine evidence from many fields (archaeology, linguistics, and history), and they acknowledge that this process is fraught with difficulties, especially when trying to discuss the issue of identity in pre- and early history. In this sense, the emphasis on science and objectivity within the DAG appears very progressive and antinationalist.

Zimmerman's emphasis on the organization of scientific knowledge, however, does raise several important questions about anthropology, race, and identity in modern German history. First, if leaders of the DAG were so skeptical about race, why were they so obsessed with it? For decades, they argued that studies focused on race would remain inconclusive, yet they did not abandon race as a meaningful concept for the study of humanity. They continued to debate the topic, collect more and more data about physical attributes, and publish studies packed with details but light on broader interpretations. Second, did the DAG's practice of discussing ancient and modern material together have the consequence (intended or not) of validating the idea that physical anthropology could connect prehistoric and present-day populations? Did it reflect the belief that ancient and modern material could be connected in a racial history of Europe once more material had been gathered?

Here, again, an emphasis on the relationship between archaeology and historical narration is helpful. For some scholars, the entrance of physical anthropology into the methodological discussions about prehistory increased the ease with which they moved back and forth between evidence from prehistory and their present day and opened the door to the idea that race, *Volk*, culture, and language were related or even overlapping categories. Virchow mocked the *Germanomanen* who advanced these ideas, but several authors—with Houston Stuart Chamberlain and Gustaf Kossinna being the best known—came to reject the work of Virchow and the DAG largely because of their reluctance to craft sweeping narratives that connected prehistory, racial studies, and modern German politics. This shift, which will be discussed at length in chapters 6 and 7, highlights the place of the DAG in between the approaches to prehistory that stressed narration (*vaterländische Altertumskunde* and nationalist versions of prehistory) and the beginning of a tradition of scientific skepticism that endured into the 1920s and then emerged again after the end of the Third Reich. Before proceeding to that history, however, we will survey two other traditions that interacted with the knowledge produced by the DAG in the German Empire. Historical associations and regional museums attempted to create comprehensive portraits of regional cultures through collections and exhibits. This work was shaped by the research questions of the DAG and was often presented at the society's annual meetings. Alongside this regional approach to prehistory, several authors found great success by telling tales set in the distant past and giving names to the anonymous creators of ancient artifacts. They did not heed the DAG's call for caution and instead created historical fiction that characterized certain virtues as "quintessentially German" and even rooted in the deep past. Virchow's vision of pre- and early history undercut these kinds of patriotic narratives, but scientists could not dictate how authors crafted their stories. Neither anthropologists nor local antiquarians nor nationalists had a monopoly on how to interpret the ancient past.

CHAPTER FIVE

Domestic Archaeology

A Preeminently Regional Discipline

In 1912, the Berlin philologist-turned-archaeologist Gustaf Kossinna declared that domestic archaeology should be promoted as "a preeminently national science."[1] Over the previous twenty years, he had advanced new methods that identified prehistoric peoples on the basis of archaeological finds and argued that "Germanic" peoples had inhabited northern and central Europe continuously since the early Bronze Age. Histories of archaeological thought rightfully connect Kossinna's fervent nationalism to the Nazi interpretations of the distant past that followed, but it is important to remember that Kossinna's call to arms in 1912 was an expression of his desire for the field, not an accurate description of contemporary practices. No institution advanced domestic archaeology as a "national science" in the late nineteenth or early twentieth century. The DAG offered a scholarly forum to representatives from across Germany, but this important network did not have authority over the field. The RGZM in Mainz and Neues Museum in Berlin housed significant collections for pre- and early history, but they existed alongside, not over and above, the numerous local and regional museums that existed in the German Empire. Indeed, Kossinna founded his own society for prehistory in 1909 as an alternative to the DAG because he felt it was not staunchly nationalist in the way he desired.

Despite the enthusiasm that accompanied national unification and the rising influence of nationalist pressure groups during the Wilhelmine era, domestic archaeology remained a preeminently regional discipline. The number of local associations, academic commissions, and learned societies dedicated to antiquities tripled from around one hundred in the 1860s to over three hundred by the 1930s. As historian Celia Applegate has shown, many of the new groups promoted local history with a missionary zeal. The Historischer Verein der Pfalz (Historical association of the Palatinate), founded in the 1830s, "had been content to preserve" historical material and discuss local history among themselves. The Pfälzerwald-Verein (Palatine forest association), founded in 1902, in contrast, "wanted to convert." It included "railroad employees, chemists, bank tellers, and city administrators," and its three hundred members developed hiking routes, erected historical markers, and maintained a museum collection. These outreach efforts helped to familiarize locals with the history that surrounded them.[2] Even as archaeology became more popular, it remained embedded in regional contexts.

The decentralized nature of archaeology was also a constitutional and administrative matter. The German Empire had national institutions for political and military affairs, but cultural policy largely remained in the hands of its constituent states. Each of the empire's four kingdoms and six grand duchies, as well as its duchies, principalities, and free cities, administered the universities, museums, and voluntary associations in its territory. Within the Kingdom of Prussia—the largest of the German Empire's states by far—policies related to domestic archaeology devolved even further. The Prussian Cultural Ministry monitored activity and provided some funding, but it was mainly the twelve provincial governments within the kingdom that supported excavations and founded cultural history museums that included collections for pre- and early history.

During the 1860s and 1870s, state and provincial governments promoted archaeology by appointing conservators and creating historical commissions. Conservators cataloged and publicized the collections held by the monarchies they served, and their work quickly extended to material held by voluntary associations, universities, and private individuals. By managing collections and facilitating communication between the government, associations, and museums, conservators continued the process that transformed royal collections into public possessions and places of education. Historical commissions were more scholarly than traditional historical associations, and they often worked together with academic institutions like universities, the DAG, and academies of sciences.

The crowning achievement of the combined efforts of state and pro-

vincial governments, conservators, commissions, and historical associations was a regional museum for cultural history. Over a dozen of these
institutions were created between the 1860s and 1910s, and each included
an archaeological collection as a major component of its display. Like the
collections of historical associations but on a larger scale, the new museums encompassed a wide array of material, including archaeological
objects, examples of medieval sculpture and woodworking, coin collections, paintings, folklore collections, and even natural history specimens.
The broad range of material was united by the fact that it was found and
usually produced in the region. In some cases, the exhibit spaces for these
museums were initially only a slight improvement over the rooms rented
by historical associations. After several years or perhaps a few decades,
though, provincial and municipal governments raised their level of commitment and allocated large sums of money for the construction of new,
opulent buildings. They, alongside the museums for art and ethnology that
also sprang up in Germany's cities around 1900, became important institutions for education and the production of scientific knowledge, as
well as prominent landmarks that exemplified civic pride.[3] Many of these
buildings still house important regional collections today. In terms of historical scholarship and tourist itineraries, these museums are overshad-
owed by the world-class collections for art and history that developed in
Berlin and other cities. Yet they are crucial institutions for understanding
the interaction between regional and national thinking, the place of the
museum in the middle-class culture of learning, and the development of
pre- and early history as a field of knowledge.

Eduard Pinder, a director of one of these museums in Kassel, explained
that central museums, like the Neues Museum (and later the Museum für
Völkerkunde) in Berlin, the BNM in Munich, and the RGZM in Mainz,
possessed comprehensive collections and offered settings where scholars
and experts could deepen their knowledge through the study of the most
important artworks and artifacts. Regional museums, on the other hand,
offered the general public an introduction to many fields, including history, archaeology, and art. Pinder emphasized that this was not a matter of
one kind of museum being more important than another, but rather that
regional museums prepared visitors for further learning. As he explained,
one first needed to understand local variations before one could appreciate general developments in fields like archaeology or applied art.[4]

Before 1909, pre- and early history was not an independent science,
and it had a minimal presence at German universities. Regional museums

therefore served as the most important setting for domestic archaeology during the German Empire, and they presented pre- and early history in three ways. First and foremost, regional museums institutionalized the connections between culture, people, and landscape that historical associations had promoted since the 1820s. By collecting everything from the region, these museums hoped to produce an all-encompassing and more scientific cultural history of their areas. Their attention to regional traditions was particularly important as a response to the processes of unification and national integration. The promoters of local and regional history presented their states, provinces, and local areas as historically coherent units and tried to preserve their legacies after being annexed into larger entities during and after the Wars of Unification.

Anthropological questions about prehistory offered a second context for archaeology. Many regional museums followed up on the DAG's questions about cultural development in eras much earlier than early history, and they developed collections of ancient bones and Stone Age implements, as well as rocks and fossils. In this sense, cultural history was viewed as related to or even as a higher form of natural history.

A third narrative connected archaeological objects to the growing interest in applied art (*Kunstgewerbe*). Regional museums displayed the development of arts and crafts—such as the production of glass, jewelry, ceramics, and porcelain—from ancient times to the nineteenth century. In this narrative, objects such as weapons, jewelry, and ceramics were not used primarily to gain information about lifeways in the distant past. Rather, they were viewed as ancient examples of applied art and compared with Roman manufactures, medieval craftsmanship, and later products.

It is important to note that these varieties of cultural history had little to do with ethnic belonging or German nationalism. The catalogs for these collections refrained from comments about race, ethnicity, or the descent from earlier peoples. As with the scholarship of the DAG, regional museums emphasized the observation and description of the objects themselves.

Archaeology collections grew rapidly through contributions from historical associations, acquisitions by the museums, and emergency and planned excavations. Ironically, the museums became overcrowded during the same years when conservators and administrators expressed their fear that so many objects were being lost because of urban expansion and the unregulated antiquities trade. Even when the museums received

larger, modern buildings, the rooms devoted to archaeology filled immediately. This crisis mirrored the problem of interpretation faced by the
DAG. The museums continued to collect as they awaited narrative frameworks that would make sense of the objects. In the meantime, displays became chaotic and held little meaning.

Pinder's comments about the relationship between central and regional museums suggest how museums came to address these problems of
organization and interpretation. Central museums offered the most outstanding artifacts for scientific study and often strove for broader, even international coverage. Regional museums, though, played the crucial role
of introducing a broad public to pre- and early history. Museum administrators believed that artifacts spoke most clearly as evidence of local and
regional cultures and that they would lose much of their meaning if they
were transferred to a distant central collection. Regional museums were
local enough to maintain the connection between objects and the places
where they were found, but they were large enough to allow scholars to
address specific research questions that were important for a given region.
In this way, regional museums lifted archaeological material out of its narrow local setting and contributed to a broader understanding of the prehistoric past. Curators, archaeologists, and historical associations enthusiastically conveyed this new knowledge to a broad public both inside and
outside the walls of their museums.

Central and Regional Museums in the German Empire

Following the Wars of Unification and the annexations that accompanied them, the Kingdom of Prussia sought an administrative structure
that integrated new territories but also preserved regional autonomy.
The *Provinzial-Ordnung* (Provinces order) of June 29, 1875, confirmed
the executive authority of provincial governments in Prussia's central
and eastern provinces, and this arrangement was applied to western territories in 1888. Each of Prussia's twelve provinces had a provincial director, executive committee, and an assembly. The provincial governments
were responsible for roadbuilding and land improvements; welfare institutions for the poor and disabled; and the advancement of art and science
through museums, libraries, voluntary associations, and historical preservation efforts.[5] Within this decentralized arrangement, pre- and early
history developed as both a scientific field of knowledge (based largely

on expert knowledge in Berlin) and a historical practice that maintained the various cultural traditions within the kingdom (housed in regional museums).

The Sammlung der nordischen Alterthümer in the Neues Museum had continued to grow through acquisitions and excavations since the 1850s, and it remained the largest collection for pre- and early history in Germany. In the early 1870s, the hall decorated with mythological murals still presented the hesitant exhibit that ordered artifacts according to type. In 1874, Leopold von Ledebur, who had managed the collection since the 1830s, retired. Adolf Bastian, who had been hired as an assistant in 1869, became the new director, but he dedicated himself increasingly to international research and focused more and more on the museum's ethnology collections. Albert Voß (1837–1906), who was technically Bastian's assistant, emerged as the de facto director of the section for pre- and early history. Both Bastian and Voß were medical doctors and active members of the BGfAEU, and they sought to bring a more scientific orientation to the collections for ethnology and archaeology in the Neues Museum.

Voß set out to transform the collection for pre- and early history into a wide-ranging display that would enable the international and comparative work envisioned by the DAG. He visited several major collections that were ordered according to chronology, including the Museum for Nordic Antiquities in Copenhagen, the Historical Museum in Stockholm, and the antiquities section of the British Museum in London. Voß became convinced of the value of a display that would shed light on questions of chronology and cultural development. In his first reorganization of the display, he brought artifacts from single sites back together, and he organized the display according to political boundaries. This was a partial shift toward a more scientific display.[6]

As part of the Prussian Königliche Museen, the Sammlung der nordischen Alterthümer operated as a central institution for pre- and early history. The DAG carried out special projects for the collection, and its staff came from the leaders of the BGfAEU. The growth of the collection was a sign of the rising value of archaeology and the activity of associations like the BGfAEU, but it also brought a crisis in the management of objects. In 1880, the very year when the DAG celebrated the all-German exhibit of anthropological and archaeological finds and envisioned a bright future for domestic archaeology, Voß lamented that conditions inside the museum made it impossible "for such a rich and instructive collection to fulfill its essential mission—to provide an understanding of the develop-

ment of our land's culture from its earliest beginnings."[7] This was partly
an issue of space: nearly eighteen thousand artifacts were crammed into
the long gallery and two adjacent rooms devoted to the collection. Voß
desperately hoped that the new ethnology museum would provide the
room needed for a clear and informative display. But the problem was
also related to the development of archaeology as an interpretive science.
The overcrowded rooms in the Neues Museum were a physical manifestation of the DAG's desire to save artifacts from destruction and to accumulate material for historical knowledge.

The Museum für Völkerkunde—the museum that Virchow announced
to the DAG in August 1880 and that Voß hoped would improve the conditions of the archaeology collection—opened on Königgrätzer Straße
in December 1886. It provided a prominent home to the young fields of
anthropology, ethnology, and prehistory. When the prehistory collection
was moved to the new building in early 1887, it had a strong collection
of German and especially Brandenburg antiquities, but the Prussian Cultural Ministry supported the idea that the new museum should become a
central institution with a national, even international collection. The Cultural Ministry gave Voß an independent budget for acquisitions, and he
traveled throughout Prussia, including visits to Holstein in 1887 and the
provinces of West and East Prussia in 1892, to assess various collections.
He then filled some of the gaps in the Berlin collection through exchanges
with other museums and the purchase of several private collections.[8] The
Cultural Ministry authorized Rudolf Virchow to examine early medieval
fortification sites in Lower Saxony, and the historical association there co-
operated with these efforts. The prehistory section also benefited directly
from an edict that gave the Cultural Ministry access to antiquities that
were brought to light during the construction of railroad lines and canals.
In the early 1890s, the Cultural and Finance Ministries created a holding
area in Münster for these finds, and representatives from the Königliche
Museen brought much of this material to Berlin.[9]

With these acquisitions, the section for pre- and early history in the
Museum für Völkerkunde became vast and varied, but it did not appear
to be getting any closer to providing answers to the questions that visitors
might have about prehistory. Despite advances in comparative studies and
relative chronology, the collection retained its geographical organization
throughout the 1890s. There were special displays for new acquisitions
and items made of silver and gold, but the majority of objects were set out
according to the province where they were found. In this sense, they still

belonged to Brandenburg, Saxony, or Posen, as opposed to being part of
an instructive display about a specific prehistoric age or culture. The labels on the display cases included only an abbreviation for the prehistoric
period (*StZ = Steinzeit*, or Stone Age; *HP = Hallstätter Periode*, or Hallstatt period), brief information about the find or excavation, and a museum inventory number.[10] This sparse commentary reflected the DAG's
general approach to prehistoric artifacts. Anthropologists posed questions
about cultural development, but they believed they were still in the data-
gathering phase. The museum, likewise, embraced the task of collecting
fragments of the past and offered only empirical observations. Historical
information about the peoples who created these fragments was withheld,
and the accumulation of objects continued.

The ability to tell stories about the artifacts was further undercut by
the continuing disorder in the new museum. Excavations and acquisitions brought the inventory of the pre- and early history section to over
a hundred thousand objects by 1906, more than a fivefold increase since
the mid-1870s. This growth was impressive, but it led to what historian
H. Glenn Penny has described as "museum chaos." Scholars hoped to provide an orderly account of the prehistoric past, but their accumulation
of material culture and lack of a clear interpretive framework created
an overwhelming wealth of artifacts.[11] Albert Voß repeatedly requested
more staff to process artifacts, and there was a constant debate about the
best use of the rooms available for the massive collection. Within six years
of the opening of the new museum building, Voß and Virchow even pe-
titioned the Prussian Cultural Ministry for the construction of another
museum so that the prehistory and ethnology collections could be separated. For museum visitors, the packed display spaces and seemingly ran-
dom layout of the museum were baffling. The *Vossische Zeitung* captured
the sad state of affairs in 1900: "As he [the visitor] moves from the am-
phorae of ancient Troy or the countless pots from the Prussian districts,
he suddenly finds himself in the middle of a hall, half of which is filled
with mummies and clay vessels of the Inca and the other half with the furs
made by the inhabitants of the Amur River valley and metalworking from
the land of the shah. And naturally all the displays are labeled as 'provisional'!"[12] In the following years, the acquisitions and the debates about
how to use the rooms continued. When Voß died in 1906, the prehistory
collection held over a hundred thousand objects, and the gap between the
wealth of the collection and the lack of an interpretive framework for the
display was greater than ever.[13]

In the Museum für Völkerkunde, it was not only the exhibits that were provisional but even the knowledge scholars had about the distant past. The section for pre- and early history was not used to bolster a sense of identity, whether Prussian or German. Rather, it was still a place where material was accumulated before any conclusions would be drawn. These conditions, which continued into the first decade of the twentieth century, provide a stark contrast to the "preeminently national science" that Gustaf Kossinna was formulating during these same decades and the images of a glorious Germanic past propagated by nationalist pressure groups like the Pan-German League. Professionals associated with the DAG and the administration of the Königliche Museen took a tentative approach to the prehistoric past. They offered questions of interpretation, not sweeping narratives about national origins and cultural greatness.

In addition to the central museum in Berlin, the Kingdom of Prussia supported numerous regional and local museums, many of which became important institutions for science and history in their own right. The Prussian Cultural Ministry and provincial governments shared the common goals of preserving antiquities and educating the public about them. The balance struck between priority for the Berlin museum and autonomy preserved for regional institutions was evident in a series of edicts that regulated excavations and the possession of artifacts. In 1887, the Cultural Ministry declared that provincial museums were responsible for excavations and artifacts in their areas, which would prevent them from competing with each other over artifacts. But the ministry also required that regional authorities submit inventories of archaeological material to the Berlin administration. This allowed authorities in Berlin to remain informed about major finds and opened the door to the possibility that significant artifacts would go to the Museum für Völkerkunde.[14]

The collecting vigor and enthusiasm for new museums spread throughout the Kingdom of Prussia. In the Rhineland, for example, the provincial government approved plans for museums in Bonn and Trier in 1874 and named a director and a nine-member commission for each museum. These leaders were called "to seek out, excavate, and preserve the antiquities in their jurisdictions" and "to advance the research and preservation of antiquities in the Rhineland and make appropriate recommendations" for this field of study.[15] This initiative revitalized an antiquarian tradition from the early nineteenth century. The new Provinzialmuseum zu Bonn (today, the LVR-LandesMuseum Bonn) combined the antiquities cabinet created by Wilhelm Dorow in the 1810s and the collection of an association that had taken up excavations in 1841 and advocated for a museum ever since. The Provinzialmuseum zu Bonn did not have a permanent home for its first fifteen years, but its collection continued to grow. Construction for a museum finally began in 1890, and the *Bonner Zeitung* celebrated this important civic event, reporting that "the property purchased for the new museum is excellent. . . . It lies directly next to the railway station and [the city's] most beautiful promenade. The front of the property is very broad and will allow for the construction of a noble building. We may anticipate that this museum will bring honor and become a jewel of the city and the entire Rhineland."[16] When it opened in 1893, the neoclassical building was indeed impressive: its facade was over 130 feet long, and the building was nearly as deep. It contained almost 12,500 square feet of exhibition space.

The Provinzialmuseum zu Bonn brought together the region's Roman heritage, the anthropological approach to prehistory, and new ideas about cultural development as seen through applied art. The oldest items in the museum included Hermann Schaaffhausen's anthropological collection and the famous Neanderthal skull and partial skeleton found near Düsseldorf in 1856. As the museum guidebook explained, these bones represented "a stage of human development that is to be seen as a predecessor of the present human being."[17] They were followed by the remains of humans who hunted mammoths and other extinct animals, along with their tools made from stones and bones.

The basement and ground floor featured an impressive number of Roman sculptures and architectural remnants from several sites, including Xanten. The second floor presented an eight-room tour through "pre-Roman antiquities"; Roman bronzes, ceramics, and glassware; weapons and jewelry from Frankish graves; and wood carvings, sculptures, and paintings from the Middle Ages and more recent times. The guidebook noted the cultural context and remarkable features of specific antiquities, but these explanations were kept sparse because, according to the guidebook, viewing the artifacts "revealed the individual character of past cultures much better than if visitors read a lengthy commentary about them."[18]

For later periods, the catalog commented on the craftsmanship and style of the objects, not the historical information they might provide. A comparison of items from Schaaffhausen's collection with well-polished clay vessels from around 2000 BCE, for example, illustrated the high level of cultural achievement during the Bronze Age. Male graves from the Migration Age contained an impressive array of weapons, and female graves included beautiful jewelry and other grave goods. The guidebook noted that "Frankish metalworking was particularly sophisticated and unique. Iron buckles often included fantastic serpent ornamentation and inlaid silver." The ceramics and sculpture from this period, however, were less developed and paled in comparison to items from Roman sites.[19]

The Provinzialmuseum in Trier (today, the Rheinisches Landesmuseum Trier) brought together antiquities owned by two historical associations and two individual collectors, as well as those held by the city and royal governments. The collection then grew through the work of the museum commission, and its building opened in 1889. Most of the museum was dedicated to the extraordinary Roman sites in and around Trier, including the thermal baths in the city and several villas nearby. The museum also included three galleries devoted to non-Roman antiquities. The commentary on Stone Age and Bronze Age artifacts provided excavation details and references to literature about the finds but offered fewer comparative comments about craftsmanship than the Bonn museum guidebook. The section devoted to the Migration Age opened with a review of the characteristic finds from row graves.[20]

Both of the Rhineland institutions reveal how archaeology was featured in cultural history museums. First, they offered excavated objects a much more prominent setting than historical associations could provide. This was especially true after the completion of the new buildings in 1889 (Trier) and 1893 (Bonn). The museums stressed the observation of artifacts, which was part of the anthropological orientation toward prehistory. Finally, nowhere did the museum guides stress a Germanic or national history. Instead, they encouraged visitors to view and compare the development of handicraft traditions and decorative arts in the Rhineland as seen through the objects.

The Kingdom of Hannover looked back on a proud history of antiquarian research. Johann Wächter's 1841 inventory of "heathen antiquities" recounted the excavations by the Historischer Verein für Niedersachsen. In the early 1850s, the historical association, along with two other associations dedicated to natural history and to art, began to advocate for a museum. With the support of the monarchy and the city, the Museum für Kunst und Wissenschaft (Museum for art and science) opened in a new building in Hannover in 1856. (Today, this ornate building on Sophienstraße is the Künstlerhaus, a center for art, cinema, and literature.) The Museum für Kunst und Wissenschaft provided a permanent meeting space for the three associations, but more important, it advanced their educational goal of sharing their collections with the public. The museum was divided into sections devoted to history (including pre- and early history), natural history, and art. The royal government signaled the importance of the archaeology collection and the close ties to the Historischer Verein für Niedersachsen by naming Dr. Johannes Müller, the head of the association, as conservator for the Kingdom of Hannover in 1866.

After the annexation of Hannover by the Kingdom of Prussia in the wake of the Austro-Prussian War of 1866, the funding and administration of the Museum für Kunst und Wissenschaft were transferred to the provincial government of Hannover, and officials began to refer to the institution as a *Provinzialmuseum*. Müller continued to serve as conservator, and he corresponded regularly with the Prussian administration about historic preservation efforts and support for cultural museums. Under his leadership, the Historischer Verein für Niedersachsen raised the public's awareness of prehistoric finds by answering inquiries, visiting sites, and excavating on behalf of the museum.[21]

As in the Museum für Völkerkunde in Berlin, museum chaos soon plagued the provincial museum in Hannover. The three associations maintained possession of their individual collections for archaeology, natural history, and art, and they continued to acquire material for the museum. Despite two additions to the building between 1856 and 1890, the museum faced a constant shortage of space. Jacobus Reimers, who succeeded Müller as leader of the Historischer Verein für Niedersachsen and became the museum's first director in 1890, complained each year that the museum lacked the space and personnel to manage the growing collections. The important collection of prehistoric artifacts, which had grown to upward of fifty thousand objects by 1890, was displayed in five rooms.[22] After visiting the museum in the spring of 1899, Gustaf Kossinna commented that the treasures in the Hannover museum were most instructive, but the exhibit was chaotic and marred by imprecise, sometimes incorrect labeling. Furthermore, he disapproved that finds from a single site were divided among different parts of the display in ways he found completely arbitrary and even unthinkable. Kossinna, knowing that the province was planning a new museum, called for changes: "Hopefully the administration will carry out a thoroughly scientific reordering of the priceless prehistoric material that has been entrusted to them when the new museum is built!"[23]

Indeed, the province was planning a new museum, and in the presence of city and regional dignitaries, the Museum für die Provinz Hannover opened to the public on February 14, 1902. The neo-Renaissance building, which serves as the home of the Niedersächsisches Landesmuseum today, immediately became a prominent landmark. It was a major piece in the development of the Maschpark area, which also featured a new city hall building. The museum's pre- and early history collection was exhibited on the ground floor, along with the ethnographic and coin collections. The second floor included the rest of the cultural history display, as well as rooms devoted to sculptures and natural history. The top floor presented the museum's art collections and more of the natural history collection. But already in the new building's first year, Director Reimers complained that there was no space for the collections to grow and that the museum would therefore acquire only exceptional antiquities and artwork. The assistant for the archaeological collection sadly explained that the Historischer Verein für Niedersachsen wanted to cultivate relationships with landowners so that they could "save archaeological sites from the plow and win them for science," but because there was so little room in the museum, the association could sponsor excavations only when objects were in immediate danger of destruction.[24] The crisis became so extreme that Reimers halted excavations on the museum's behalf in 1906 and expressed his exasperation that there was simply no room in the museum for the effective display of large items like church altars. "The need for more space has become a burning issue and can no longer remain unaddressed without seriously diminishing the museum."[25]

Conditions in the Museum für die Provinz Hannover capture the mixed status of domestic archaeology around 1900. On the one hand, the province had a wealth of excellent material and a long tradition of support from groups like the Historischer Verein für Niedersachsen. Members of the aristocracy and middle class, as well as city and provincial officials, wanted to promote Hannover's cultural history, as evidenced by the building of the Museum für Kunst und Wissenschaft in 1856 and the Provinzialmuseum forty-six years later. Strong and steady visitor numbers throughout these decades showed that the museum's educational mission resonated with the public. Yet the museum faced some basic limitations. The three associations kept their collections separate for several decades, making it impossible for the museum to organize thematic exhibits or exchange parts of the collection with other museums to create a more rational presentation of the region's cultural heritage. The collections kept growing, and even when the provincial government built additions to the initial building and then supplied the museum with a new building in 1902, the space was not enough. Museum officials who so fervently wanted to "save archaeological sites from the plow and win them for science" were oversupplied with the kinds of artifacts that they described as scarce, endangered, and irreplaceable. Displays that were supposed to excite and inspire visitors became overcrowded and chaotic.

The solution to this problem was to make sense of the objects in a new way. Conditions in cultural history museums forced directors to abandon the idea of displaying everything the museum possessed. New displays after 1900 would feature fewer objects and place a much greater emphasis on the interpretation of prehistory. They explained objects in greater detail and used them to illustrate the chronological development of prehistory. In Hannover, this shift from preservation to interpretation was accompanied by a new focus on a nationalist version of prehistory, but this was not the case in most other museums. The efforts to make museums more engaging and scientific after 1900 will be covered in chapter 7.

Voluntary associations in Berlin were critical for the advancement of domestic archaeology in the province of Brandenburg. Members of the Verein für die Geschichte Berlins (Association for the history of Berlin, founded in 1864) and the BGfAEU, with the support of the city government, founded the Märkisches Provinzial-Museum in 1874. Initially a small display in the rotunda and a corridor of Berlin's city hall, the museum enjoyed a warm reception from Berliners and averaged around a hundred visitors per day. In January 1876, it received a larger and more suitable display space on the second floor of Klosterstraße 68. In this new location, the museum had eight rooms at its disposal, two of which displayed prehistoric artifacts from Brandenburg and other German regions. In the summer of 1880, the relocation of various city offices forced the Märkisches Provinzial-Museum to move again, and it received a larger home on the ground floor of the Köllnisches Rathaus at Breite Straße 20a, where the museum remained for the next nineteen years.[26] These addresses are significant because they are in the oldest part of Berlin, near the site of the twelfth-century settlement of Kölln. This was an ideal setting for a museum that documented the region's cultural development. (The building constructed for the museum in 1909, which still serves as its home, is also located in this central part of Berlin.)

Ernst Friedel served as the first director of the Märkisches Provinzial-Museum. As a member of the steering committee for the BGfAEU and a city councilor, Friedel garnered support from scientific, government, and

middle-class circles. In its first years, the museum received donations from
the kaiser and crown prince; imperial and Prussian officials; municipal
governments, churches, and *Gymnasien* in Brandenburg; various voluntary associations and guilds; and over 470 individuals. With these gifts, the
museum's collections grew rapidly from 14,200 items at the end of 1875
to close to 30,000 items at the end of 1879.[27] In 1887, the museum for-
malized its relationship with supporters by creating a *Pflegschaft*, a core
group of patrons and cultural administrators throughout Brandenburg
who pledged "to add to the museum's various collections whenever possible and to be ready with word and deed to advise the director about acquisitions and other matters." The *Pfleger* served on a voluntary basis, and
the museum honored their efforts with three levels of distinction. By 1901,
240 people had received a recognition certificate; 77, a silver emblem; and
11, a gold emblem, which was the highest distinction.[28]

   In 1892, the circle supporting the museum grew even wider with the
founding of the Gesellschaft für die Heimatkunde der Provinz Brandenburg (Local history society for the province of Brandenburg). This group,
also known as the Brandenburgia, quickly became the largest historical
association in Berlin and the first group of its kind to include women.
Legal councilors, commercial councilors, and city councilors and their
wives joined the Brandenburgia.[29] They came together to socialize and to
share their interest in the culture and history of Brandenburg. Monthly
meetings were held in various civic and cultural institutions in Berlin, where members heard presentations about topics ranging from geology to heraldry to art history. Weekend excursions allowed members
to cultivate their knowledge of art, architecture, and history with visits
to churches, museums, and other attractions in Brandenburg's towns and
villages. Other outings fed an interest in science and technology. For example, members admired the modern cooling system at Julius Bötzow's
brewery and the new machines at Commercial Councilor Spindler's textile factory.[30]

   A main goal of the Brandenburgia society was to promote the
Märkisches Provinzial-Museum and its presentation of Brandenburg's
cultural heritage. The museum had an important integrative function, as
the imperial capital became a major manufacturing center in the 1870s
and many of its residents were recent migrants from nearby towns. When
the museum was housed in the Köllnisches Rathaus, the museum's entrance hall was filled with a curious mixture of sacred and profane works
from Brandenburg, including an elaborate medieval altar from Reinicken-

dorf, a collection of ornate pieces of wrought iron used to decorate residential stoves, and other pieces and architectural details created by various guilds. From here, visitors proceeded through the cultural history of Brandenburg. A large room displayed glassworks, masonry, ceramic beer mugs, jewelry, and clothing, as well as implements for fishing, hunting, and farming from the nineteenth century. These items were followed by porcelain pieces, medallions, coins, and religious objects that illustrated the history of the region from the sixteenth to the eighteenth centuries. A medieval section featured not only weapons, armor, and implements of torture but also Bibles, chandeliers, and money boxes (*Gotteskasten*). The next three rooms were dedicated to archaeology. One room was devoted to antiquities found in Brandenburg. A second displayed artifacts from other German regions and abroad for comparison. Finally, a long hall displayed the prehistoric bones of the anthropological collection. After 1885, this room also included a natural history display.[31]

The archaeological objects, the paintings and sculptures from medieval churches, and the copperplate engravings and other elaborate handicrafts from more recent periods revealed Brandenburg's unique culture. This wide range of artifacts also displayed the unifying theme of cultural progress. By looking at the changes in craftsmanship and the variety of materials used by artisans and artists from prehistory to the present day, one could see the development of human creativity and view current practices as the culmination of local traditions.

Archaeology was central to the Märkisches Provinzial-Museum and to Berlin's associations. Since its founding in 1869, members of the BGfAEU had excavated throughout Brandenburg and discussed local finds at their meetings. The society built its own collection, but it also donated many artifacts to Berlin's museums. Ernst Friedel published several articles about Brandenburg's botany, natural history, and pre- and early history, and he prepared an inventory of Brandenburg's antiquities for the DAG's all-German archaeological exhibit in 1880.[32]

Excavation outings were always part adventure, and sometimes they had more to do with sharing the excitement of archaeology with a wider audience than with the discovery of important finds. On the occasion of the 1880 congress, members of the BGfAEU scouted out and excavated sites in advance and then brought prominent visitors to these sites, where shiny objects appeared relatively quickly. The staged nature of the events did not diminish the interest of the participants, however. One reporter covering an excursion to the Spreewald comically noted that despite the obvious reburial of objects (the area was covered in evenly raked sand), the crowd watched with fascination as several perfectly preserved urns were lifted from the ground. The ladies of the expedition then gently spooned out the urns' gritty contents. A few days later, the Prussian crown prince and princess joined an outing to the Römerschanze, a prehistoric defense wall near Potsdam. They were so enthralled that they asked the leader of the excavation "to dig a bit more because they speculated that more finds were just a bit deeper." The request kept the company standing in the rain for several minutes and did not yield any more treasures.[33]

Archaeology was a regular activity for the Brandenburgia society. Friedel organized outings with members interested in archaeology, folklore, natural history, and geology. They set out at least twice a month from April to October, but sometimes in winter too, to seek out prehistoric settlements or graveyards, to document the unique architecture of a church building, or to examine an unusual plant or animal that had been the subject of a debate at a recent meeting. One participant recalled the group's enthusiasm: "Oh how we investigated and explored, even in rough conditions brought by rain, storms, blowing snow, or July's burning heat! Once we were out east in a blowing wind and 16 degrees Fahrenheit to observe a curious boulder that had recently been discovered. Naturally we couldn't take it back with us, but we were able to photograph it."[34] On other occasions, the group sought out movable objects, and they discussed them at the Brandenburgia's monthly meetings and presented them to the Märkisches Provinzial-Museum. One example of this came at the society's very first meeting in April 1892, when Friedel presented several bronze finds the group had recently excavated near Köpenick.[35]

Both the archaeological outings and the monthly meetings of the Brandenburgia society were festive. Their "digging song," written by Friedel, captured the merry spirit associated with archaeology:

In Berlin höchst wunderbar	In most wonderful Berlin
Juchheidi, juchheida,	Juchheidi, juchheida,
Buddelt man das ganze Jahr,	we dig all year long,
Juchheidi, heida,	Juchheidi, heida,
Wenn's auch kostet Heidengeld,	even if it costs lots of money,
Uns das Buddeln sehr gefällt.	we really love to dig.
Juchheidi, heidi, heida,	Juchheidi, heidi, heida,
Juchheidi, juchheida,	Juchheidi, juchheida,
Juchheidi, heidi, heida,	Juchheidi, heidi, heida,

Juchheidi, heida!	Juchheidi, heida!
Diesem Beispiel folgen wir,	We follow this model:
Juchheidi, juchheida,	Juchheidi, juchheida,
Mit der Wissenschaft Begier,	with scientific curiosity,
Juchheidi, heida,	Juchheidi, heida,
Scherben, Knochen, Glas und Holz —	shards, bones, glass, and wood—
Alles buddeln wir mit Stolz.	everything we dig up with pride.
Juchheidi heidi, heida.[36]	Juchheidi heidi, heida.

As with other cultural history museums, the results from frequent excavations soon filled the archaeological section of the Märkisches Provinzial-Museum. Photographs of the display in the Köllnisches Rathaus show chaotic rooms packed with pieces of Brandenburg's heritage.[37]

A high point for both Friedel's group and the Märkisches Provinzial-Museum came in 1899 with the discovery of the "royal grave" of Seddin. A local newspaper announced that workers had hit upon a chamber grave on September 19. Heinemann, one of the museum's *Pfleger*, went to the site immediately, and Friedel and two others from Berlin joined him the next day to investigate the site. Friedel gave a preliminary report about his findings at the November Brandenburgia meeting and a longer lecture about the excavations in December. The royal grave of Seddin turned out to be a magnificent early Iron Age tomb, and it was quickly recognized as the largest grave of its kind in all of Germany. The Brandenburgia celebrated the site with a poem by Carl Bolle, and even *Die Gartenlaube* publicized it nationally with a set of images.[38]

The urns and swords from the royal grave of Seddin did not make it into a display in the Köllnisches Rathaus, however, because the building was torn down in 1899 to widen the streets in central Berlin. The city had already approved funds for a new building dedicated solely to the museum and the closure was to be temporary. There were lengthy construction delays, however, and parts of the collection were put on display in a temporary location after 1904. The new building, beside Köllnischer Park along the Spree River, finally opened in 1908. (This location still serves as the home of the Märkisches Museum.) It was hailed as an architectural landmark and demonstrated the city's commitment to the museum. The new building also marked a turning point for the museum as a scientific institution, as it received a more professional staff in 1909 that included an archaeologist.

Archaeological engagement through collecting, display, and outreach

FIGURE 5. Members of the Brandenburgia association acquired artifacts for the Märkisches Provinzial-Museum through their excavations. This outing to an urnfield in Brandenburg resulted in several finds for the museum. (Photo by Herbert Mauerer, Grabungen der Brandenburgia, Urnenfeld bei Fürstenberg i/M, nahe der brandenburg. Grenze, September 22, 1901. Courtesy of the Stiftung Stadtmuseum Berlin.)

occurred in Prussia's other provinces. Schleswig-Holstein, created as a province after the German-Danish war of 1864, received a provincial museum in 1873. The Museum vaterländischer Alterthümer combined two older collections and became affiliated with the university in Kiel. Johanna Mestorf (1828–1909) served as the custodian for the collection and was named director in 1891. (She was the first woman to become the director of a German museum.) Mestorf was an active member of the DAG, and she was admired in professional circles for translations of several important works on Scandinavian archaeology. These changes raised the status of the Kiel museum to that of a central museum for the province.[39] In Westphalia, the Museum vaterländischer Alterthümer zu Münster served as the official provincial museum. The historical association founded this collection in 1844, and it was moved to the town hall in 1886.[40] Hesse-Nassau had important cultural history museums in both Wiesbaden and Kassel because of the independent traditions of the territories that had been combined to form this province in 1868. In the province of Saxony, the Thüringisch-Sächsischer Verein, whose dedication to domestic archaeology went back to Friedrich Kruse's efforts in the 1820s, called on the provincial government to create a historical commission and a museum. The Museum für heimatliche Geschichte und Altertumskunde opened in Halle in 1884, and the commission began working on a map of prehistoric finds in the province in 1886.[41] Pomerania, Posen, Silesia, West Prussia, and East Prussia all established provincial museums in their capital cities. Even the special territory of Hohenzollern had an antiquities collection that opened to the public in 1868.

States beyond the Kingdom of Prussia developed institutions for domestic archaeology also. In the Kingdom of Württemberg, the Württembergischer Alterthums-Verein in Stuttgart and the Verein für Kunst und Altertum in Ulm und Oberschwaben had carried out excavations since the 1830s and 1840s and amassed substantial collections of archaeological objects. In 1858, the kingdom created the position of conservator for art and antiquities, and Konrad Dietrich Haßler, the investigator of the Alemanni graveyard in Ulm, was the first person to hold this office. He was charged with the tasks of creating an inventory of the kingdom's cultural treasures and crafting policies for their preservation. From the beginning, Haßler stressed that the state needed to do more than register finds, and he advocated for the creation of a central collection. In 1862, the royal house established the Königliche Staatssammlung vaterländischer Kunst- und Altertumsdenkmäler in Stuttgart "to prevent the destruction, division, and export [of the kingdom's artworks and antiquities] and to bring them to the attention of the public through display." A museum commission received six thousand to eight thousand *Gulden* each year for preservation work and acquisitions. It purchased several major private collections (including Haßler's in 1863), and the collection grew through donations from historical associations and emergency excavations that happened during railroad construction.[42]

The treatment of domestic archaeology in Württemberg displayed a blend of Romantic enthusiasm and new scholarship. Eduard Paulus (1837–1907), the son of Eduard Paulus the Elder, who had worked for the kingdom's topographical bureau and produced an archaeological map that plotted Roman settlements and defense walls in Württemberg in 1860, succeeded Haßler as conservator in 1873. He published an updated and expanded version of his father's map in 1878 and executed the state's first planned excavations, but his career-long passion was an inventory of Württemberg's art and antiquities. The four volumes of *Die Kunst- und Altertums-Denkmale im Königreich Württemberg* were organized by district, and they cataloged examples of architecture, church altars, archaeological finds, and many other kinds of cultural material. The entries expressed so much of Paulus's love for his homeland that a conservator from Württemberg recently described the catalog as part reference work and part fairy tale. Throughout, Paulus moved seamlessly from the specialized terminology of geology, archaeology, and geography to poetic waxing about the beauty of the landscape and the fascinating history that it contained.[43]

In 1871, a branch of the DAG was founded in Stuttgart. Eugen von Tröltsch (1828–1901), a retired military officer and well-respected cartographer, led this regional society and emerged as the driving force in the DAG's national committee for prehistoric maps. He helped to develop the legend symbols and naming conventions that were to be used for all the maps sanctioned by the DAG, and he advised scholars in other regions on the production of their maps.[44] One of the first products of this effort was Tröltsch's 1874 map for Württemberg. In contrast to the maps by Paulus and his father, which focused on the Württemberg landscape and especially the remnants of Roman defense walls, Tröltsch's work reflected the new interest in prehistory and specifically his interest in the dwellings and Bronze Age finds around Lake Constance. Tröltsch represented how the research agenda of the DAG entered into regional contexts. He contributed to the standardization of archaeological cartography and to the DAG's general goal of collecting data about prehistoric finds from all of Germany.

The Königliche Staatssammlung vaterländischer Kunst- und Altertumsdenkmäler did not have a permanent home until it was displayed in the Landesbibliothek in 1886. In this improved location, the archaeology collection was enhanced by major donations from the Württembergischer Alterthums-Verein and the Cannstatter Altertumsverein. Peter Goeßler, a classical archaeologist, was hired in 1905 to oversee the archaeology and coin collections, and he became Württemberg's conservator in 1909. Goeßler's approach to pre- and early history was fundamentally different from Paulus's and his folksy inventory. He implemented a chronological exhibit that moved from the Stone Age to the Bronze Age and to the

Frankish and Alemannic antiquities of early history. The wing of the library that held the prehistory display also included rooms with Greek and Roman artifacts and ironwork from the seventeenth and eighteenth centuries, encouraging viewers to compare the evolution of metalworking techniques over the centuries.[45] Under Goeßler's direction, the primary meaning associated with the archaeology collection shifted from a love of Württemberg to lessons in prehistoric chronology and the history of applied art. The collection is now part of the Landesmuseum Württemberg, which is housed in Stuttgart's historic castle.

The BNM, as the central museum for Bavaria's cultural heritage, had been created to strengthen the sense of a Bavarian identity. During the 1870s and early 1880s, its collection of archaeological objects was growing rapidly, especially because of finds made during the construction of railway lines from Munich to Nuremberg and Bayreuth and between Munich and Passau. Yet the museum did not provide a very compelling setting for its collection of archaeological objects. Jakob Heinrich von Hefner-Alteneck, who served as the museum's director from 1868 to 1885, was a specialist in medieval and Renaissance culture, and he had a strong interest in applied art. In the early 1870s, he divided the museum's collection between two displays. One remained a cultural history display based on chronology, but it no longer included prehistoric epochs. The other focused on particular craft traditions, like metalworking, ceramics, textiles, and wood carving, and on unique collections of *Trachten* (traditional Bavarian costume) and weapons. Some of the prehistoric items were included in these specialized displays as early examples of metalworking and ceramics. In this display, the golden hat from Schifferstadt served as an example of decorative innovation, not as a representative of the Bronze Age. The 1882 guidebook (incorrectly) explained that it was "a golden boss . . . that covered the protrusion of a prominent man's shield." No date was given for the magnificent object, and it was displayed alongside a golden armband from Hungary from the Carolingian era, a bronze helmet from the early Germanic period, and an unpolished, pear-shaped emerald that came from a bishop's crown.[46] Other items of stone and iron were arranged in the garden (and exposed to the outdoor elements) or removed from view. This version of the museum, which treated artifacts as part of the development of the applied arts and not as evidence of the history of the land, remained in place until the mid-1880s.

Johannes Ranke, the nephew of the famous historian Leopold von Ranke and the leader of the Munich branch of the DAG, painfully recalled the conditions inside the museum: "For those of us who value and understand these objects, it made our hearts bleed to see them ... without documentation, damaged, and not properly preserved, all their scientific value gone, some even bought, sold, and traded at the local flea market."[47] In an attempt to address what he considered the unscientific approach to archaeology in Bavaria, Ranke joined with like-minded scholars to found the Munich branch of the DAG in 1870, and he emerged as an important organizer for the DAG's annual meetings, including the 1875 congress in Munich that featured the impressive display of Bavarian antiquities. As a professor of anatomy, Ranke brought the anthropological approach to artifacts from pre- and early history. The Munich DAG established *Beiträge zur Anthropologie und Urgeschichte Bayerns*, a new publication for Bavarian prehistory that reflected Ranke's interest in both physical anthropology and prehistoric archaeology. The journal brought together announcements of new finds, detailed studies of archaeological sites, and anatomical research in order to "reconstruct the ethnographic conditions among the prehistoric population of Bavaria."[48] It also contained the maps of prehistoric finds in Bavaria that Friedrich Ohlenschlager produced as part of the DAG's nationwide project.

Three developments in the mid-1880s further demonstrated a shift toward a more scientific approach to prehistory. First, the Bavarian state created the Kommission für Erforschung der Urgeschichte Bayerns (Commission for research into the prehistory of Bavaria) as part of the Bavarian Academy of Sciences in 1886. The initial membership of the commission had the highest scientific credentials and strong connections to other institutions in Munich. It included Wilhelm Heinrich Riehl, who became the director of the BNM in 1885; Karl von Zittel, a cofounder of the Munich branch of the DAG and director of Bavaria's natural history museum; the anatomist Karl von Kupffer; the classicists Heinrich Brunn and Wilhelm von Christ; and Joseph Würdinger, an active member of the Historischer Verein von Oberbayern and the DAG who had excavated throughout Bavaria. Much of this group's work related to the Roman walls in Bavaria, but it also oversaw the excavation of pre-Roman grave mounds near Murnau and premedieval cemeteries near Thalmässig. Julius Naue began working as part of the commission in 1890 and became well known for his excavations, as well as for his artistic renderings of prehistoric finds.[49]

Second, changes in the BNM's display reflected the growing understanding of archaeology as an anthropological and historical science. The museum's new director, Wilhelm Heinrich Riehl, restored the prehistory collection as the beginning of a chronological display, and scholars began working on a comprehensive catalog for the museum's collections. The volume on the archaeological objects, which appeared in 1892, was a clear advance over earlier guides to the museum. It explained the provenance of the items and described them as representatives of specific chronological periods.[50] Editions of the museum guidebook now contained enthusiastic comments about this section of the museum and explained how the display was divided into prehistoric finds, Roman finds, and Merovingian antiquities. Continued acquisitions, however, led to the overcrowding that became common in other cultural history museums. The 1894 guidebook lamented that "limitations of space in the museum did not allow for the desired chronological display."[51] Construction for a new building began in the late 1890s as part of the development of Prinzregentenstraße. When the museum opened in 1900, the prehistory collection remained at the beginning of a chronological cultural history display. The greatly expanded guidebook from 1908, over three hundred pages in length, included more coverage of life in prehistoric times and details about the objects. In this iteration of the display, the golden hat from Schifferstadt stood in its own display case and was correctly identified. The guidebook explained that "it is formed completely out of gold foil and was presumably part of the headdress, similar to the tiaras of Assyrian kings. Hallstatt period."[52]

Third, Johannes Ranke created a stand-alone museum for prehistory called the Prähistorische Sammlung in October 1885. Initially, it consisted of objects donated by Ranke from his own excavations and purchases, and it was part of the kingdom's paleontology collection, located in the former Academy of Sciences building known as the Wilhelminum.

Other states within the German Empire maintained regional traditions through similar associations and institutions for prehistory research. The Kingdom of Saxony established a royal society that combined the efforts of voluntary associations with state efforts for historical preservation. The Grand Duchy of Baden created a more scientific collection in Karlsruhe, and the state of Hesse had one in Darmstadt. Robert Beltz created a map of prehistoric finds in Mecklenburg-Schwerin. Even smaller states like Oldenburg and Saxe-Weimar-Eisenach established museums that featured archaeological collections.[53]

Saving Artifacts through Public Awareness

Communication with landowners, teachers, and pastors had been a part
of antiquarian research since Friedrich Kruse's travels through Silesia
and the questionnaires Leopold von Ledebur sent to historical associations in Brandenburg. The nature of this information was changing in the
1870s and 1880s, though, as regional museums and historical commissions
covered larger areas and as the anthropological orientation in prehistory
emphasized the production of scientific knowledge. Increasingly, domestic archaeology was viewed as something more than a curiosity about the
local landscape. Scholars wanted more systematic information about archaeological sites so that they could substantiate chronologies and better
explain long arcs of settlement and cultural development. They therefore
turned to the public for information about finds and, along with provincial governments, museum administrators, and historical associations, promoted a greater awareness of domestic antiquities.

These efforts were urgent as the urbanization that accompanied the
Second Industrial Revolution brought more construction in and on the
outskirts of many towns and cities. Eduard Krause, an assistant at the Museum für Völkerkunde in Berlin, reported in 1899 that plans for a new gas-
works in Tegel threatened three sets of grave mounds that the BGfAEU
had found. He published a notice about these mounds because they would
"soon disappear owing to the continuing development of this terrain and
because their existence could perhaps be overlooked because, as far as I
can tell, they are not mentioned in any other literature."[54] Farming, roadbuilding, and the development of railroads and canals also disturbed new
tracts of land and exposed ancient material.

In addition to threats brought by modernization came the competition
of a growing market for antiquities. State authorities and museum personnel constantly mentioned the prevalence of middlemen who acquired antiquities for private collectors and other museums. Konrad Haßler had denounced the traffic by "sleazy antiquities dealers" during his excavations
near Ulm. In 1865, the Prussian Cultural Ministry circulated an impas-
sioned plea from Ignaz von Olfers, the director of the Königliche Museen
in Berlin, to the provincial governments. Olfers was alarmed by the awful
circumstances surrounding domestic antiquities, explaining that

> it is often the case (and with improving communication it is becoming ever
> more frequently the case) that coins and other items that need to be preserved

because of their historical value are scattered or sold to the nearest dealer immediately after they are found. Those who find these items deceive themselves into thinking that they will get a higher price this way or they fear that the landowners will make a claim on the objects. The antiquities then end up in the smelting oven and thereby not only cannot be added to a museum collection but are also robbed of the scientific value they have for the study of archaeology and the fatherland.[55]

To combat these circumstances, the Königliche Museen offered immediate payment for objects sent to the central museum for pre- and early history. Olfers promised that the museum would compensate finders with the value of the precious metal that the objects contained or the value of the objects warranted by their rarity or significance, whichever was greater.

In other cases, roving agents bid up the prices that states and museums had to pay and undermined the scientific study of archaeological finds. H. Runde, the assistant director of the Provinzialmuseum in Hannover, expressed his frustration that local publications praised the owners of some private collections as "trailblazers in the science of archaeology." He conceded that they had accumulated large collections but excoriated them as part of a major problem. Private collectors, he charged, sponsored excavations "for sport" and offered bonuses for certain kinds of objects. This practice inspired "a widespread rampage of unsystematic excavations that amounted to thievery." Because they kept only what they considered valuable and discarded everything else, these treasure hunters tore apart finds that belonged together and destroyed their scientific value. Runde therefore acknowledged the value of only finds that were well documented and reported to the museum. He refused to list "the numerous additional finds that had been intentionally removed from the public eye by private collectors and dealers" in the museum's inventory for the region.[56]

Museum directors addressed these problems by encouraging greater communication with schoolteachers, pastors, and local officials and by raising their awareness of the value of ancient artifacts. After the opening of the provincial museum in Kiel, Johanna Mestorf appealed to the residents of Schleswig-Holstein for their support of prehistory research. She noted that experts cannot be everywhere and see everything. They therefore needed the help of landowners and field workers who had daily contact with the land and might hear about new finds. Mestorf then offered advice on how to recognize ancient settlements, *Pfahlbauten*, and graves and urged her readers to notify the Kiel institute of any finds. She hoped that a local notable (a doctor, pastor, teacher, estate owner, or mayor) in each town would serve as a contact person who would inspect sites for the museum and ideally send items when possible. (The museum would pay the shipping costs.) All this information would be logged in the museum's topographical-statistical record and help experts recognize larger patterns in the distribution of finds.[57]

The most widely distributed official guidance concerning domestic antiquities came from Albert Voß, the curator of the pre- and early history section of the Museum für Völkerkunde in Berlin. In 1888, he published his *Merkbuch, Alterthümer aufzugraben und aufzubewahren*, a seventy-page manual about the excavation and preservation of antiquities. It was printed in a pocket-sized format so that it could be brought into the field by excavators. The Prussian Cultural Ministry distributed it to the presidents of all the provincial governments, and the Kingdom of Bavaria sent a version of the manual to local officials there.[58]

By the end of the nineteenth century, many museum administrators felt that it was not enough to educate laypeople in the basics of excavation. As Voß explained, he hoped more and more people would realize "that collecting represents only one aspect of research. It is crucial that it be supplemented by the much more difficult task of investigating antiquities at the excavation site."[59] And a reviewer in the *Literarisches Centralblatt für Deutschland* echoed Johannes Ranke's insistence that the time of amateurs had passed. It was not enough to remove objects from the ground; the science of archaeology required much more information about the context and contents of archaeological sites that only an expert could provide.[60]

Museum administrators and scholars associated with the DAG not only campaigned for greater awareness but also wanted to educate the public about prehistory. In 1889, Eugen von Tröltsch, the cartographer and military officer involved in the DAG's creation of maps of prehistoric finds, produced a poster that depicted weapons, vessels, coins, and decorative items common in the areas along the Rhine and Danube Rivers, grouped according to the "pre-Roman," "Roman," and "Alemanni-Frankish" periods in Württemberg's history. He hoped his "Altertümer aus unserer Heimat" (Antiquities from our homeland) would contribute to the protection of antiquities by teaching people what to look for. The Württemberg Cultural Ministry ordered three thousand copies of Tröltsch's poster to be distributed to the kingdom's schools. The DAG happily reported that the eight-color, seventy- by ninety-centimeter sheets cost less than a mark and recommended their display in town halls, schools, and other public spaces.[61]

Other regions followed Tröltsch's lead. The Museum für heimatliche Geschichte und Altertumskunde in Halle produced a similar poster for the Prussian province of Saxony in 1898, and six thousand copies were distributed to local schools. Hugo Conwentz, the director of the West-preußisches Provinzial-Museum in Danzig, created a set of six color images with artifacts from his region. His work was celebrated in the *Naturwissenschaftliche Wochenschrift* in August 1899.[62]

These outreach efforts extended to other publications that reached wide sections of the middle class. The nationally distributed *Illustrirte Zeitung* from Leipzig featured a weekly section called "cultural history news." Along with discussion of theater events, music concerts, and museum exhibits, the column announced archaeological excavations in Germany, as well as in Italy and the Middle East.[63] The *Augsburger Abendzeitung* contained a regular column during the 1880s called Der Sammler (The collector) that provided details about discoveries and excavations throughout Swabia and Bavaria. This kind of news was so popular that one contributor to the *Pfälzisches Museum*, a historical magazine for the Palatinate region, claimed that the archaeology section kept the entire magazine financially afloat.[64]

This press coverage intersected with the outdoor pursuits of the growing number of voluntary associations in the German Empire. Archaeological sites and historical places became destinations for hiking clubs and casual strollers. For many, a walk in the woods was not only a way to enjoy the contemporary landscape but also a chance to see early medieval fortifications and prehistoric burial mounds. In 1888, for example, citizens in Bad Urach founded the Schwäbischer Albverein (Swabian Jura association), and it soon became the largest of Germany's hiking clubs, with over six hundred branches in Württemberg's cities and towns. The group published maps and trail guides that included commentaries on geological features and sites from pre- and early history that dotted the forests and low mountains of Swabia.[65]

Historical commissions and museums also worked with schools to promote pre- and early history. Mestorf viewed teachers as an essential partner in her outreach efforts in Kiel. "They possess the secret weapon: they can train a thousand eyes to look out for antiquities, a thousand ears to listen for news of finds, and a thousand tongues to spread the word about

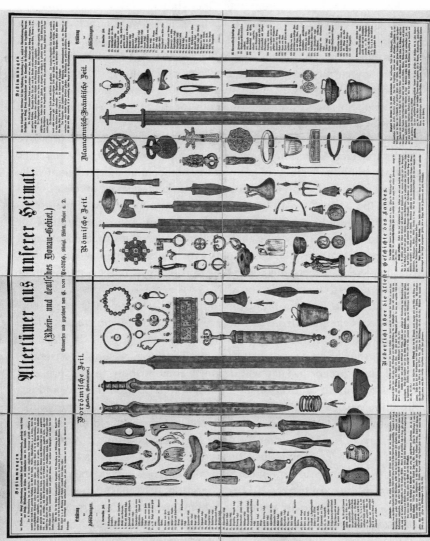

FIGURE 6. Eugen von Tröltsch's "Altertümer aus unserer Heimat" helped residents and students recognize and gain a greater appreciation of local antiquities. The upper corners of the poster offer advice for those who might find or hear of antiquities: the left-hand text explains that the Kingdom of Württemberg will pay to acquire local items, and the right-hand text urges care and caution when handling fragile material. The text along the bottom of the poster provides a basic timeline for pre- and early history. (*Source:* Bayerische Staatsbibliothek. 2 Arch. 257 to.)

the value of local history." Mestorf was certain that students would be fascinated by prehistory and gain a new respect for ancient objects. Once teachers talked about "how our land looked in ancient times and how we can still find traces of successive cultural periods . . . students will stop smashing clay pots for fun, as has recently happened, and they will learn to admire the skills of the ancient potter."[66]

Prehistory was also presented to older students. Over fifty *Gymnasien* in Germany had archaeological collections. These were largely the result of teachers like Eduard Roese and Albert Baum in Dortmund, who filled two rooms of a schoolhouse with the products of local excavations.[67] Classroom learning was enhanced by local and regional museums. Hugo Conwentz regularly welcomed school groups to the Westpreußisches Provinzial-Museum. After his visual aids for the region's antiquities were distributed, he visited each district in the province to speak with teachers and local officials.[68] And a teacher in Württemberg noted that twelve of the fourteen *Gymnasien* and *Realgymnasien* in his state were in cities that had public collections of antiquities. He recommended frequent visits to these collections and excursions to historic places and archaeological sites as the perfect way to teach science, history, and a love of the homeland. These would not be visits to "pretty places" but opportunities for pupils to "take in everything that *Heimatkunde* values." He encouraged teachers to decorate their classrooms with Tröltsch's map and use books that depicted artifacts, including Ludwig Lindenschmit's *Handbuch der deutschen Alterthumskunde*.[69]

Bavaria, too, brought prehistory into the classroom. In January 1890, Riehl recommended Tröltsch's depiction of domestic antiquities to the Bavarian Interior Ministry as an important pedagogical tool. He described "the distribution of wall hangings like these in schools, town halls, etc. as most practical. They will do much to contribute to the spread of knowledge about prehistoric antiquities and to prevent the loss and destruction of such finds."[70] The following month, the ministry wrote to district governments that "hundreds of pieces of the prehistoric past" were lost every year. "The reason for these regrettable losses . . . is without a doubt . . . ignorance about the value of these items. In addition to their odd form, these objects are often covered with corrosion, rust, and dirt. Most ceramic pieces are broken. In this condition, the objects appear worthless. . . . Those who do not understand them throw them away, or they send bronze pieces to money mints to be smelted or place them in the hands of traveling antiquities dealers." Greater recognition would counter these circumstances and encourage the wider public to preserve the region's archaeological heritage.[71]

The desire to reform museum displays was a natural extension of the desire to teach the public about pre- and early history. Eduard Pinder, the museum director in Kassel, explained that regional museums were stuck in a debate about the best organizing principle for an archaeological display. Should objects be grouped according to type and material or according to where they were found? Both of these approaches had advantages, but Pinder urged museum administrators to move beyond this either-or debate and place more emphasis on what a guest might learn from a museum visit. He recommended a hybrid exhibit that displayed objects along two axes whereby a vertical line of objects and information would relate to a specific location and a horizontal line of objects would demonstrate the evolution of the ways that a certain material had been used over time. Explanatory texts and regional maps would be used to give the display coherence. Pinder also emphasized the need for better museum catalogs that would connect the specific finds in a regional museum to broader insights into pre- and early history.[72] By the last third of the nineteenth century, museum administrators were beginning to focus on the explanatory power of a few objects as opposed to feeling the need to display everything that a museum possessed. This idea would gain even greater traction after 1900, as more and more museums began to explain how artifacts were part of a larger story.

Local Objects between Regional Museums and Central Museums

State governments allowed regional institutions for cultural history a great deal of autonomy. Provincial directors in Prussia requested subsidies for their museums and special funding for specific acquisitions and renovations. The Cultural Ministry received reports from historical associations, provincial museums, and local museums from Cologne to Königsberg about their collections and excavations. The vast majority of these reports received little more than an acknowledgment of receipt and some praise for the shared work of maintaining the state's cultural heritage.[73] This hands-off approach created a set of overlapping and sometimes-conflicting responsibilities with regard to domestic archaeology. On the one hand, officials from the Prussian and Bavarian Cultural Ministries wanted to promote central museums, like the Museum für Völkerkunde in Berlin and the BNM in Munich, as leading scientific institutions. On the other hand, regional museums emerged as the agencies that oversaw most excavations and promoted archaeology in the public's eye. They benefited from state resources and consultation with central museums, but they wanted to maintain their scientific status and sometimes had to rebuff what they saw as attempts by Berlin and Munich to claim archaeological objects for the central museums. This dynamic repeated itself on the regional level. Local museums and historical associations turned to regional museums for assistance, but they also viewed larger provincial museums with suspicion. These rivalries reinforced the local and regional nature of domestic archaeology even as the nation-state sought to forge a national culture.

The most important find in Brandenburg in the late nineteenth century, the royal grave of Seddin, showed that lines of authority and ownership rights were sometimes unclear. In September 1899, Ernst Friedel received two telegrams from Dr. Heinemann, a *Pfleger* for the Märkisches Provinzial-Museum who lived in Perleberg. The first message alerted Friedel to a grave mound in Seddin that locals referred to as a "royal grave." "Full of urns and bronze vessels. Can you come tomorrow?" The matter was urgent because workers were taking the contents of the grave, but Heinemann's next message contained a second reason for Friedel to hurry: "Dr. Brunner of Royal Museum says he's coming to Seddin tomorrow. What should I do?" Friedel arrived early the next day and arranged for a payment of 120 marks to four workers for the items they had already removed. His action prevented them from selling the items to someone else and preempted the Berlin museum's access to the site.[74]

When Friedel presented the finds from Seddin at the meetings of the Brandenburgia society in late 1899, it seemed that the circumstances surrounding the finds were settled. But Traugott von Jagow, a local notable from Perleberg, started to advocate on the workers' behalf, calling for a higher payment. The town of Perleberg also created its own antiquities museum after the sensational finds at Seddin. In 1912, Jagow requested copies of the Seddin finds for the local *Heimatmuseum*, but the Märkisches Provinzial-Museum did not provide them. The issue was still unresolved when Max Viereck, a notable from Perleberg, again asked the Märkisches Provinzial-Museum for copies in 1927. When he was told that they would cost 400 marks, he asked if the price could not be lowered since "the museum basically came into possession of the [Seddin] finds for free and even with the support of people of Perleberg." Viereck then gave up because his small museum could not afford this price. In the end, copies of the Seddin antiquities were made and loaned to the Perleberg museum for a temporary exhibit.[75]

Disputes between local, regional, and central collections also arose in the Kingdom of Bavaria. When Johannes Ranke began plans for his Prähistorische Sammlung in Munich in 1885, the Historischer Verein von Oberbayern responded with indignation. The association argued that King Ludwig I had given the historical associations the right to preserve local items in the 1830s. The association acknowledged the value of a central Bavarian collection, but it believed that this should be created with replicas from local collections, not from originals found in the domains of the local associations. The group therefore refused to turn over its collection and accused Ranke's museum of encroaching on its traditional responsibilities. The organization associated with Ranke's museum tried to defuse the situation and presented itself as a partner that stood between the Munich branch of the DAG and the local associations.[76] The parties initially did not find a resolution. The Historischer Verein von Oberbayern retained its collection, and Ranke's group continued to develop its museum using the material that they excavated. After 1900, though, several groups recognized the scientific value of a modern and more comprehensive collection. The Historischer Verein von Oberbayern and the BNM turned their archaeological collections over to the Prähistorische Sammlung, and over time it developed into Bavaria's main scientific institution and museum for archaeology.[77] (Today, the museum is called the Archäologische Staatssammlung.)

The desire of local actors to assert their independence also appeared in complaints about the BNM. In the Bavarian parliament in 1900, Representative Zimmern from the Palatinate expressed his concern that so much emphasis was being placed on the centralization of Bavaria's museums. He complained that when one read about an archaeological find in the newspaper, "a Herr Director from an outside institution swoops down on the object like a bird of prey [wie ein Stossvogel], and we have to watch this happen or pay more [in order to keep the object in the Palatinate]." He called upon the BNM and the GNM to stay out of local affairs. Their competition only inflated the prices paid for antiquities or robbed local institutions of what was rightfully theirs. Zimmern then cited a nagging example. The golden hat of Schifferstadt was purchased by the Bavarian monarchy after its discovery in 1835 and then went to the BNM. Zimmern lamented its removal from his homeland and asked that it be returned.[78]

Today, the golden hat of Schifferstadt is back in the Palatinate and on display at the Historisches Museum der Pfalz in Speyer.

In other cases, regional museums appealed to central authorities to uphold their status vis-à-vis smaller, local museums. The historical commission in Kiel, for example, explained to the Prussian Cultural Ministry in 1892 that Schleswig-Holstein should be regarded as "an archaeological province" with a unique prehistory and unique finds. According to this argument, the Kiel museum was the best institution to research the prehistory of this area, and it therefore required a comprehensive collection of the region's antiquities. The historical commission used this logic not only to fend off the Berlin museum's first right of refusal for artifacts uncovered in northern Germany but also to assert its authority over local collections in Ditmar and Flensburg. The commission noted that a Prussian ordinance from 1878 established the principle that the provinces should not compete with each other for archaeological objects, but it did not clarify the relationship between provincial and local museums. In its correspondence with Berlin officials, the historical commission in Kiel recognized local museums as wonderful ambassadors for archaeology and valued their ability to inspire greater interest in local finds, but it asked the Cultural Ministry to categorize these local collections as branches of the provincial museum in Kiel, not as independent institutions. The commission argued that it was critical that Berlin support Kiel's provincial museum because it had a more professional staff (which included Johanna Mestorf) and the ability to ensure that all excavations were carried out in a professional manner.[79] This exchange was more than a turf battle between Kiel and Berlin. The commission in Kiel was also asserting its authority over local *Heimatmuseen* and presenting the provincial museum as the best custodian of Schleswig-Holstein's cultural heritage.

The proliferation of local museums was also an issue in the province of Hannover. In 1901, Jacobus Reimers, the director of the provincial museum, complained about the "incessant creation of small and even tiny museums in the province." These institutions had little scientific value, according to Reimers, because they prevented larger museums like his from attaining a more comprehensive collection.[80] By 1908, however, the relationship between the regional museum and local museums in Hannover was changing. The annual report of the provincial museum spoke of the need for greater cooperation, which would arise from leadership provided by the provincial museum and put an end to dilettantism.[81]

Even the project that seemed most obviously suited for a nationalist interpretation remained hampered by scientific uncertainty and local allegiances. In 1875, Ernst von Bandel completed the Hermannsdenkmal atop the Grotenburg just outside Detmold. Already at that time, though, scholars and local associations proposed several different locations as the site where Arminius had ambushed Varus and his Roman legions in 9 CE, and they hoped to use the growing knowledge about early history and archaeology to determine the location of the battle. (Today, archaeologists place the Battle of the Teutoburg Forest at Kalkriese, about twenty kilometers north of Osnabrück and eighty kilometers northwest of the Hermannsdenkmal.)[82] The two most famous scholars in the German Empire to weigh in on the topic were Theodor Mommsen and Hans Delbrück. Mommsen, the leading scholar in Roman history, was most interested in sites to the east and northeast of Osnabrück between Haltern and Barenau where concentrations of Roman coins contemporary with the battle had been found. (Barenau is very close to the correct location at Kalkriese.) And Delbrück, a military historian and editor of the *Preußische Jahrbücher*, argued for a site outside Rehme (fifty kilometers east of Kalkriese on the Weser River).

In *Die Örtlichkeit der Varusschlacht* (The location of the Varus battle, 1885), Mommsen had expressed his conviction that finds of Roman coins would prove central to solving this historical question, and he called for the careful and patient documentation of finds across central and northern Germany. Mommsen thought that local antiquarians could be ideal partners in this project, but only if they would stop "filling the small and large gossip pages with their beloved patriotic-topographical squabbles and firing up naive observers with their parochial controversies."[83] Over fifteen years later, the search for the battle site continued, and according to an assessment sent to the Prussian Cultural Ministry in 1901, it was still marked by a proliferation of theories and local allegiances: "Unusually diligent research has spun a thick web of investigations and speculations around this quite paltry core [of sources], and dilettantism and local pride have done their ample part to increase the darkness [that shrouds the Varus question]. Just as the use of the same sources has not prevented the philologists from reaching conclusions that vary widely, so too has the investigation of the landscape not led to agreement. . . . Hypotheses crisscross through a wide area, but they have not led to a solution."[84]

Despite the uncertainty surrounding the location of the battle, archaeology's growing importance was evident in the assessment of several sites. As the report to the Cultural Ministry explained, interpretations had to make use of the Roman sources, but it was also impossible "to make an assertion without studying the terrain in person. . . . Moreover, [scholars] are now beginning to apply ground-surveying methods developed recently during work on the *limes*." Specifically, scholars were able to eliminate many sites that were touted as "Varus encampments" because closer observation showed trenches and walls to be modern constructions. Yet the report noted that the question was still unresolved, and it even referred to growing interest in locations closer to the Hermannsdenkmal than Kalkriese, noting, "Increasingly investigators are again turning to the forests around Lippe, the place that Melanchthon had already described as the site of Arminius's deed and that researchers have never completely abandoned during the centuries-long debate."[85]

The ability to distinguish between ancient sites and modern alterations to the landscape was based in part on new research into fortification structures throughout Lower Saxony. General August von Oppermann, with the support of the Historischer Verein für Niedersachsen, began this project in the 1880s, and he was joined by Carl Schuchhardt in 1888. Their work was published as the *Atlas vorgeschichtlicher Befestigungen in Niedersachsen* (Atlas of prehistoric fortifications in Lower Saxony) in several volumes between 1887 and 1916. Trained as a classical archaeologist, Schuchhardt had investigated Roman fortification ruins in Romania and participated in several large-scale excavations in Greece and Turkey. He also authored a widely read publication about Heinrich Schliemann's famous discovery of Troy. In 1888, Schuchhardt became the director of the Kestner-Museum, a cultural history museum in Hannover with an impressive collection of classical artifacts, and he joined the Historischer Verein für Niedersachsen. By 1901, scholars were using Schuchhardt's work to refute various hypotheses about supposed Roman encampments and the location of the Battle of the Teutoburg Forest. Friedrich Koepp, the director of classical antiquities research in Westphalia, for example, urged the Cultural Ministry to deny a funding request by E. A. Mueller. According to Koepp, Mueller based his "rash conclusion" that the Dietrichsburg near Melle was the battle site on the presence of a single Roman inscription. Koepp noted that this site was near Mueller's hometown and that his publication was the epitome of "the combination of dilettantism and local pride that appears in so many investigations of the *Varusschlacht*." Koepp then recommended the consideration of sites that Schuchhardt had determined to be truly ancient, like the Grotenburg outside Detmold.[86]

Throughout this discussion, there were scholars who tried to steer the

discussion away from emotional expressions of local pride and rash conclusions. A teacher from Reuthen named Hartmann described what was
considered to be the remains of an encampment near Erle in the *Westfälischer Merkur*. He, along with other members of the local historical
association, excavated in the area in the early months of 1907 and was
alarmed when several newspapers jumped to the conclusion that this was
definitely a Roman camp even before more thorough excavations had
taken place. Hartmann noted that "even the gathered experts did not yet
know, or even could not yet know on the basis of the status of the current excavations, what the Erle site contained." Although he then cited
other evidence about the location that suggested that it could be Roman,
he stressed that "this could be determined only on the basis of the views
enabled by the planned excavations. May they be successful!"[87] And in
evaluating another site nearby Friedrich Koepp urged patience and careful excavations, explaining in the *Osnabrücker Zeitung*: "The fact is that
it is still fashionable even today simply to claim the remains of every wall
as Roman and to push aside all contradicting evidence based on serious
science. One must accept that this evidence sometimes cannot be brought
to bear until after much labor and the commitment of substantial financial resources."[88]

As these disputes over objects and competing interpretations demonstrate, the connection between local landscapes and archaeological objects remained strong, even within the German Empire. The directors of
regional museums based their outreach and professionalization efforts on
the belief that objects were closely related to their immediate archaeological context and on their desire to ensure their jurisdiction over their
local areas. They also exhibited a clear awareness of the limitations of
their knowledge about the archaeological record and about prehistory in
general. Their museums overflowed with objects, but this supply caused
more chaos than confident interpretations. They therefore focused on limited geographical areas as opposed to writing a sweeping national cultural
history. Even in the case of the central moment in German history—the
battle that stopped Roman influence among ancient Germans—the professionals were hesitant in the face of local claims about the site of Arminius's deed, and they called for further investigations.

Museum administrators performed a delicate balancing act. They
wanted to provide a fuller portrait of the distant past, and they encouraged
archaeology as a way to create ties between residents and the landscape.
They hoped their cultural mission would benefit from local and national

patriotism. At the same time, though, they were primarily concerned with the rising scientific status of domestic archaeology. They understood the difficulty of narrating the past, despite the growing number of excavations and improved methods. Their careful interpretations were based on empirical observations and on the immediate context of archaeological sites. Archaeologists connected their findings to regional cultural history narratives, but they were hesitant to draw sweeping national conclusions from local sites. Even as the process of unification was celebrated and a national culture was forged in the German Empire, domestic archaeology flourished as a local activity and in regional displays. Professionals at regional museums, for the most part, did not take the step toward the narration of the nation's distant past.

Outside the walls of archaeology museums, other actors combined an interest in prehistory with nationalist politics more readily. As chapter 6 will show, authors of historical fiction, in particular, were not held back by the limits of the archaeological record. They grasped the enthusiasm for domestic archaeology and steered it in a new direction by overriding the cautious interpretations that archaeologists prepared for regional museums. Novelists filled the gaps in the prehistoric record with action and constructed sweeping stories that connected the present day with the distant past.

CHAPTER SIX

Narrating the National Past

The German Empire did not gradually develop out of the cultural nationalism that intellectuals had promoted since the late eighteenth century, nor did it embody the integrating ideas of liberal nationalism that had inspired many members of the middle class since the 1840s. The political and military events of the 1860s and early 1870s created the German Empire, and after 1871, this new geographical and administrative entity needed a symbolic language that could fuse the concepts of state and nation together. The narration of ancient events played an important role in this process, mitigating the newness of the German Empire by connecting the modern state to the premedieval past.[1]

This new culture of memory found expression in the realm of historical fiction. As the previous two chapters explained, the actual practitioners of domestic archaeology thought largely in anthropological and regional terms that ran counter to a nationalist interpretation of the prehistoric past. Authors of popular history and literature, however, abandoned this cautious and patient stance. The year 1871 brought a sense of closure and transformed the new nation-state into a historical subject whose story needed to be told. In the place of inventories and descriptions, novelists produced narratives that had a clear beginning, middle, and end. The most popular works sketched out sweeping arcs of history and presented the German Empire as the fulfillment of a long process. While Rudolf Virchow argued that the nation disappeared as one went farther back in time, his contemporaries in the field of literature made the opposite claim. They grounded present-day values and national characteristics in the ancient past. This kind of narrative filled pre- and early history with a national content that was largely absent in the work of historical associations, regional museums, and the German Empire's network of anthropologists, but it became a very powerful reading of the distant past.

The literary works discussed below by David Friedrich Weinland, Gustav Freytag, and Felix Dahn appeared in the decade immediately following the creation of the German Empire. They coincided with the founding of the DAG and came just before the enhancements made to regional cultural history museums throughout Germany, but they offered a dramatically different approach to pre- and early history. The authors emphasized narration instead of description, the story instead of the inventory. Taken together, these books prompted the German reading audience to contemplate their relationship to "ancient Germans." Weinland showed his readers how ancient peoples used artifacts and how specific sites in Swabia witnessed decisive battles and the arrival of new peoples. Instead of knives, battle-axes, or urns being static artifacts in a museum, they became *the* items used in warfare and burial rituals. Archaeological sites were not merely the locations of finds but *the* places where specific action took place. Freytag and Dahn presented what they claimed to be the timeless virtues of the German people. Their novels taught lessons about German productivity and valor. And all these works did something that anthropologists and museum directors would never do in their scientific studies: they filled the anonymous past with sympathetic characters and encouraged readers to identify with the victories and struggles of earlier peoples.

Although the presentation of the distant past as a national story was new, it would be a mistake to view these novels separately from the other approaches to prehistory. The desire for a national past, the continued development of domestic archaeology in regional and local settings, and the ascendancy of a bourgeois culture that valued science and history were inseparable in the reading public's mind. These approaches to the distant past mutually reinforced one another. Domestic archaeology made historical fiction about "German ancestors" more alluring and concrete, and the increasingly professional nature of museums in the German Empire gave narratives about the ancient past a greater sense of plausibility. Certainly, the immense popularity of these books was related to the growth and vitality of historical associations in the late nineteenth century and the investments made in institutions for cultural history.

In the climate of rising nationalism before World War I, the desire for a national story anchored in the deep past crossed over from the realm of fiction to the field of domestic archaeology itself. Gustaf Kossinna, the most influential example of this nationalist turn, went beyond traditional archaeological descriptions and argued for a national history that was rooted in prehistory. More measured interpretations continued along-side Kossinna's position, but nationalist authors were impatient with inventories and museum collections that did not venture ethnic interpretations. The crucial step toward a nationalistic telling of the German past required a willingness to push the interpretation of archaeological material well beyond the inventory or description and confidently assign cultural, even national labels to ancient finds. Throughout the eighteenth and nineteenth centuries, antiquarians had suggested this as a future outcome of their efforts, but they respected the fragmentary nature of the archaeological record. Nationalists around the turn of the twentieth century overrode this hesitation and used prehistory as the starting point for their nation's history.

Bringing the Past to Life in the Swabian Landscape

The literary path from inventories to narratives was already present in the work of the Bavarian ethnographer Wilhelm Heinrich Riehl. In addition to his nonfiction surveys of customs and folklore, Riehl produced dozens of novellas from the 1850s through the 1880s that traced his four *S*s of folklore (*Stamm*, *Sprache*, *Sitte*, and *Siedlung*) throughout history. As literary scholar Mary Beth Stein has explained, these novellas reworked Riehl's ethnographic observations into narrative form and provided a contrast to the approaches taken by early nineteenth-century collectors. Riehl

> used the materials of folklore toward different ends than his Romantic precursors. Whereas the latter published collections of folksongs and folktales out of their belief in the intrinsic value in the study of language, history, and folklore, Riehl produced *Darstellungen*, ethnographic and literary sketches of folk life. What had been a primarily unconscious process of folklore recreation in the hands of the Grimms becomes a self-conscious and intentional act in Riehl's works. While basing his ethnographies and novellas on empirical fact, his primary aim is the poetic realization of the intrinsic truth of his data.[2]

Most of Riehl's works were set in the Middle Ages or early modern period, but they offered a model for authors who delved into a deeper past. A key point was Riehl's combination of historical accuracy (his raw material was ethnographic data) and accessibility (his plotlines were easy to read and filled with colorful characters). Indeed, Riehl admired authors who straddled the line between scholarship and storytelling. Considering the popularity of the English historian T. B. Macaulay, he explained that "the novels and historiography of an era complement one another. . . . The most popular historian of our day, Macaulay, must attribute half of his popularity among the people to the art of capturing . . . the inner connections between the customs and circumstances of an era. At the same time he knows how to paint the colorful details of these circumstances with the brush of an artist. That's why one can say of the third chapter of his *History of England*: it reads like a novel."[3]

For Riehl, the task did not end at "capturing . . . the inner connections between the customs and circumstances" of early modern Europe. He wanted to show that present-day conditions grew out of this earlier era and that the "connections" were timeless. Specifically, he wanted to explain the positions of the middle class and the landed nobility in the nineteenth century as positive outcomes of long-term historical developments.[4] In order to realize the instructive potential of ethnography and folklore studies, he had to connect the past to the present in narrative form. This could happen as either fiction or nonfiction, but historical fiction was more accessible and had fewer restraints.

Several authors achieved enormous success by integrating historical research into their novels. Joseph Scheffel, who published *Ekkehard* in 1851, had spent years studying legal documents from the Middle Ages, and the original version of his story about a tenth-century king had over 250 footnotes.[5] David Friedrich Weinland based his stories from prehistory on the work of Württemberg historical associations and the DAG. And Gustav Freytag and Felix Dahn, two of the most successful novelists of the second half of the nineteenth century, also composed cultural histories that reflected their engagement with a wide range of historical sources. All these authors claimed that their literary output was something more than pure invention; they brought research and creativity together and thereby created new possibilities for the evidence that domestic archaeology had uncovered.

During the 1870s, David Friedrich Weinland (1829–1915) composed two novels for young readers that were set in prehistoric Swabia: *Rula-*

man: Erzählung aus der Zeit des Höhlenmenschen und des Höhlenbären
(Rulaman: A tale from the time of cavemen and cave bears, 1876) and
*Kuning Hartfest: Ein Lebensbild aus der Geschichte unserer deutschen
Ahnen, als sie noch Wuodan und Duonar opferten* (Kuning Hartfest: A
portrait from the history of our German ancestors from the time when
they still sacrificed to Wuodan and Duonar, 1878). Both books were in-
stant successes among children, parents, and teachers. Over fifty editions
of *Rulaman* have appeared, and it is still a favorite today. New editions of
Kuning Hartfest appeared through 1936.[6]

Weinland was born in Grabenstetten, a village in southwestern Germany near Reutlingen, in 1829. He was the son of August Johann Weinland, a pastor and advocate for technological advancements in farming.
David Friedrich enrolled at the renowned Tübingen seminary where both
Hegel and Hölderlin had spent some of their student years, but he eventually moved from the study of theology to an interest in animals and
nature. He worked at the Berlin zoo in the early 1850s before an extended
stay in the "New World" as a natural scientist. Harvard University invited
him as a research scholar, and he later traveled to Haiti to investigate the
growth of coral reefs. After his return to Germany in 1859, he worked for
the Frankfurt zoo. In 1863, after suffering a relapse of yellow fever that
he had initially contracted in Haiti, Weinland moved back to Swabia and
lived a quiet life. While raising four sons with his wife Sophie, he told tales
from prehistory that would become his best-selling works of juvenile literature.[7]

Sympathetic heroes and action-packed plotlines turned Weinland's
books into classics, but they were also part of a middle-class culture that
placed a high value on scientific and historical knowledge. Weinland advocated for zoos and botanical gardens because he believed that these kinds
of public institutions contributed to the intellectual and moral improvement of individuals.[8] His novels were related to this commitment, for they
provided a popular forum where Weinland conveyed the current thinking
about archaeology, anthropology, and natural history.

Yet Weinland's works were very different from the archaeological reports that appeared in journals for regional history or anthropology in the
1870s. He drew connections between the worlds of his prehistoric characters and the nineteenth-century landscape. He emphasized that the beautiful hills of the Swabian Jura were the same ones where Stone Age cave
dwellers had built their homes and that this terrain had also witnessed the
Germanic settlements of the early Common Era. Readers sought to ex-

perience these historical developments for themselves. The Schwäbischer
Albverein sponsored "Rulaman hikes," and trail guides for the region still
recommend tracing Rulaman's steps through the caves, cliffs, and lakes
that Weinland described.[9]

Rulaman, published in 1876, takes place during the Neolithic period
in the Swabian Jura, the range of low mountains southeast of Stuttgart. It
relates the decline of the Tulkas, a tribe of the ancient Aimats who lived
in the caves of prehistoric Europe. Weinland describes the Aimats as the
"original Europeans" (*Ureuropäer*). The story follows Rulaman, a young
Tulka, who learns the ways of his people and develops a close relationship with a revered female soothsayer. The drama of the story comes with
the arrival of the Kalats from the east. They are a materialistic people
who possess weapons made of bronze. (The Aimats have only stone implements.) The Kalats destroy the Tulkas, but not before the soothsayer
curses the spiritual leader of the Kalats. She tackles the threatening
Druid, and they both fall over a cliff to their death. Rulaman escapes as
the sole Tulka survivor.

The tremendous success of *Rulaman* set the stage for *Kuning Hartfest*'s
publication in 1878. Weinland invited his young readers to return to the
same geographical setting as *Rulaman*, but at a different moment in time:
"May I lead you once again to our beautiful *Sueben-Alb*. . . . Although
about a millennium may lie between [these two stories] . . . mountain and
valley, cliff and caves are indeed the same ancient ones."[10] *Kuning Hartfest* is set in 8 CE among the Suebi, the tribe that Julius Caesar had battled
in Gaul and drove east across the Rhine. The king of the Suebi, Kuning
Hartfest, is now old and tired, but he is still respected by his people and
seen as the embodiment of all the positive Germanic values. Like *Rulaman*, this is an action-packed novel: the Suebi are always off to hunt or
fight, and their religious practices and social customs are portrayed as
dramatic and profound. The threat of invasion again plays an important
role. Hartfest's warriors begin to encounter Romans in their territory,
and the appearance of transportation roads east of the Rhine foreshad-
ows the "strangulation" of Germanic freedom.[11] Weinland emphasizes a
larger Germanic victory over the Romans, however, by closing the novel
with Berchta, the maiden granddaughter of Hartfest, and her search for
Agilolf, a Suebi warrior and her destined lover. In the story, she arrives in
the Teutoburg Forest just after Arminius and the Germanic tribes have
defeated Varus and the Roman legions in 9 CE. Agilolf was killed in the
battle, but Berchta's love resurrects him so that they may experience the
Germans' good fortune together.

A critical aspect of both novels was Weinland's claim that they were grounded in archaeological and historical knowledge. Weinland was aware that the scarcity of textual and material sources made it impossible to reconstruct life in prehistoric times. He explained that he could not write the "truth" about ancient epochs, but he promised his readers that he would not include anything that was scientifically impossible.[12] He then filled his narratives with references to antiquarian and anthropological literature. In *Rulaman*, for example, Weinland placed the Tulkas near Lake Constance because this is where prehistoric lake dwellings had been found in the previous decade. Antiquarians and archaeologists believed these sites to be the oldest settlements in central Europe, and Weinland found it appropriate to connect the "original Europeans" with this kind of dwelling. He based his description of the cave where Rulaman's family lived on excavations in Hohlenfels (eight miles west of Ulm) carried out by anthropologist Oscar Fraas in the early 1870s. And he drew on Carl Vogt's *Vorlesungen über den Menschen* (Lectures on mankind, 1863) for his discussion of prehistoric animals and the physiology of the Aimats. Editions of *Rulaman* after 1905 included an increasing number of references to the work of antiquarians, folklorists, and anthropologists. These citations brought the expanding literature about prehistory and archaeology to the attention of a large reading public, and they imbued Weinland's tale with scientific authority.

Weinland's treatment of the Kalats was also based on recent debates and research. His descriptions of the Kalats' appearance drew on a presentation about the physical anthropology of the Celts at the DAG's 1881 meeting in Regensburg. And he enhanced his claim that the Kalats introduced bronze into the area with references to the growing literature about the Hallstatt finds in Austria. This site had been excavated from the 1840s through the 1860s, and publications about the extraordinary bronze finds strengthened the nineteenth-century association of copper and bronze with the Celts. Weinland went beyond the inventories and the description of finds, though, by speculating about the manufacture and use of ancient items. His book included an illustration of a prehistoric foundry, and it portrayed bronze swords as sacred objects in the cultic practices of the Druidic priesthood.

Weinland's attention to historical and archaeological sources was even clearer in *Kuning Hartfest*. In order "to provide a *complete* look . . . into the *vigorous life* of this ancient people [the Suebi]," Weinland gathered information from Greek and Roman sources, antiquarian finds, and that "which still lives on in our culture from ancient times."[13] In the first edi-

FIGURE 7. The illustrations in David Friedrich Weinland's *Rulaman* helped his young readers imagine life in prehistoric times. Here the Kalats manufacture copper swords in a foundry. (Drawing by H. Leutemann. Weinland, *Rulaman*, 190. Reproduced courtesy of the Zach S. Henderson Library, Georgia Southern University.)

tion, footnotes contained information about research into German linguistics, mythology, and ancient place-names. These references became even more extensive in editions after 1904.

One scene among Hartfest's people appears to draw directly on an archaeological discovery that created a sensation in the mid-nineteenth century. Late in the book, Germanic warriors, adorned in shining armor, marched in mourning to a cemetery in a forest of oak and fir trees. There, they constructed a funeral pyre and burned the bodies of their fallen comrades. The bones of most of the men were placed in a row grave along with their swords, spears, and shields. One of the warriors, however, was placed in an oak coffin, and his horse was buried with him "to take his master to Walhalla." The mourners constructed a burial mound over the grave containing the horse and rider.[14] Weinland does not cite a specific inspiration for this scene, but the description of grave goods and the details about the horse skeleton recall the excitement about the initial excavations at Nordendorf in the 1840s and 1850s. Rudolf Marggraff, an art professor in Munich, recounted the most fascinating details in the *Allgemeine Zeitung*, which included the burial of several horses next to men's bodies.[15] In Weinland's telling, this archaeological site became the result of dramatic action. Readers were encouraged to imagine the warriors at the funeral scene and the bonds between horsemen and their equine companions.

Weinland's footnotes also reviewed the investigations into the battle between Arminius and Varus. He noted the spectacular find of silverware near Hildesheim in 1868, which some people took to be Varus's table service, and later editions of *Kuning Hartfest* referred to research into the battle's location by archaeologists and classicists, including Eduard Paulus, Carl Schuchhardt, and Theodor Mommsen.[16] Weinland's main goal, however, was to bring this battle to life through Agilolf's participation and have his readers identify with the combatants.

In scene after scene, Weinland's novels cast archaeological objects in a new way. Regional museums exhibited bowls, swords, and jewelry and explained what they were made of and where they were found, but Weinland put these items into action. In *Kuning Hartfest*, Germanic heroes drank ale from earthen bowls, sounded horns before the hunt, and drew their swords in battle. The maiden Berchta wore a golden brooch and arm ring like the ones found in Merovingian row graves. Suebi warriors recreated the rituals that Tacitus had described in the *Germania*: they convened in the tribe's assembly hall to discuss their military strategy and to celebrate victories, and the community celebrated a spring festival, the summer solstice, a harvest festival, and the appearance of the mother goddess in winter. These stories put the wide range of materials that historical associations and regional museums had amassed into motion and invited readers to reenact the ancient past.

Weinland's works also introduced the reading public to the emerging field of anthropology. In *Rulaman*, the Aimats, whom Weinland described as Europe's earliest inhabitants, lived as hunters and gatherers in harmony with nature. The women had dark complexions and dark hair. They dressed in reindeer skins and spent most of their time at home in the community's cave. The men, introduced as they returned from a five-day hunt, were sturdy and alert. They used stone tools but had no knowledge of bronze or iron. The seasons punctuated the life of the community, as the Tulka tribe traveled to Lake Constance every summer for a celebration in nature. This idyll breaks down as the Kalats enter the area. Weinland described them as also having dark complexions, but these invaders had metal swords and were able to defeat and subjugate the Aimats completely (except for Rulaman's escape). The invading culture replaced the primitive life of the Aimats with colorful clothing, shiny swords, and new religious practices, including live sacrifices.[17]

Several aspects of this encounter between the Aimats and Kalats are striking when read against the rise of anthropological knowledge in the last third of the nineteenth century. The frequent references to skin color and hair color represented the metrics used by anthropologists in racial surveys of present-day European populations. Although anthropologists grappled with the problem of defining racial categories among peoples and acknowledged that all peoples exhibited wide variations, Weinland presented each cultural group in his story as having clearly defined physical attributes.

The plotline in *Rulaman* stressed that the settlement of central Europe was the story of one people replacing another. Peace and tranquility existed within the Aimat community, but the arrival of another group represented a challenge that resulted in the triumph of new technology and the dispersal or enslavement of the earlier inhabitants. The Aimats used stone tools, whereas the metal weapons wielded by the Kalats helped them to defeat the Aimats and destroy their culture. This version of prehistory—a primitive people being replaced by one with greater technological skills—created a clear succession of well-defined communities. New settlements were a matter of conquest, not co-optation or assimilation.

Sharp distinctions between cultural groups shaped the narrative of *Kuning Hartfest* as well. In the introduction to his second novel, Weinland explained that "the simple Aimats and weak Kalats [from *Rulaman*] do not live here; now it is the Germanic peoples. They migrated here and made slaves of the Celts." The conquest by the stronger people was complete, but history did not stop there. Newcomers again threatened a settled people. In *Hartfest*, the story shifted from the Aimats' fight for survival to "the struggle between the German and Roman worlds."[18] Weinland associated physical appearance and cultural values with the outcome of the territorial struggle. The enslaved Kalats were weak and darkhaired, and the Romans who encroached upon the Swabian hills were "the weak, bland people" who were dwarfed in comparison to the large stature of Germanic men.[19] For these images, Weinland referred to Tacitus's description of the Germans' "fierce blue eyes, tawny hair, bodies that are big but strong only in attack"[20] and to the passage from Julius Caesar that said several Roman officers fled the Battle of Wasgaubergen because they feared their giant adversaries in Gaul.

The emphasis on successive cultures had important consequences for the way that Weinland framed prehistory. By emphasizing sharp distinctions between Romans and Germans (and drawing a similar distinction between the Aimats and Kalats in *Rulaman*), Weinland could portray each people as a separate epoch. The arrival of a new people (and the defeat of an old one) provided an ethnic baseline that erased what had come before. Time and culture started over again with each wave of settlement.

Many anthropologists criticized this simple schema from the 1860s through the 1890s. They pointed to the mixture of peoples and cultures that occurred over time and the way that the decorative arts were transferred among prehistoric groups. At one point in the notes to *Kuning Hartfest*, Weinland acknowledged the high degree of mixing among peoples that must have occurred. Yet he quickly moved on from this, and the overwhelming impression he left—indeed, the basic structure of both novels—was that different peoples possessed distinctive physical traits, cultures, and even values and that each invasion brought a complete replacement of earlier practices by the new people and culture.[21]

The replacement of one people by another also allowed Weinland to attach specific virtues to ancient peoples in ways that were impossible for an archaeologist or museum director. The Kalats intrude upon the Aimats in *Rulaman*, and the Romans threaten the Suebi in *Kuning Hartfest*. Weinland depicted the settled peoples, the Aimats and Suebi, in sympathetic terms. They live in harmony with the landscape. They are simple, yet noble, and they possess a clear code of honor. The invaders, by contrast, are marked by their avarice. The Kalats seek to exploit the Aimats, and the Romans are bent on strangling the freedom of the Germans. Conveniently, the two books skip over the period when the Suebi encroached upon the Celts. *Kuning Hartfest* begins after this has happened, and the author dispenses with this transition by mentioning the weakness of the Celts and the ability of the Germanic peoples to enslave them. Despite the previous occupation of the land by Aimats and Kalats, the model of one people replacing another allowed Weinland to describe the Suebi as an indigenous people. They were fully at home in the Swabian landscape.

Weinland's books framed the prehistoric past in another important way. His compelling stories added personalities and human action to the skeletal knowledge that researchers had about prehistoric epochs. The anthropological literature on *Pfahlbauten* was completely depopulated because it could identify only the remains of dwellings, food scraps, and other fragments of domestic life that were preserved in peat bogs and alpine lakes. Weinland filled these sites with people and activity. He gave the anonymous "original Europeans" a name (the Aimats) and went even further to identify characters within the Tulka tribe. In this way, *Rulaman* encouraged readers to identify with the young protagonist as he learned the ways of his people.

Weinland filled the gaps in knowledge about the distant past with family scenes that felt quite contemporary. Despite the thousands of years that lie between the settings of the two stories, the family structures in the prehistoric caves and among the Suebi look very similar, and they affirmed the nineteenth-century notion of separate spheres in the bourgeois family. As the women in *Rulaman* sewed pelts and the daughters combed their long hair at home, the tribe's men related stories from the hunt. Sons, including young Rulaman, eagerly listened to their fathers. An early climax in the book charts Rulaman's entrance into his manly role within the community. On his first hunting expedition, he embarrasses himself by shooting his arrow too soon and scaring off a deer, but he later proves himself a hero by rescuing his father from an attacking lion. In *Kuning Hartfest*, Weinland again portrays the women taking care of work at home while the men are off to fight, and much of the book revolves around the protection of young Berchta's honor and the attempt to find her a suitable husband.[22] Such descriptions not only made the family practices of the Aimats and Suebi recognizable to nineteenth-century readers but also made current notions of family appear ancient. The German family appeared to have emerged out of these hunting and warring societies fully intact.

Weinland relied on archaeological descriptions and anthropological literature, but he suggested new possibilities for the antiquarian imagination. By setting his stories in distant times but in familiar locales, he brought his readers closer to Germany's ancient past. Visitors to the Swabian hills that Rulaman and Hartfest had known (or to any of the places where antiquarians had uncovered ancient artifacts) could identify with these characters and imagine that they shared the same traits and values as the ancient inhabitants of central Europe. Fiction, more so than inventories or museum displays, could put ancient objects in motion and create bonds between nineteenth-century readers and their supposed ancient ancestors.

Weinland's approach to antiquarian literature was not the only option for authors who wanted to bring the past to life. Theodor Fontane, for example, did not fill Germany's ancient past with warriors or heroes. Instead, he celebrated the curiosity of the antiquarian. Born in the town of Neuruppin in 1819, Fontane developed a deep love for regional history, and many of his best-known works delved into Brandenburg's past. Between 1862 and 1892, he published a five-volume survey of the landscapes and historical monuments of his native area. In terms of historical narration, the *Wanderungen durch die Mark Brandenburg* (Strolls through the Mark Brandenburg) is something more than an archaeological description but yet very different from the dramatic action of *Rulaman* and *Kuning Hartfest*.

Fontane was an honorary member of the Brandenburgia society in Berlin, and he corresponded with Leopold von Ledebur about the archaeological collection in the Neues Museum.[23] Fontane's *Wanderungen*, however, went beyond the spare entries of an inventory. His volumes cataloged legends, customs, noteworthy architecture, and antiquarian research in a way that was much more inviting to readers. Fontane filled the *Wanderungen* with history, personal memories, and quotations from local figures. In many ways, his audience was reading not a narrative but rather the landscape itself.

A striking example of the interaction between the distant past, recent events, and personal memories appeared in Fontane's treatment of his hometown. Fontane began with comments on the proud history of Neuruppin's *Gymnasium* and the rectors from the late eighteenth and nineteenth centuries who had made it a serious place of learning. He then turned to a rich collection of natural history objects and local antiquities that had been donated to the *Gymnasium* by a local count. There were over two hundred artifacts from the Stone and Bronze Ages, and Fontane gave special attention to two items—a battle-ax covered with six spikes known as the *Kommandostab* and a bronze wagon with three wheels that was decorated with small bird figures. Both of these items were uncovered in 1848 and had generated lively interpretive debates.[24] While Fontane had his own opinions about these items, he presented alternative interpretations in an evenhanded way. For him, they represented the ancient past that the landscape contained and influenced the ways that antiquarians, local residents, and schoolchildren engaged with this past. As Fontane explained:

> there's something exciting in the fact that the best objects the collection has to offer were found either in the immediate area of the local province or even on the count's personal property. A battle-ax, like the one detailed above, is gener-

ally interesting, but it is twice or three times as interesting when it was dug up in one of my neighbor's fields. It's precisely that point [the proximity of the object] that awakens the otherwise-dead landscape, Elsengrund woods, or peat bog and conjures up a world full of life even in the most barren stretch of heath.[25]

With Fontane as a guide, artifacts remained artifacts. They did not need to be plugged into exciting plotlines to be evocative.

Gustav Freytag: Prehistory as National Background

Other authors focused less on specific objects and more on German characteristics that supposedly developed out of the distant past. In novels and historical works, Gustav Freytag (1816–95) crafted broad stories of national development that extended back to the beginning of the Common Era. His literary success helped to make prehistory an integral part of the historical narrative. Freytag was born in Silesia in 1816 and moved to Berlin during the 1840s. He became a celebrated figure in German-speaking Europe through his editorial work for *Die Grenzboten* (Dispatches from the border, 1841–61 and 1867–70) and through his extremely successful novel *Soll und Haben* (1855; translated in 1857 as *Debit and Credit*). He admired the economic development Prussia had brought to his native Silesia, and his political position followed that of many National Liberals during the 1860s and 1870s: he celebrated Prussia's military victories, supported unification, and shared the national pride of this era; however, he had misgivings about the authoritarian political order in the Second Empire and the power of Chancellor Otto von Bismarck.[26]

Beginning in the late 1850s, Freytag turned to the distant past with *Bilder aus der deutschen Vergangenheit* (Scenes from the German past, 1859–66), a work that followed the history of the German people from the first century BCE to the middle of the nineteenth century. A little over a decade later, Freytag traversed this same expanse of time in a cycle of novels called *Die Ahnen* (The ancestors, 1872–80). This project traced multiple generations of a single family from the fourth century CE to the nineteenth century. As literary scholar Lynne Tatlock has argued, *Bilder aus der deutschen Vergangenheit* valorized the middle class. The multivolume work covered several centuries in order to portray the "ascendancy of the bourgeois subject . . . as the goal of history."[27] A closer look at several passages from *Bilder* and *Die Ahnen*, along with a consideration of the development of domestic archaeology in the late nineteenth century, pushes Tatlock's insight even further. Freytag's works suggested that contemporary values developed out of the distant past. These included not only the productivity of the middle class but also a set of national characteristics that could supposedly be found in the premedieval period.

Freytag's *Bilder aus der deutschen Vergangenheit* purported to be a work of history. It moved forward in chronological order, as one would expect of a historical survey, and it was based on an array of primary sources. Yet the author's references to Roman commentators, Byzantine documents, German philologists, and archaeological excavations were infrequent, and more often than not, they were used to add a colorful detail, not to document a specific event. For example, while describing the constant warfare in central Europe during the fifth and sixth centuries, Freytag cited an excavation near Verdun to illustrate the dedication of vassals who walked ahead of their riding lords and the presence of women among troops on the move.[28] The purpose of this note was more literary than evidentiary; it illuminated wartime loyalty rather than a particular battle or army. As history, *Bilder* was very different from a political narrative devoted to diplomacy and statecraft. Instead of covering battles and the succession of rulers, it highlighted the presence of timeless virtues and cultural traits.

Freytag's creative approach to history was made clear in his comments on the most important source for central Europe's prehistory, Tacitus's *Germania*. Freytag acknowledged the centrality of this text, but he noted that it was written from the perspective of a Roman official, and it contained a rhetorical overlay that was meant more for Roman readers than as a rendering of life north of the empire. Freytag's concerns were different: "The historiography of the ancient world did not know the rich registration of small but telling details—the aspect [of historical writing] that has become so dear to us since the blossoming of the novel. . . . In the end, Tacitus was no painter of details."[29] Although the Roman author made valuable observations about the cultural practices of the Germans, these were "the impressions of a respected traveler." Freytag imagined how the text would have looked if it had been written by a merchant in a German village,[30] and he set out to read between the lines of Tacitus's observations to re-create the daily life of the ancient Germans.

Freytag's treatment of pre- and early history emphasized two main ideas: first, that the premedieval past exercised a formative influence on the later course of German history; and second, that clear parallels existed between ancient events and nineteenth-century conditions. Both of these ideas compressed the time between the early Common Era and the present day and encouraged the association of modern values with the ancient past. The first theme was clear from the first chapter of *Bilder*. Freytag began in the centuries just before the beginning of the Common Era and described the sixteen hundred years between the invasion of the Cimbri and the life of Martin Luther as "the German nation's first adolescence." It was "a long political history filled with blood and the destruction of peoples, with enormous deeds and incalculable suffering, with the cheerful growth of the *Volk*'s spirit and with the corrupting turbulence that caused young blossoms [of this spirit] to fade."[31] This kind of stilted language pervades *Bilder* and shows how Freytag connected mundane developments to what he described as national characteristics. The development of farming communities in the distant past, for example, was not just an economic or social transformation. Instead, it was the moment when ancient Germans abandoned "defiant egotism . . . in favor of more noble ideals," which led to a sense of community that was then passed down to later Germans.[32]

The emphasis on agriculture was part of Freytag's attempt to revise the perception of the Germanic peoples. Both Strabo and Caesar had described them as nomadic, and Tacitus reported that "every year they cultivate new ploughlands. . . . For they do not struggle with the richness and extent of the soil in order to plant orchards, enclose meadows, and irrigate gardens."[33] In contrast, Freytag viewed the early Germans as settled, and he linked landholding patterns in the early Common Era to the Germans' strong attachment to the land and their sense of community. He explained that individual families owned their houses, gardens, and animals, but that they shared the administration of fields, lakes, and forests. This arrangement formed the backbone of the German community (*Gemeinde*), which was the basic unit of larger geographic regions (*Gaue*).[34] For Freytag, agriculture was central to the development of civic institutions. This idea was not unique to Freytag's writings. Wilhelm Heinrich Riehl and Friedrich Ratzel wrote extensively about the place of the German peasantry in German society, expressing both praise for the virtues that German farmers embodied and the need for reforms that would allow for economic modernization and the preservation of the peasantry.[35] Freytag's discussion was unique, though, because it anchored these contemporary concerns in the deep past.

Ancient landholding patterns were also at the heart of a contradiction Freytag hoped to explain. On the one hand, Freytag noted, Germans loved the land, and their development was intimately tied to the territory they inhabited. At the same time, though, the Germanic peoples had a *Wanderlust* like no other group, and emigration and expansion were essential parts of their history.[36] Freytag resolved this contradiction by pointing to the productivity of the Germanic peoples. They were successful farmers, and the growth of their communities created a demographic pressure that sent young sons off to find new lands. This did not mean that ancient Germans did not love their homeland. Instead, it was a matter of transplanting their productivity to new places. In Freytag's narrative, the ancient pairing of the love of the land with the need for expansion provided a historical precedent for later movements, including the medieval conquests by the Teutonic Knights throughout eastern Europe and the expansion of the Kingdom of Prussia since the seventeenth century. The pairing also contained a clear message in the nineteenth century. Leaving the land for the city or shifting from agriculture to industrial enterprises did not suggest a breakdown. Instead, these two environments remained connected to one another, as the land continued to feed urbanization and industrialization. And in a period when tens of thousands of Germans left Europe for the Americas, this model presented emigration as a long-standing practice and one that did not disrupt the cultural connections between Germans abroad and at home. The Germans, in the ancient past and in the present day, could appreciate their rootedness and *Wanderlust* at the same time and treat both as valuable national characteristics.

The spread of Germanic productivity had a shadow side, however. Freytag described those whom he viewed as standing in the way of this expansion in negative terms. The presence of Slavs in Bohemia since the early centuries of the Common Era "had become an injury of German history that we still feel today."[37] To nineteenth-century readers, the contrast between Germans as the bearers of progress and Slavs as a less advanced people affirmed the benefits that German influence brought to east-central Europe and justified the Germanization policies that were implemented in Prussia's eastern provinces. This hierarchical thinking was present among archaeologists of this period also, who claimed to see the difference between "German productivity" and "Slavic indolence" in the relative cultural levels of Germanic and Slavic finds.[38] On this issue, archaeology and Freytag's nationalist narrative worked together to present the ancient Germans as brave warriors and settlers who brought prosperity to eastern Europe.

In the early 1870s, Freytag began to publish *Die Ahnen*, a series of novels that began in the year 357 CE. The author claimed that this extended family history had little to do with the realm of history. The characters and plotlines were invented, and the first two books in the cycle, *Ingo* and *Ingraban*, "lead into a period of time that the poet can understand better than the historian."[39] Yet a comparison of *Bilder* and *Die Ahnen* reveals how the line between history and literature was unclear. The early books of *Die Ahnen* contain many of the themes Freytag had developed a decade earlier in his cultural history. In the novels, however, Freytag turned from the collective story of the Germans to a story devoted to a single family.

Book 1 of *Die Ahnen* revolves around Ingo, who grows up among his people but sets out on his own as a youth. He encounters wild animals and fights several rivals. He falls in love with a girl who is his social superior, but he is able to win acceptance for their marriage with a series of heroic deeds. Their bliss is short-lived, however, because rivals kill both Ingo and his wife. With a device that David Friedrich Weinland would repeat in *Rulaman*, the series of novels could continue because the woman who cares for the doomed couple's son escapes with their little boy.

Ingo's life combines the love of home and *Wanderlust* that Freytag identified in *Bilder*. There are several scenes where young sons set off from their village in search of better land. They travel to the edge of Thuringia and create a new village, confirming their productive capacity and explaining the expansion of the Germanic peoples through central Europe. Freytag added two layers to this scene that suggested connections to his present day. First, he described the land-seeking characters as if they were settlers moving west in North America. The men led caravans of covered wagons and yoked oxen, while the women and children followed with their herds. Sparsely populated areas in central Europe during the early Common Era invited the same kind of adventure and opportunity that the American West provided in the nineteenth century. Second, Freytag drew attention to the ways that ancient landholding patterns had created nobles, free peasants, and serfs in the early modern era. One character explained how free peasants and nobles were useful to each other, while the serfs appeared as an economic hindrance.[40] Freytag used this conversation to highlight freedom as a necessary ingredient for prosperity, whether in the distant past or the present day.

The theme of German productivity replacing Slavic backwardness appeared in *Ingraban*, the second of the novels in the series. This story is set in the early ninth century in the border area between Saxony and Thuringia. The Slavic chief Ratiz, who is depicted as deceitful and inert, provides a stark contrast to the Germanic hero Ingraban.[41] Here again territorial expansion was the outcome of productivity, not conquest. Freytag presented Ingraban's people as more deserving of the frontier land. Freedom and productivity appear as national virtues in the story and provide a rationale for German expansion in the east.

Freytag's forays into historical narration were fundamentally different from the details of an archaeological description. The general survey in *Bilder* and the individual characters in *Die Ahnen* conveyed messages of social cohesion and historical continuity. Both works inserted virtues and values into the early Common Era and invited readers to view these values as enduring features of German history.

Die Ahnen remained very popular throughout the 1870s and 1880s, although some commentators expressed distaste with the way that the ancient past was treated as a prologue for later German history. In a review essay from 1891, Munich scholar Leo Gregorovius criticized historical novels that asserted superficial connections between distant ancestors and contemporary Germans. He discussed several works but devoted much of his essay to *Die Ahnen*. Gregorovius described the series of novels as fictional creations that masqueraded as history. Perhaps this would not have been a problem if the works were merely entertainment, but Gregorovius explained that the books were read and even used as works of cultural history.[42] This raised the stakes in the reception of the novels and drove Gregorovius to a fuller analysis of the way the books presented the past. For Gregorovius, the problem was more than the presence of anachronisms or the archaic language that Freytag developed for the novel. (Gregorovius was certain that early medieval Europeans did not speak the elevated language that Freytag used.) The main issue concerned the meanings generated by the author's inventions. For example, Gregorovius took issue with the overarching theme of *Die Ahnen*—that the German bourgeoisie and the value of freedom had grown out of the distant past and now flourished in the present day. By transporting social and political values from the nineteenth century to the premedieval period, Freytag gave present-day conditions a historical justification that they did not deserve.[43] Furthermore, Gregorovius claimed that the noble and courageous actions of Ingo were too moral and unrealistic. If this were simply a fictional account of a hero's life, the story could be accepted as entertainment. But Ingo's actions were seen as expressions of German national characteristics and *Die Ahnen* more broadly as the "psychology of a people."[44] This was the dominant interpretation of *Ingo* and *Ingraban* that had emerged in the twenty years since their publication, and Gregorovius found this use of historical material unsettling.

Gregorovius's review did not discuss the works of another author who had achieved great fame by the 1880s with his novels set in the early Common Era. Felix Dahn was even more explicit than Freytag about his desire to ground the "psychology of a people" in the distant past, and Gregorovius's concerns would have applied to his works as well.

Felix Dahn's Monumental Nationalism

Born in Hamburg in 1834, Felix Dahn studied law and philosophy in Munich and Berlin from 1849 to 1853, and he enjoyed a successful career as a law professor in Munich, Würzburg, Königsberg, and Breslau. He was tremendously prolific in the fields of legal theory and early German history and wrote several novels, including *Ein Kampf um Rom* (A struggle for Rome, 1876), one of the most popular novels of the late nineteenth century. Dahn was an ardent nationalist, and his worldview shaped his legal, historical, and fictional works and his own actions. In 1870, for example, he took leave from his professorship in Würzburg to volunteer for the war against France and hoped to die a hero's death.[45] In legal theory, Dahn began from the premise that the law, society, and the economy operated for the benefit of the nation. He therefore rejected the concept of a universal natural law that protected individual rights and notions of international law that would place any limits on national sovereignty.[46]

Although Dahn's expertise was legal history, professional publications in Germany and abroad praised his works on Germanic history. The British historian Thomas Hodgkin relied on Dahn's *Die Könige der Germanen* (The Germanic kings, 1861–70) in his classic work *Italy and Her Invaders* (1879), and he wrote a glowing review of *Geschichte der deutschen Urzeit* (History of German prehistory, 1883–88) in the *English Historical Review*.[47] The 1911 edition of the *Encyclopaedia Britannica* praised the way Dahn's nonfiction works transformed the dark, gray past into narratives about exploits and conquests. The article noted that Dahn's cultural histories contained "a wealth of picturesque detail [that] has been worked over and resolved into history with such imaginative insight and critical skill as to make real and present the indistinct beginnings of German society."[48] The combination of "imaginative insight and critical skill" marked both Dahn's histories and his novels, blurring the line between history and fiction and inviting readers to imagine his plotlines as part of their history.

In *Urgeschichte der germanischen und romanischen Völker* (Prehistory of the Germanic and Romanic peoples, 1881–90), Dahn attempted to answer several basic questions. He explained the origins of the Germanic peoples, what Europe looked like before they arrived, and the characteristics that bound them together. This history began with the Aryans, "from whom so many [nations] developed," and their movement westward from their original homeland east of the Caspian Sea. The Aryans included the ancestors of the inhabitants of India, who arrived in southern Asia between 2500 and 2000 BCE; the Celts, who had migrated to the westernmost end of Europe by 2000 BCE; the forerunners to the Greeks, who reached the Mediterranean also around 2000 BCE; and the Germanic peoples (*die Germanen*), who had entered the land along the Vistula and Oder Rivers by 700 BCE.[49] Dahn claimed that the Celts had defeated and replaced an earlier people who lived at the *Pfahlbau* sites and that they came to dominate all of western Europe and reached their climax by 400 BCE. Many rivers, creeks, mountains, hills, and forests therefore still had Celtic names because of this long period of settlement. Dahn explained that the Germanic peoples later moved west and defeated the Celts and that later battles with the Romans established the Rhine River as the border between the Germanic peoples and Romans for the next five hundred years. Like Weinland (and unlike most anthropologists associated with the DAG), Dahn explained the distant past as a series of peoples replacing others through migration and conflict. For Dahn, each cultural group maintained its own characteristics, and there was little room for encounters or exchanges that would have shaped the practices of ancient peoples over time. Indeed, Dahn felt that writers had obscured the identities of ancient peoples, and he sought to clarify several definitions. The term *Germani*, for example, was a Celtic word and described "all the branches of our *Volk*" from England to Scandinavia. The word *Deutsche*, according to Dahn, came into general use between the ninth and tenth centuries to describe a linguistic group and had its origins in the name of the Germanic god Thiod.[50]

Dahn drew the sharpest contrast between the Germanic peoples and Slavs. In lines that clearly expressed his prejudices, he explained that

> as the still mighty Roman Empire prevented a further expansion of the Germanic peoples in the west along the Rhine and in the south along the Danube

and Alps, masses of rougher tribes pushed on the Germanic rear guard (the Goths, Lugii, and other Ostrogoths). They came along a wide front from the Danube in Hungary to the Düna as a threatening wave: it was the *Slavs*. They came so close on the heels of the Germanic peoples that Tacitus had trouble distinguishing them [from the Germanic tribes], despite their lower cultural development, their dirtiness, and their dull idleness. . . . The Germans named them "*Wenden*" (meaning those who pasture their animals), and their nomadic lives as shepherds distinguished them from the Germans, who were forming farms at that time.[51]

This account contains several features similar to the coverage in Freytag's *Bilder*. Dahn associated farming with settlements and a higher level of culture and claimed that the Slavs possessed neither. He made clear distinctions between peoples, even though antiquarians were still debating decorative styles and burial types that could be associated with Germanic and Slavic peoples. Leopold von Ledebur, who had spent over forty years working with the archaeological evidence from Prussia's provinces, posed questions about the eastern borders of Germanic settlements, but he did not venture answers to them. Dahn overrode this hesitation and treated the migrations, conflicts, settlements, and village life of early history as the formative elements of a German national character.

Dahn's attempt to explain the coexistence of German unity and regional loyalties as characteristics with a deep history echoed Freytag's attention to the interaction between a love for home and *Wanderlust*. According to Dahn, ancient Germans respected the authority of their kings, but they also had a strong desire for freedom. The interplay between these two forces explained the expansion of the Germanic tribes in central Europe. As kings established their rule and amassed territory, groups would eventually break off to seek their fortune in new territories. Monarchs could maintain their rule and the loyalty of their subjects only through fair governance and military success.[52] Both Freytag's description of love of home and *Wanderlust* and Dahn's narration of the conflict between allegiance to the king and the allure of new adventures emphasized stability and loyalty, on the one hand, and movement and conquest, on the other. Both models included a strong connection to a German homeland in the ancient past and a more dynamic element that explained territorial acquisition.

Dahn's version of the ancient past was more monumental than Freytag's, however. *Bilder aus der deutschen Vergangenheit* focused on the growing prosperity and self-reliance of the ancient, then medieval Germans, and *Die Ahnen* illustrated these themes with details from individual lives. In contrast, Dahn presented pre- and early history in an epic way, valorizing kings, castigating villains, and depicting war and conquest as moments when all could be gained or lost. *Ein Kampf um Rom*, for example, did not merely describe the Ostrogoths in the sixth century; it portrayed the fate of the Gothic kingdom squeezed between Roman insurrections and the specter of Byzantine intervention. The story, set in Ravenna, Italy, between 526 and 553 CE, is packed with intrigues, dramatic decisions, displays of strength, and tragedy.

Throughout *Ein Kampf um Rom*, Dahn recommended self-reliance and strong leadership as keys to survival. In the first chapter, the royal adviser Hildebrand implores Ostrogoth warriors to avoid intrigues or personal ambition and reminds them that they are fighting to protect their most deeply held traditions.[53] Furthermore, Hildebrand rebukes those who were willing to adopt the customs of other peoples, arguing that the Goths' strength came through the maintenance of their distinct nationality. Even though they ruled in Italy, they would not become Romans or a hybrid people.[54] Dahn describes these strategies with great urgency, presenting the Ostrogoths as facing a downfall that can be prevented only by a strong commitment to the *Volk*.

The fate of the people, according to Dahn, was closely tied to the institution of monarchy. In his multivolume *Die Könige der Germanen*, Dahn had depicted the concept of the king as the natural outcome of a long historical process. The monarch—the office and not just the specific individual who held that position—represented the strength and endurance of the people.[55] This theme creates a dramatic scene near the close of the fourth book of *Ein Kampf um Rom*. The Gothic assembly removed King Theodahad from power and condemned him for betraying the people. Yet his removal did not diminish the *Volk* or the monarchy. In one of the most famous lines in the book, the elder Haduswinth proclaimed to the thousands of Goths assembled before him, "As far back as our legends and sayings reach, our ancestors have raised one man up on their shields as the living symbol of the power, fame, and fortune of the noble Goths. As long as there are Goths, they will have kings; and as long as there is a king, the *Volk* will persist."[56] It was even possible for a defeated king to appear as a hero, as was the case with Teja, the final king depicted in the book. He is praised for his willingness to die as he tries to defend the Italian peninsula from the invading Byzantine army. He is not judged in terms of his success or failure but rather in terms of the noble sacrifice he made in a fight against a superior force.[57] Dahn elevated the monarchy and the people to normative concepts, even eternal entities. The people could cast out an individual bad king and withstand a military setback, but such short-term events did not tarnish the *Volk* or the legitimacy of the monarchy.

The contrast between freedom and occupation was a matter of both personal and collective virtue for Dahn. In *Ein Kampf um Rom*, the Roman desire for independence from the Goths was aligned with deception, intrigue, and plans for aggrandizement. This is best represented by Cethegus, who claims to speak for the Roman interest but makes several asides about his own rise to power.[58] The most negative lines in the novel criticize the imperial ambitions of foreigners who want to intervene. The Byzantine emperor Justinian has an insatiable appetite for expansion, and the representatives of the papacy, including Silverius, will cut any deal to maintain their power. In contrast, Dahn depicted the Goths in Italy as noble and self-confident. They are occupiers (and Dahn allowed expressions of legitimate unrest against this occupation) but are firm and honest in their diplomatic dealings. This contrast presents honesty and loyalty as Germanic virtues and associates avarice and deceit with other actors.

Dahn, like Freytag, was explicit about reading recent history as the final outcome of ancient patterns. Germanist Kurt Frech has argued that *Ein Kampf um Rom* can be seen as a commentary on Italian and German unification in the nineteenth century. Dahn started the novel in the early 1860s, amid the diplomacy and war that marked these two political processes. In this reading, the Gothic kingdom represented the Austrian Empire's territory on the northern Italian peninsula; the Romans and their insurrection mirrored contemporary Italians and their national demands; Silverius, a Catholic archdeacon and later pope, stood for the Catholic Church and its suspicion of the Italian nationalist movement; and the Byzantine emperor Justinian, the major villain of the story, clearly represented the French emperor Napoleon III and his intervention in Italian affairs.[59] These parallel conditions allowed Dahn to show how personal virtues and flaws, as well as national strengths and weaknesses, shaped history and continued to affect present-day international relations.

In *Urgeschichte der germanischen und romanischen Völker*, Dahn celebrated the nineteenth century as the long-awaited resolution to the deep historical forces of centralization and particularism. The defeat of Napoleon, Dahn argued, had brought a synthesis of these two forces, and "the glorious establishment of our empire [in 1871] was the victorious outcome of this process."[60] Yet *Ein Kampf um Rom* ended with the defeat of the Goths in Italy, and this perhaps reflected the apprehension that Dahn felt about international politics even after Prussia's success in the Wars of Unification and the creation of the German Empire. In a world of military rivalries and power politics, the novel reminded readers to remain vigilant and committed to the preservation of the nation.

Dahn's view of early history as a grand stage for heroes and villains, epic victories, and devastating defeats provided a stark contrast to the archaeological guidebooks of his day. References to Ludwig Lindenschmit's *Die Alterthümer unserer heidnischen Vorzeit* (The antiquities from our heathen prehistory) appeared throughout Dahn's *Urgeschichte*. This catalog provided hundreds of images of artifact types and became a reference work for regional museums throughout central Europe. Besides an introduction, though, the curator at the RGZM included almost no text beyond brief explanations of the objects and notes about where they were found and the museum that currently held them. Dahn's narratives, in contrast, brought these objects to life with his sweeping account of German history and his tales of conquest and settlement. His stories invited readers to view objects in much more dynamic ways than Lindenschmit's accurate depictions allowed.

Action scenes from the ancient past extended beyond the pages of historical novels in the German Empire. Images had decorated the pages of *Die Gartenlaube*, one of the most influential middle-class journals of the nineteenth century, since its publication began in 1851. The subject matter in the 1850s and 1860s was mostly subdued and realistic in nature. Beginning in the 1880s, however, dramatic images of Germanic kings and especially coronation ceremonies began to appear in *Die Gartenlaube*. Germanist Kirsten Belgum has described the appearance of these images as a shift toward a mythic "monumentalism" in popular middle-class home journals that expressed a "contemporary self-assurance, even boastfulness, of the German nation."[61] These images were different from the mythological sagas depicted in the Neues Museum in Berlin. They were done in a highly idealized way, but they purported to depict actual figures from the Germanic past and suggested the period of tribal migrations and early kingdoms as the beginning point of German history.

Some museums even blurred the line between fiction and history. The Prähistorisches Museum in Cologne, which opened in 1907, displayed an impressive collection of artifacts in one of the towers of the city's medieval fortifications. The museum's walls were decorated with eighteen historical paintings, including works by the Bavarian artist and archaeologist Julius Naue. It also featured a portrait of the "Bavarian hero" Hortari based on a scene from *Felicitas* (1882), Felix Dahn's novel that told how Hortari took Salzburg from the Romans in 476 CE.[62]

National narratives about the distant past remained popular throughout the German Empire. Dahn's heroic vision of this past spoke to Germans, who were witnessing an export-driven economy that produced great wealth, the acquisition of colonies, and new campaigns to build up the German army and navy. Cultural history and historical fiction anchored this rising strength and vitality in history. By the 1890s, this pride in the past erupted in the field of archaeology itself.

Prehistory as a Preeminently National Narrative

Regional museums and the anthropologists involved with the DAG pursued pre- and early history research in the German Empire, but their work contained nothing of the timeless national characteristics described by Freytag and Dahn. By the last decade of the nineteenth century, however, some scholars began to use the growing accumulation of archaeological knowledge to narrate Germany's ancient past in confident and scientific terms. Their efforts attempted to overturn a decades-long tradition of antiquarianism that postponed formulating broader conclusions about the German past and to counteract the anthropological orientation in archaeology that was devoted to comparative studies and European questions. The author who most forcibly advanced these views within the field of domestic archaeology was Gustaf Kossinna.

Gustaf Kossinna (1858–1931) was born in Tilsit, East Prussia, into a well-educated and financially stable household. He attended secondary school during the first years of the German Empire and absorbed the national enthusiasm of his day. Kossinna attended several German universities, but it was Karl Müllenhof's lectures in Berlin on German and Indo-European linguistics that inspired Kossinna's first thoughts about the original homeland of what he termed "the Indo-Germanic peoples." In 1881, Kossinna defended a dissertation on the "written monuments of ancient Upper Franconia," a topic in linguistics, not archaeology. Unable to secure a university teaching position, Kossinna became a librarian and worked in Bonn and Berlin.[63]

During the 1880s and 1890s, Kossinna read the growing literature on pre- and early history, joined the Berlin branch of the DAG as well as the Brandenburgia society, and traveled throughout the German Empire (and beyond) to view archaeological collections held by provincial and local museums and by private individuals. Although his efforts were somewhat limited by poor health and inadequate funding, he visited over three hundred collections between 1896 and 1931 and received modest support from the Prussian Academy of Sciences and the Prussian Cultural Ministry. This background made Kossinna an expert on domestic archaeological finds, and in 1902, he received an untenured appointment at the Friedrich Wilhelm University in Berlin (known as the Humboldt University since 1949) as a professor of German archaeology. This position allowed time for his research, and he trained a cohort of students who became enthusiastic ambassadors for the fledgling field of pre- and early history at provincial museums throughout the German Empire.

Kossinna first articulated his vision for domestic archaeology at the DAG's annual meeting in Kassel in 1895. His lecture, titled "Die vorgeschichtliche Ausbreitung der Germanen in Deutschland" (The prehistoric expansion of the Germanic peoples in Germany), emphasized three themes. First, he called for a greater emphasis on Germanic finds, in contrast to what he viewed to be the DAG's overly broad interest in the Celtic, Germanic, Slavic, and Roman pasts of central Europe. Second, he argued for the primacy of archaeology in studying ancient peoples: it was through the presence of real artifacts, not the study of linguistics, place-names, or folklore, that one could document the location of ancient cultures. And third, he boldly asserted an expansive view of the Germanic peoples that was temporally deeper and geographically wider than what most archaeologists of his era claimed.

Kossinna opened his 1895 presentation with a sarcastic jab at Rudolf Virchow, the seventy-three-year-old father of anthropology in the German Empire. Kossinna declared:

> When I venture to attempt to connect patriotic archaeology with history and thereby restore to their owners the rich finds from native soil that have been collected through the efforts of our century, I am in no small part prompted by the words that Rudolf Virchow spoke on the occasion of the anniversary of the Berlin Society for Anthropology: "We must again turn ourselves energetically to the question of the Celts, a question that has remained dormant in archaeological circles for the past quarter century." The flip side of the Celtic question for Germany is the Germanic question.[64]

Kossinna rejected Virchow's call for more research into Celtic sites as a betrayal of the "native soil," and he proceeded to dismiss the work of philologists who attempted to trace European languages back to the ancient cultures of India, central Asia, and the Middle East, claiming that too much time and cultural change stood between an original language and modern word usages. In this "linguistic paleontology," Kossinna said, "all the steps in between [an Indo-Germanic language and a modern language] are completely unknown."[65] In place of a focus on the Celts and linguistic research, Kossinna called for the study of the Germanic peoples based on archaeological finds.

Several scholars have documented Kossinna's complaint that "foreign" cultures received too much attention and all the funding in German scholarship. This was part of the long-running animosity between Roman scholarship and barbarian research, and it certainly contributed to Kossinna's later rivalry with the classical archaeologist Carl Schuchhardt.[66] But the Kassel lecture makes clear that Kossinna was also criticizing the way the DAG framed anthropological questions in European or even wider human terms. In contrast, Kossinna endeavored to transform domestic archaeology into German archaeology—a discipline that would focus primarily on the Germanic past and inspire pride and loyalty in his fellow Germans.

Kossinna's main topic in the Kassel lecture was the extent of the geographic areas inhabited by Germanic peoples. He began by describing the movement of the Cimbri from northern Europe across the Main River in the second century BCE and further migrations southward into the Rhineland and Alsace and to Lake Constance. This part of Kossinna's summary followed the basic outline established by earlier antiquarians who had linked archaeological finds in northern Germany and Denmark to the Cimbri. But Kossinna also commented on Saxony, Thuringia, Bohemia, Moravia, Silesia, West Prussia, and the Baltic coastline, noting that all these eastern lands were Germanic during Tacitus's time. (And this area included almost all the land in the Second Empire and a few Habsburg lands.) Kossinna then went much farther back in temporal terms. He suggested that an "original" Germanic culture that had spread from northern Europe was vastly older than the turn of the Common Era, perhaps dating as far back as 1500 BCE.[67] This proposal added another one to two thousand years to "Germanic" history and implied that one could go much earlier than the Cimbri to search for the beginning of German history.

Kossinna based this extended timeline on the work of Oscar Montelius

(1843–1921). After decades of study, this Swedish scholar had concluded that the prehistoric material of Denmark, Sweden, and Norway revealed the continuous development of a single cultural group from the Bronze Age through the early Iron Age and into historical times. Although Montelius did not make much of this sweeping conclusion, in Kossinna's eyes it suggested the possibility that the "original" (Indo-Germanic) and historically recorded (Germanic) peoples, known to researchers through two distinct sets of evidence, were related.[68] The attempt to fill in the temporal gap between these two peoples (and between prehistory and early history) became Kossinna's lifework. One of his primary motivations derived from his frustration with the work of the DAG. By viewing artifacts as small pieces of evidence related to cultural questions and by refusing to apply ethnic labels to archaeological material, Kossinna claimed, the DAG denied the national story. Kossinna's advocacy for his alternative vision of archaeology will be examined further in the following chapter.

A second assault on the DAG's version of pre- and early history came with the publication of Houston Stuart Chamberlain's *Foundations of the Nineteenth Century* in 1899. Chamberlain (1855–1927) was born in Britain but became an ardent admirer of German culture. During the 1890s, he wrote several studies and commentaries on cultural issues, including a widely read veneration of Richard Wagner. In 1908, he married Wagner's daughter and lived in Bayreuth, the home of the annual Wagner festival, for the rest of this life. *Foundations of the Nineteenth Century* was a blockbuster and became a core text for radical nationalists in the German Empire. In this work, Chamberlain claimed to reveal the incredible creativity of the Teutonic peoples through an analysis of major achievements in art, history, and politics in the roughly two thousand years that preceded 1800. Part 1 surveyed the ancient world (Greece, Rome, and the coming of Christianity), the "heirs" to this world (focusing mostly on the entrance of the Jews and Germanic peoples into European history at the end of the Roman Empire), and the ensuing struggle whereby Germanic influences fought against the power of the Catholic Church. Part 2 was then devoted to the new world that the Teutonic peoples created after 1200 (their discoveries, industry, statecraft, philosophy, and art). Throughout this discussion, Chamberlain emphasized that the creative capacity of the Teutonic peoples derived from their race, and he had no patience with the DAG's research into this issue.

Chamberlain admired Gobineau's *Essai sur l'inégalité des races humaines* as "rich" and "intuitive," but he thought this earlier book contained too many incorrect suppositions and therefore belonged to that "hybrid class of scientific phantasmagorias."[69] In contrast to Gobineau's fixation on racial purity, Chamberlain viewed race as an innate capacity that could be bred toward nobility. Historical advances, in his view, did not come when the Teutonic race was kept pure but rather when the right proportion of Teutonic blood was infused into a population stock.

Like earlier antiquarians, Chamberlain sought to revise history's view of the barbarian peoples. He abhorred presentations that described a glorious Roman Empire followed by the Dark Ages. His alternative vision described the late Roman Empire as a "raceless and nationless chaos" that was redeemed by the "one ray of light [that] shone over that degenerate world. It came from the north." In this reading, the Migration Age did not bring a "chaos of peoples" to Europe. Rather, it brought "the salvation of a great inheritance from unworthy hands, the dawn of a new day." Chamberlain continued with a hyperbolic, racialized narrative about the rise of the early monarchies of central Europe. He spoke of "the great Teuton Princes" whose nobility developed gradually in the first centuries CE and was on full display by the fourth century.[70]

Claims of scientific merit pervade Chamberlain's book, but a quick reading of any extended passage reveals twisted logic and slippery concepts. He claimed, for example, that knowledge about the Aryans was uncertain, but that did not stop him from confidently treating them as the central actors in history.[71] And his definition of the Germanic peoples must have confused researchers who had been trying to distinguish between ancient Germans, Celts, and Slavs for decades. Chamberlain explained that

> in this book I understand by "Teutonic peoples" the different North-European races, which appear in history as Celts, Teutons (Germanen) and Slavs, and from whom—mostly by indeterminable mingling—the peoples of modern Europe are descended. It is certain that they belonged originally to a single family . . . ; but the Teuton in the narrower Tacitean sense of the word has proved himself so intellectually, morally and physically pre-eminent among his kinsmen, that we are entitled to make his name summarily represent the whole family. The Teuton is the soul of our culture.[72]

How did Chamberlain view the racial background of this indeterminately mingled but, to his eyes, clearly preeminent people? He simultaneously invoked and mocked the work of the DAG, claiming that the hesitancy of these scientists blinded them to the most obvious truths.

Chamberlain's *Foundations* contains over fifty references to prominent members of the DAG. The author cited the DAG's anthropological surveys of European populations but applied their conclusions in warped ways. He explained, for example, that measurements by Rudolf Virchow and Johannes Ranke, as well as by American anthropologists, attested to the purity of the Jewish race. (Virchow's surveys did the opposite.) And he reminded his readers several times that Jews and Slavs could have blond hair and fair complexions, "as Virchow has proved by countless statistics."[73] In this instance, the DAG had emphasized the great variety within racial types to cast doubt on the racial types themselves. Chamberlain wanted to use the scientific authority of the DAG for the opposite claim: that racial categories were so fundamental that even the variety of physical attributes did not disprove their existence.

Most of Chamberlain's references to the DAG accused its members of naïveté and recklessness, and these charges revolved around the issue of narration. In words that seemed to echo Alexander Ecker's uncertainty about the Celts from the DAG's 1875 meeting, Chamberlain expressed frustration that, in his day, the study of the Aryan race seemed far from conclusive: "the more we study the specialists, the less certain we become." Even as he repeated that these men were knowledgeable (and he claimed his authority was based in part on their science), Chamberlain mocked their science as unreliable and ever changing. The study of the Aryan race, he claimed, was a farcical parade. First came the linguists, then the anatomical anthropologists, followed by ethnographers who traveled abroad and studied other peoples. All this led to a most confusing situation, and the layman could not help but be perplexed. The field of anthropology was lost at sea and therefore useless, Chamberlain concluded.[74]

Chamberlain also condemned anthropologists for spreading dangerous ideas that derived from their liberal politics. Citing a speech Rudolf Virchow gave as rector of the Friedrich Wilhelm University, Chamberlain claimed that the leader of the DAG contradicted himself by speaking about the equality of humanity and the need to develop "beautiful self-dependent personalities." Chamberlain asked how one could be so naive as to believe that all were equal but still promote the advancement of some individuals. He castigated "our phrase-making authorities [who] preach amalgamation, when ennoblement is clearly the road of history." In all of this, "the school of Virchow" was drawing the wrong conclusions from their facts.[75] In Chamberlain's view, the DAG wholly ignored the fundamental significance of race. He noted that scholars did not take notice of Gobineau's "brilliant work" when it appeared. "Like poor Virchow men stood puzzled before the riddle."[76] And in the 1890s, the "empiricists" were no more helpful. They had reached a dead end, and their gibberish blocked everyone from understanding the most basic questions about culture and humanity. Mocking the outcome of the DAG's 1892 annual meeting, Chamberlain decried:

> upon what a dangerous path these worthies take us, when they suddenly interrupt their discussion of "chameprosopic, platyrrhinous, mesoconchic, prognathic, proophryocephalous, ooidic, brachyklitometopic, hypsistegobregmatic Dolichocephali" in order to link on to it general remarks about history and culture. The layman understands little or nothing of the remainder; he wades hopelessly about in this barbaric jargon of neoscholastic natural science; only the one point is printed in all the newspapers of Europe as the visible result of such a congress: that the most learned gentlemen in Europe have solemnly protocolled the fact that all the races bear an equal share in the development of culture; there have never been Greeks, Romans, Germanic peoples, Jews, but from time immemorial there have lived peacefully side by side, or it may be, devouring each other, leptoprosopic Dolichocephali, chameprosopic Dolichocephali, leptoprosopic Brachycephali and chameprosopic Brachycephali, "all working unitedly at the furtherance of culture" (sic!). It provokes a smile! But crimes against history are really too serious to be punished merely by being laughed at; the sound common sense of all intelligent men must step vigorously in and put a stop to this: we must say to these worthies, "Cobbler, stick to your last!"[77]

In a stunning statement, Chamberlain recommended a way out of the anthropological confusion. Instead of the skepticism and reservation that marked the work of the DAG, he recommended that his readers simply skip over science and the search for hidden origins and causes. Life, after all, "is something different from systematic knowledge, something much more stable, more firmly founded, more comprehensive." Instead of careful study, people should embrace the "automatic" knowledge that nature provides.[78]

Kossinna's exclusive focus on the Germanic past and Chamberlain's claims about Germanic creativity intersected with a broader, *völkisch* nationalism that arose in Wilhelmine Germany. As historian Uwe Puschner has explained, the *völkisch* movement was committed to a virulent defense of what it viewed as Germany's national interests and to the veneration of a pure and vital Germanic past.[79] Political pressure groups like the Pan-German League and radical cultural societies like the Reichshammerbund and Thule-Gesellschaft identified closely with an exclusionary vision of national belonging based on Germanic or Aryan origins. After 1900, these groups published numerous studies about Germanic religion and customs and emphasized the extent of Germanic expansion in the early medieval era. Germanic motifs appeared frequently in their journals and newspapers.[80]

Readers of a *völkisch* persuasion latched on to two concepts that were central to Chamberlain. First, Chamberlain asserted that cultural capacity was determined by race and that the ancient Germans were preeminent in their cultural contributions. By taking this as a fundamental given, *völkisch* thought saw something very different in domestic archaeology. It was no longer primarily about documenting who lived where in ancient times, which had been the main antiquarian concern for the entire nineteenth century. Instead, archaeology showcased the spread of cultural greatness that occurred through the expansion of the Germanic peoples. Antiquarians admired their ancestors, but they did not go this far.

Second, the *völkisch* movement shared Chamberlain's sense of urgency about contemporary politics and his idea of "the creation of race by nation-building." Chamberlain explained that race was "an organic living thing" that could rise or fall, become ennobled or degenerate. "But the firm national union is the surest protection against going astray: it signifies common memory, common hope, common intellectual nourishment; it fixes firmly the existing bond of blood and impels us to make it ever closer."[81] If state polices did not place the proper emphasis on racial matters, the state and civilization were doomed. In this mind-set, the caution and uncertainty associated with regional museums and the DAG were unacceptable. It was better to create inspiring plotlines that confirmed Germanic greatness than to wallow in the growing amount of archaeological material amassed by museums.

The novels of Freytag and Dahn, reissued throughout the late nineteenth and early twentieth centuries, were popular in this milieu. Nationalists valued Freytag's emphasis on German creativity, the contrast between productive settlers and backward Slavs, and the image of eastern Europe as land to be colonized. The reception of Dahn was even more closely linked to the intensification of German nationalism because of his use of heroic images from the distant past. It is worth noting, though, that the creation of novels like the ones by Weinland, Freytag, and Dahn belonged largely to the 1870s and 1880s. By the 1890s, nationalist scholars wanted to bring the inspiration developed in fiction to the realm of nonfiction.

Gustaf Kossinna had direct ties to *völkisch* groups,[82] and he endeavored to present a heroic, nationalist prehistory as scientifically verifiable. By 1902, he was confidently repeating his claims about the expanse of prehistoric German settlements as facts. He even went so far as to recommend that scholars dispense with the label "pre-Germanic" and refer to the vast sweep of time from the third century BCE to later migrations as Germanic.[83] Kossinna was explicit about the political utility of his conclusions, and his method allowed Germans to imagine a connection between Indo-Germanic and Germanic peoples that seemed to anchor their national history in prehistory.

Kossinna certainly would have vociferously contested the inclusion of his work in a chapter largely about historical fiction. And archaeologists who investigate questions of ethnicity in late antiquity may find that the coverage underestimates the contributions made by Kossinna's line of questioning. This organization, though, allows us to dig deeper than the conclusion that Kossinna was a nationalist so his archaeology was political. By placing his interpretations alongside narratives that traced German families and German virtues from the early Common Era to the nineteenth century, we see the desire for a coherent story about the distant past and the interpretive leap that was required to address it.

It is also important to place these narratives in the broader context of the German Empire. A national story did not exercise a monopoly on the interpretation of the ancient past. Regional and anthropological approaches to prehistory had other stories to tell, and they largely rejected the notion that they were writing the prehistory of nation-states. The interaction between these perspectives continued into the twentieth century. The desire for a national story grew stronger during and after World War I and with the rise of Nazi ideology. But as pre- and early history became an independent scientific field, it continued to stress a European story of human development. The first half of the twentieth century was marked by this issue: Was the growing knowledge about pre- and early history leading to scientific knowledge about an ethnic or racial past? Or did this knowledge continue to complicate the understanding of pre- and early history in ways that made it difficult to connect ancient peoples to modern nation-states?

PART III

Between Science and Ideology

CHAPTER SEVEN

Professionalization and Nationalism in Domestic Archaeology

After 1900, pre- and early history attained the hallmarks of a scientific discipline. Armed with advances in comparative methods and better dating techniques, archaeologists moved from the work of collecting toward the task of interpretation. Scholars described archaeological sites in a more systematic way, and they produced surveys that explained Europe's cultural development from the Stone Age to the Middle Ages. New journals and monographs presented specialized research and reviewed the findings of archaeologists throughout Europe, and universities began to train students in the methods and questions of the field. Graduates brought this specialized knowledge to new positions as researchers and administrators at regional museums throughout Germany. This wave of hiring transformed these museums into important research institutes that communicated new insights to a wider public.

The rising status of pre- and early history had important implications for the question of historical narration. As the field developed, it promised to deliver more confident conclusions about the ancient past. But what kinds of narratives could pre- and early history provide? The literature on archaeology in Germany has stressed the rise of nationalist, even chauvinistic interpretations during the first three decades of the twentieth century, focusing especially on Gustaf Kossinna's ethnic interpretations and his emphasis on the study of prehistory as a patriotic duty. This perspective resonated in nationalist circles, and it gained Kossinna many adherents. Although most museum directors and professionals, including several of Kossinna's former students, shared his frustration with the disadvantaged position of prehistory, they did not blindly follow his call for "a preeminently national science." Instead, they noted the limits of what their science could tell them about questions of ethnicity and identity in the distant past.

Viewing the period between 1900 and 1933 exclusively as the time when a nationalist version of the ancient past triumphed conceals as much as it reveals. Gustaf Kossinna did, in part, prepare the way for National Socialist ideology by emphasizing Germanic greatness and the political utility of the study of the ancient past, but his outlook did not dominate the field of pre- and early history before 1933. A better model for capturing developments in this period emphasizes the interaction between rising nationalism and other traditions in domestic archaeology. As the practitioners of pre- and early history made better sense of chaotic museum collections and as they synthesized information about the ancient past for a general audience, they continued to work within local and regional contexts, and like the early scholarship of the DAG, they investigated all epochs and often placed prehistory in a broad, European framework. Prehistorians were patriotic, and they were drawn to domestic archaeology because they wanted to research their homeland. Yet most distanced themselves from the simple conclusions of *völkisch* thinkers, who viewed prehistory exclusively as the story of Germany's ethnic or racial development. They wrote about what had happened on German territory in ancient times, but they did not present this as the prehistory of the German people.

The Development of Scientific Interpretations

In the late nineteenth century, prehistory was not offered as an independent field of study at any German university. A handful of students pursued prehistory topics in related disciplines. Gustaf Kossinna had completed a dissertation in German linguistics in 1881; Alfred Götze, who became an assistant in the prehistory section of Berlin's Museum für Völkerkunde, earned a doctorate in art history at the University of Jena in 1890 with a study of Neolithic pottery. Other scholars interested in the field studied medicine or classical archaeology. In 1896, the year after his Kassel lecture, Kossinna promoted the idea of a professorship for German archaeology, but this idea did not find immediate resonance.[1]

In the first decade of the twentieth century, seven German universities offered lectures on prehistory to supplement programs in geology, ethnology, cultural or physical anthropology, and Germanic languages.[2] And in 1902, Kossinna became a professor for German archaeology in the Department of Germanic Studies at the Friedrich Wilhelm University in Berlin, but this was a nontenured position that did not include the traditional privileges of a German professorship. The position gave Kossinna time for research and publishing, and he trained a cohort of students in the decade before World War I, but he was not given adequate resources to establish the field. Professors in other disciplines did not look upon prehistory favorably. They did not allow their students to include prehistory in their programs of study, and they disputed the evaluations Kossinna gave his students.[3] Despite these obstacles, the circle of students around Kossinna became dedicated to the field of prehistory. Their motivation came from a nationalist desire to advance what they perceived to be a neglected field of study, and they practiced Kossinna's method of settlement archaeology, which attempted to draw conclusions from the detailed study of the distribution of finds.

Several of Kossinna's students carried their commitments into new positions for archaeology at regional museums. Albert Kiekebusch, who in 1908 completed the first dissertation under Kossinna's direction, dramatically improved the scientific status of the pre- and early history section of the Märkisches Museum in Berlin,[4] and Hans Hahne, who studied under Kossinna from 1905 to 1907, became the director of a new museum in Halle, which had been an important center for domestic archaeology since the early nineteenth century. Erich Blume produced a catalog for the collection in Prenzlau in 1908 and then headed the pre- and early history section at the museum in Poznań (Posen).[5] Kossinna's students benefited from their teacher's knowledge and passion for the field, but several of them, including Albert Kiekebusch, distanced themselves from Kossinna's chauvinistic formulations.

In addition to Kossinna's students in Berlin, others graduated from programs that combined the study of prehistory with anthropology, history, or linguistics. Max Ebert (1879–1929) studied German literature and history and would later work at the Museum für Völkerkunde in Berlin. He became a professor for prehistory in Königsberg in 1921 and took over Kossinna's chair in Berlin in 1927. He is best known as the editor of the *Reallexikon der Vorgeschichte*, the day's most comprehensive encyclopedia for prehistory. Karl Hermann Jacob-Friesen (1886–1960) was an assistant in the ethnology museum in Leipzig and later became an assistant and then director of the provincial museum in Hannover. These career paths show that the generation who brought the science of prehistory into regional museums had more training in archaeology, but they still worked within a tradition that viewed prehistory as closely tied to anthropology, ethnology, linguistics, and history. For these scholars, the influence of Kossinna's nationalist outlook and archaeological method was combined with approaches that had shaped domestic archaeology since the 1860s.

A clear sign that prehistory was not developing exclusively along nationalistic lines was the choice for the director of the pre- and early history section at Berlin's Museum für Völkerkunde. Albert Voß, who had worked alongside Rudolf Virchow in the BGfAEU and had served as the director of the section since the 1870s, died in 1906. The general director of the Königliche Museen, Wilhelm von Bode, hoped to fill the director's position with someone with substantial excavation experience but also a dedication to Germanic studies. Kossinna expressed strong interest in the job but was not considered. Bode first turned to Rudolf Much, the son of one of the founding figures in domestic archaeology in Austria and a scholar of German and Scandinavian philology. Much declined the position, and Bode then turned to Carl Schuchhardt, the director of the Kestner-Museum in Hannover, who was known to the Prussian Cultural Ministry through his investigations into the location of Arminius's victory over Varus. Bode was initially concerned that Schuchhardt was too much of a classicist and would view "our pre- and early history culture with prejudice." But after interviewing Schuchhardt and reviewing several of his works, Bode saw that his concerns were unjustified. He came to realize that Schuchhardt worked from the assumption that "the ancient Germanic culture was also the mother of the ancient Greek and Italian ones and that it was only later influenced in various ways by the latter."[6]

In this very important personnel decision, the Berlin cultural establishment chose Schuchhardt, not Kossinna. This was not simply a decision between science and nationalism. After all, Rudolf Much was an ardent German nationalist in Vienna. The major difference between Schuchhardt and Kossinna lay in their approach to archaeology and the field of prehistory. Schuchhardt, while also a German nationalist, was quite cautious in his formulations about ethnic claims, and he framed research questions in a broad, European context. Kossinna, on the other hand, focused on what he viewed as Germanic and German finds, and he pursued domestic archaeology as a way to promote national pride.

Upon being called to Berlin, Schuchhardt created a new exhibit for the prehistory section of the Museum für Völkerkunde that presented the prehistoric development of Europe.[7] As Hubert Schmidt, one of Schuchhardt's museum assistants, explained, "The main question is not the development of specific areas, provinces, or countries, as it has been in the organization of local or provincial museums. It is not even about the development of larger areas like northern, central, or southern Europe. Instead, the layout should be determined by questions about Europe's prehistoric development that are broadly conceived and *general*."[8]

This European orientation was fundamentally new. Since Ledebur's installation in the Neues Museum, the largest collection for domestic archaeology in Berlin had been organized by geography. Groups of objects represented various German provinces and specific international locations. The new display was ordered chronologically and set out to convey a clear understanding of the Stone, Bronze, and Iron Ages based on material from throughout Europe. This organizational principle contrasted with the local and regional approaches that valued objects as evidence of nearby events. And it offered a much clearer narrative than the crowded display that resulted from the DAG's seemingly endless tasks of describing, cataloging, and preserving objects. The museum was no longer a warehouse for the numerous items that were found or acquired each year. It now told a clear story of cultural development organized by chronology. Schmidt noted that there was a national story in the display because such a large portion of the material came from Germany, but the main goal of the exhibit was to "place the prehistoric development of Germany within a larger framework of European prehistory."[9]

The Museum für Völkerkunde also became more directly involved with excavations, which Carl Schuchhardt directed. During the late nineteenth century, the Reichs-Limes-Kommission systematically mapped the Roman walls and outposts that stretched throughout western and southern Germany. With this work, archaeologists developed techniques for using the course of the walls and constructions along the walls as evidence for relations between the Roman Empire and barbarians. They investigated these sites as sources for social and military history and even treated these structures from the perspective of architectural history. This new way of reading archaeological sites was evident in the first excavations Schuchhardt undertook after arriving in Berlin. A team of archaeologists investigated the Römerschanze near Potsdam. Despite this name, the site had nothing to do with Romans. As Schuchhardt explained, the fortified settlement's main features were constructed during the Bronze Age and had been modified during later periods of habitation. Excavations revealed an outer wall made of earth and wood that stood three meters high, and one-meter-wide postholes were clearly recognizable at the front and rear of the construction. Two large gates provided access into the redoubt. The discovery of numerous postholes inside the walls suggested that the interior included several structures and was densely populated.[10]

Excavations at the Römerschanze caused quite a sensation. It was located just a few miles from Sanssouci, the Hohenzollern summer palace, and Emperor Wilhelm II followed the work closely. Schuchhardt reported that "the fame of this excavation spread. . . . Fifty members of the Anthropological Society came, as well as representatives from the university, museums, and archaeological society. On the 17th of September [1908], His Excellence Cultural Minister Dr. Holle came, and on the 2nd of November His Majesty the kaiser and the empress appeared, as well as Princess Victoria Louise with a retinue of around twelve ladies and gentlemen."[11]

Like the royal participation in the DAG's archaeological outings in 1880, the visit to the Römerschanze was a celebration of archaeology, not German nationalism. The excavation site was not just a repository of objects that could be removed and displayed in a museum. It was a historical source that allowed observers to reconstruct the floor plan of this ancient site and speculate about the inhabitants' architectural achievements. Schuchhardt did apply ethnic labels to the different periods of occupation. He referred to the original construction as "a large Germanic fortress" from the eighth to fifth centuries BCE, to a later period of Slavic occupation, and finally, on the basis of pottery shards and modifications to the site, to the "re-Germanization" of the area. But he also compared the site to constructions found along the Roman *limes* and to the layout of the walls and houses uncovered at Troy. These exciting parallels among the construction practices of early peoples were as important as an ethnic attribution at the site.[12]

The changing nature of the field of prehistory was also signaled by the appearance of a major encyclopedia dedicated to the study of the distant past in Germany. The *Reallexikon der germanischen Altertumskunde*, edited by Heidelberg professor Johannes Hoops, combined the efforts of more than eighty scholars from the fields of linguistics, literary studies, and archaeology and was published in four volumes between 1911 and 1919. As the preface to the first volume explained, the *Reallexikon* presented "a general depiction of the culture of the Germanic peoples from the most ancient times . . . to the eleventh century." In covering this sweep of time, Hoops hoped that the work would establish "connections between prehistory and history, on the one hand, and between archaeology and linguistics, on the other hand."[13] The encyclopedia sought to explain developments on German territory, but it did not focus exclusively on Germanic finds and it certainly did not project the existence of the German nation into the distant past. Hoops clearly stated that the study of pre- and early history was still in its early stages and incomplete. Writing ninety years after Leopold von Ledebur, he noted that archaeology still could not answer questions of ethnography:

> Prehistoric typologies are first of all ordered according to a purely archaeological standpoint and depicted in a line of progressive development. The next main task of archaeology—the determination of objects according to their ethnography and the identification of that which is specifically Germanic—can be approached in two ways: one way is to investigate which types of prehistoric material in northern and central Europe appear to be developmentally related to Germanic finds from the historical period; and, second, one can combine archaeological facts with the evidence from Germanic and Indo-Germanic linguistics and information from ancient and medieval authors. Admirable progress has been made in both of these directions in recent years. But in general the ethnographic problem in prehistory is still far from being solved.[14]

The contributors to the *Reallexikon* hoped to answer ethnographic questions, but their reserved approach showed that they understood the limits of narrating a national past.

The encyclopedia's coverage did not display an eagerness to draw connections between ancient peoples and contemporary Germans. The lengthy article "Vorgeschichtliches deutsches Siedlungswesen" (Prehistoric German settlements) by the anthropologist Alfred Schliz explained conditions on German territory, not the origins or development of a German people. The author preserved the anonymous nature of Stone Age finds and stressed that lifestyles and the choice of dwelling locations responded to conditions dictated by physical geography and changes in the climate. Instead of claiming a unique prehistory for Germany, Schliz placed developments in a European context by comparing sites in Germany with places throughout Europe.[15]

A Nationalist Reading of the Ancient Past

Nationalists criticized the developments in prehistory research in the national capital, claiming that the changes did not adequately respect what they viewed as the pressing need for more research into the Germanic past. Gustaf Kossinna had hoped to become the director of the prehistory section of the Museum für Völkerkunde, and the hiring of Schuchhardt, "the classicist," rankled him.[16] When Schuchhardt created a new journal for prehistory research, the *Prähistorische Zeitschrift*, Kossinna founded a separate learned society, the Deutsche Gesellschaft für Vorgeschichte (German society for prehistory), that embraced a specific focus on the Germanic past and rejected the general orientation of prehistory research. To further emphasize its German-national orientation, the society's journal and book series were called *Mannus*. This name came from a passage in Tacitus's *Germania* that related the claim that the Germanic peoples descended from Mannus, a son of Tuisto, who the ancient Germans supposedly believed was "the first man and father of all men."[17] In 1913, the society made its national emphasis even more explicit by changing its name from the Deutsche Gesellschaft für Vorgeschichte to the Gesellschaft für deutsche Vorgeschichte (Society for German prehistory).

Curiously, Kossinna visited few archaeological sites, and he did not lead any excavations himself. His research focused almost exclusively on museum objects. He took detailed notes about the ornamental details of finds and their provenance during each museum visit. He then plotted this information on maps to register the distribution of different types of finds.[18] The transformation of museum inventories into maps became the basis for Kossinna's *Siedlungsarchäologie* (settlement archaeology). This method did not focus on the contents of settlements (households or villages) as the name seems to imply. Rather, it mapped the actual settlement areas of prehistoric peoples on the basis of archaeological finds. Kossinna explained this procedure in *Die Herkunft der Germanen* (The origins of the Germanic peoples, 1911) with characteristic hyperbole: "The leading idea, whose correctness for the period of early history has been tested a thousand times and is constantly affirmed by new studies and therefore must also have validity for the periods closely related to early history as well as for the prehistoric periods that lie much farther back in time, is the following: *sharply defined archaeological cultural provinces correspond in all time periods to specific peoples or tribes.*"[19] It was the task of the archaeologist to reveal these ancient borders and connect them to ancient peoples.

Kossinna's use of the term "cultural provinces" was important. He found the notion of "culture" unsatisfactory because it could be transferred among different groups of peoples. For example, if a people produced a certain style of ornamental work, it is conceivable that another group could learn this form of production through contact and trade. This kind of cultural exchange was a key concept among linguists who traced Indo-European and Celtic words and among anthropologists who explained cultural patterns. Kossinna rejected the idea of a free-floating culture and argued instead that culture traveled with the migration of peoples. His concept of "cultural provinces" therefore enabled him to attach specific peoples to defined territories when recognizable cultural artifacts of those peoples appeared.

Kossinna was explicit about his ideological goals for archaeology. In his best-known work, a 1912 essay intended for a lay audience, Kossinna described prehistory as "a preeminently national science" that must be performed by Germans for Germans. The study of the ancient German past, Kossinna wrote, is "for *our* antiquity, for Germandom and for that which is truly German, as *our* eyes see it." He assured his readers that this national project was "scientific." It was not just a matter of feelings or preferences, for domestic archaeology rested "on the deep, reliable, and unshakable foundation of historical and natural scientific knowledge."[20] This "science," however, was about validation, not measured empiricism. Kossinna stressed that the practitioners of "a preeminently national science" could not lose sight of their field's higher purpose: "We must always keep present in our minds that the emphasis of the science of prehistory lies not in the various archaeological objects ... but rather in the way that the science of prehistory is rooted in the entire lives of our ancestors."[21]

Contemporaries had several criticisms of settlement archaeology. Archaeologists interested in the history of trade relations found Kossinna's conclusions much too static. They emphasized the cross-fertilization of cultures and the blurry line between the creation and modification of cultural practices. Others considered Kossinna's method and conclusions too simplistic. Neither of these charges deterred Kossinna. In fact, he argued that the clarity of his method was one of its primary strengths, boiling his method down to simple statements. In the introduction to *Die Herkunft der Germanen*, he declared, "Our principle is: territories of cultures are territories of peoples," and "My equation is: a cultural group = a people."[22]

Although Kossinna applied this method in a chauvinistic way, documenting what he saw as the wide diffusion of Germanic peoples and their culture, scholars acknowledged settlement archaeology as an important innovation. In a review of a later monograph by Kossinna, archaeologist V. Gordon Childe criticized the "polemic style of his [Kossinna's] writings and certain nationalistic idiosyncrasies in his speculations," but he also noted that the work contained "a detailed analysis of the archaeological remains and an exact plotting of their distributions. Few will dispute the general correctness of the identification of the 'Germans' . . . as far back as the second period of the Bronze Age."[23] More recently, archaeologist Bruce Trigger denounced the nationalist prejudices exhibited by Kossinna and others during his day but noted that Kossinna's settlement archaeology helped to shift the focus of archaeological interpretation from a comparison of finds to a much more serious discussion of chronology.[24]

Although many recognized Kossinna's knowledge of archaeological literature, his polemical style and exclusive focus on documenting Germanic areas made him an outsider to the major developments in prehistory research. Several of his students contributed entries to Hoops's *Reallexikon*, but Kossinna was not invited to participate. Moreover, Kossinna's exclusive focus on the Germanic past caused him to ignore or devalue important innovations in his field. As biographer Heinz Grünert has noted, Schuchhardt's excavations at the Römerschanze and Kiekebusch's at Buch (discussed below) were the two most important excavations near Berlin around 1910, but Kossinna paid little attention to them.[25]

Prehistory and Education at Regional Museums

New developments in the field of prehistory were evident in Germany's regional museums after 1900. These institutions were able to hire new directors with university training and excavation experience, and this new staff had three main goals. They hoped to convey a clear understanding of the prehistoric past to a wider audience by transforming their museums into settings for research and teaching. They wanted to maintain the engagement with the local landscape that had been so important during the founding phase of these museums and use this relationship to advance the field of prehistory. Finally, they hoped that regional museums would strengthen preservation efforts and become recognized as the proper custodians of archaeological material.

Three of the museum directors discussed below demonstrate the various ways that archaeological traditions interacted with nationalist thinking. Albert Kiekebusch, Karl Hermann Jacob-Friesen, and Hans Hahne were part of the new generation of museum staff, and they were committed to bringing the science of archaeology to a wider public. They shared nationalist views that were shaped by the experience of World War I, and all three (to varying degrees) expressed enthusiasm for National Socialist ideas after 1933. Yet their professional outlooks varied considerably during the period between 1909 and 1933. Kiekebusch emphasized archaeology's regional context in his work at the Märkisches Museum in Berlin. Jacob-Friesen was also devoted to regionalism, but he brought a special focus on the anthropological orientation of prehistory to his work in Lower Saxony. And Hahne, first in Hannover and then in Halle, mostly clearly presented local objects as sources for a story of national greatness.

The Märkisches Museum in Berlin provided an influential model for the combination of professionalization and public outreach. In 1908, the museum received a new building in the heart of the city. The structure, which still houses the museum today, connected the province and the city by using distinctive red brick from Brandenburg and a neo-Gothic style that imitated the design of many well-known Brandenburg churches. As part of this transition, Ernst Friedel retired, and pre- and early history became an independent section directed by a professional archaeologist named Albert Kiekebusch.

Born the son of a farmer, Kiekebusch (1870–1935) attended school in Waßmannsdorf (near Berlin) and then became a schoolteacher. At the age of thirty-four, he enrolled at the Friedrich Wilhelm University in Berlin to study prehistory, history, German literature, and philosophy. In 1907, he completed a dissertation under the direction of Gustaf Kossinna about Roman influences on Germanic burial customs. He then began his career at the Märkisches Museum as an assistant in the department for pre- and early history. In 1922, he was promoted to section director, a title he held until his retirement in 1934.[26]

Kiekebusch immediately addressed the problem of museum chaos in the archaeology display at the Märkisches Museum. He criticized museums that served only as warehouses for artifacts, complaining that "the endless repetition of clay pots and stone axes serves no purpose. . . . Most people are either exhausted or bored by such a display and therefore ignore the prehistory exhibit."[27] These comments echoed a wider discussion about the organization and conception of German museums.

Crowded displays overwhelmed visitors and hindered the ability of museums to convey specialized knowledge. To address these concerns, museum
directors made displays more accessible and placed a greater emphasis
on pedagogy.[28] As a former teacher, Kiekebusch embraced this approach.
He divided the museum's collection into three parts. He selected a small
number of items for the permanent exhibit (*die Schausammlung*). Each
room contained only a few display cases so that visitors could focus on
specific objects and read the explanations provided by museum staff.

The items not on display were kept in a research collection or in storage. The research collection contained multiple examples of different
types of artifacts. It was available for specialists who wanted to make comparisons or view all the objects from a single excavation site. According
to Kiekebusch, the research collection was essential to the functioning of
the museum. The scholarship produced there would be integrated into the
permanent exhibit and thereby present the public with the most current
research. A storage area contained the rest of the objects owned by the
museum. These were mostly fragments, and they were arranged by excavation site.[29]

Above all, Kiekebusch wanted the museum to provide visitors with
a clear message. He emphasized that the *Schausammlung* would be the
only place where most visitors encountered the topic of prehistory, and
it therefore needed to be instructive and entertaining. Kiekebusch hoped
the new display would give the public more than a simple awareness that
the land around Berlin had been inhabited for centuries. He wanted them
to understand the basic contours of Brandenburg's prehistory. For those
who wished to know more or who could not visit the museum, Kiekebusch prepared two brief works about local archaeology. Both were filled
with images of objects from the museum.[30]

The Märkisches Museum also advanced archaeology's scientific agenda
through excavations. Kiekebusch's first major investigation was a Bronze
Age site near Buch (twelve miles north of Berlin). Like Carl Schuchhardt's work at the Römerschanze, Kiekebusch's efforts at Buch from
1910 to 1914 attempted to do more than retrieve artifacts for the museum. Kiekebusch was interested in many questions, including the kinds
of handicrafts produced at this place, the trade relations that the inhabitants engaged in, and their domestication of animals.[31] His findings were
part of a special display at the Märkisches Museum that focused on daily
life in the Bronze Age. The exhibit, Kiekebusch hoped, would address
what he saw as one of the biggest mistakes in prehistory research. He

FIGURE 8. In contrast to the crowded and overwhelming displays in earlier versions of the museum, the archaeology display in the new Märkisches Museum, which opened in 1908, encouraged visitors to focus on a few objects and gain detailed information about the material culture from specific time periods. Here the urns from the royal grave of Seddin are at the center of the display devoted to the Late Bronze Age. (Photo by Ernst von Brauchitsch. Neubau des Märkischen Museums am Köllnischen Park. Erdgeschoß Raum 6. Prähistorische Abteilung. Jüngere Bronzezeit. Berlin, 1908. Inv.-Nr.: GE 2007/574 VF. Courtesy of the Stiftung Stadtmuseum Berlin. Reproduction by Friedhelm Hoffmann, Berlin.)

noted that up to that point exhibits focused only on the contents of graves but neglected settlements. This gave the impression that archaeologists appreciated only glamorous and shiny items and had nothing to say about daily life in ancient times.[32]

Other museums implemented similar changes. The Museum für die Provinz Hannover, which opened in a new building in 1902 and possessed wide-ranging collections for archaeology, history, natural history, and art, hired Hans Hahne (1875–1935) in 1907. Like Kiekebusch, Hahne had changed career paths to pursue his love of prehistory. After studying medicine and practicing as a doctor for several years, he came to Berlin to study under Gustaf Kossinna. At the Hannover museum, Hahne's main task was to address the overcrowded pre- and early history display. He also strengthened ties to sixteen smaller museums throughout the province and offered them his expert advice. This communication was not only about the science of prehistory, though. Hahne infused his instruction with a much stronger sense of nationalism than had been the case under Jacobus Reimers, and he hosted Kossinna's Deutsche Gesellschaft für Vorgeschichte in Hannover in the summer of 1909 when his new display opened.[33]

After Hahne left Hannover for a position in Halle, the museum hired Karl Hermann Jacob-Friesen (1886–1960). Previously, Jacob-Friesen had served as an assistant at the Museum für Völkerkunde in Leipzig, where he advocated for the inclusion of European prehistory in this prestigious museum. He also spent a semester in Stockholm, Sweden, in 1908, studying with Oscar Montelius, the archaeologist who greatly refined the Three Age System so that types of objects could be used to draw better conclusions about relative chronology.[34]

When Jacob-Friesen arrived in Hannover in 1913, he advocated for a regionally based, scientific prehistory that engaged the public. Like Kiekebusch, Jacob-Friesen saw that this would require a break from past practices whereby regional museums served primarily as storehouses for excavations. The most pressing task for prehistory museums, according to Jacob-Friesen, was "to evaluate the extensive and valuable material that they already possess," most of which came from excavations carried out by historical associations in the previous seventy-five years. The museum would then need to present this material to the public in a way that reflected modern, scientific knowledge. This was not an easy task. The public, Jacob-Friesen explained, did not understand the significance of prehistoric objects, and most museum directors assumed too much knowledge among their visitors. Regional museums needed to "drape the cold and rough material in warm and inviting clothing." In this way, they would "serve as the intermediary between science and the public."[35]

What did this look like in practice? Jacob-Friesen wanted his museum's display to function like a textbook that drew on regional artifacts but also provided a general overview of European prehistory. Introductory texts and charts first explained the periodization of prehistory. Individual rooms then functioned as chapters that were organized by a clear theme, and objects on display served as illustrations of the theme. Clarifying texts and timelines throughout the museum helped the visitor connect the specific artifacts to the general overview.[36] To illustrate this approach, Jacob-Friesen explained how the museum could shape a visitor's understanding of a Neolithic blade made of flint:

> With this object, we can study, first of all, the primitive culture from five thousand years ago, how humans wrestled to create a useful form out of the rough material nature provided. Furthermore, we can derive a timeline from the improving techniques and development of forms, that is, from typology, and learn about the settlement history of entire cultures [by studying] geographic distributions. Finally, aesthetic considerations come into play also. For example, is it not amazing to see the blade's perfectly elegant lines, how cleanly chip after chip has been removed, and how someone was able to give ornamentation to a material as rough as flint through variable chiseling?[37]

The emphasis here is on understanding the prehistoric past and increasing the visitor's powers of observation. Jacob-Friesen was not teaching nationalism or connecting artifacts to ancient peoples.

The Provinzialmuseum für Vorgeschichte zu Halle, which celebrated its grand opening in October 1918, presented the new emphasis on archaeological interpretation in Saxony. The museum's holdings included the items excavated in the early nineteenth century by the Thüringisch-Sächsischer Verein, as well as an even greater number of items uncovered since the establishment of Saxony's historical commission in 1876. Regional authorities decided to build a museum dedicated to prehistory (in contrast to the other regional museums, which held a wide variety of collections) because of these valuable holdings and because they planned for the museum to operate as the primary agency for evaluating finds that the museum would acquire as a result of the Prussian excavation law of 1914. The state of Saxony paid 740,000 marks for the construction of the museum, and the dedication to "Unserer Vorzeit" (Our prehistory) that crowned the building's main entrance referred to the region. In addition to plentiful display space, the new building included a one-hundred-seat lecture hall, a library, and a workshop where artifacts were cleaned and photographed and casts were made.[38]

The director of the new museum was Hans Hahne, and his work in Halle provided a stark contrast to the more measured interpretations offered by Kiekebusch in Berlin and Jacob-Friesen in Hannover. Under Hahne's direction, the Halle museum emphasized not only regional historiography and the increasingly scientific nature of prehistory but also

Kossinna's vision of archaeology as a preeminently national science. Hahne shared his adviser's frustration that philological and historical research had dominated the study of the Germanic past and that the study of artifacts (*Sachforschung*) had remained "a stepchild of German science until our present day."[39] He was certain that new developments in archaeology would yield insights into "the origins and roots of the German *Volk* and its culture" and that the new museum would help to ground national feeling in the prehistoric past. Specifically, Hahne extolled Saxony's status as an important borderland and area of conflict where German culture had developed. He claimed that finds from the Bronze Age revealed the migrations from the Germanic culture in the Baltic Sea area. Later finds showed renewed migrations from the Germanic north and mutual influences between Germanic and Celtic peoples. Under Hahne's direction, the museum delivered an overtly nationalist conclusion to visitors: "Germanization" came from northern Europe and spread across all of Germany, to the Vistula River in the east and into France in the west. Saxony had experienced a temporary "Slavicization" in the early Middle Ages, but this was overturned by recurring migrations from the north.[40] Hahne's emphasis on northern origins and influences was part of a broader trend in nationalist thought that sought to counter the established linguistic model that connected the origins of European peoples to migrations from the Caucasus and other lands to the east.[41]

The directors of regional museums viewed their institutions as the linchpins of preservation efforts. Jacob-Friesen, for example, supported the Prussian excavation law of 1914 but claimed that this legislation was not very effective because the state did not have a complete inventory of prehistoric sites and monuments. Jacob-Friesen viewed the staff at regional museums as the ideal agents for carrying out land surveys that would record topographical information and prehistoric sites. The staff possessed the necessary knowledge about archaeological sites, and they were already involved in outreach as educators of the public. They could also encourage landowners to register significant monuments on their property and turn archaeological finds over to the state. This communication network would benefit the museums greatly, for they would receive artifacts from recent, well-documented excavations, and the artifacts would be preserved for future generations.[42] This was a critical step in transforming museums from warehouses for nineteenth-century discoveries to outreach and research centers that advanced the field of prehistory.

In Berlin, Kiekebusch engaged the public in excavation and preservation work by creating a seminar dedicated to settlement archaeology in 1915. The seminar was directly affiliated with the Märkisches Museum, and it gave individuals, especially teachers, the opportunity to study prehistory more intensively. It began with around thirty individuals and had over four hundred participants between 1915 and 1955.[43] Kiekebusch described the endeavor as a way to convey a scientific understanding of archaeology and to enrich the museum: "I do not have in mind the creation of a group of 'trustees' that is made up of several private collectors. Such an arrangement could harm the entire project. Instead, I have in mind an enthusiastic circle of knowledgeable individuals who work in the service of science. They should work in harmony with the museum as a scientific institute to advance the research in all areas of settlement archaeology."[44] Kiekebusch's use of the word *Pflegschaft* (the group of trustees) was important, for that was the designation Ernst Friedel gave to the enthusiasts who helped found the Märkisches Museum in the 1870s. The contrast between a nineteenth-century historical association full of enthusiasts and Kiekebusch's scientific seminar demonstrates the formalization of prehistory that occurred after the turn of the twentieth century.

Museum directors also sought to bring archaeology into Germany's schools. The rector for Berlin's primary schools had noted in 1902 that the curriculum did not include local studies (*Heimatkunde*), and Kiekebusch, a former teacher, took up this issue after coming to the Märkisches Museum. In 1912, he organized the first of several traveling exhibits for Berlin's schools, and alongside the Berliner Lehrerverein (Berlin association of teachers), he advocated for reforms in the primary school curriculum. Their success came with a resolution dated April 1, 1915, that explicitly named "Berlin's prehistory" as a topic of instruction. Of the forty hours in the curriculum devoted to local studies, Kiekebusch calculated that thirteen to fourteen hours could be used specifically for pre- and early history each year. He hoped that school field trips would include visits to museums, excavation sites, or other historically significant places.[45]

To ensure the proper teaching of prehistory, Kiekebusch published a small manual in 1915 entitled *Die heimische Altertumskunde in der Schule* (Local antiquities in the school) that listed several lesson plans and possible destinations for school excursions. The booklet illustrated Kiekebusch's belief in the integrative power of prehistory. Through an understanding of the ancient past, people who felt "alienated by all kinds of modern circumstances" could "reconnect with their home."[46] On a trip to the Müggelsee, for example, one could tour the following points of interest:

"Travel with the Lower Silesian–Mark Brandenburg Rail . . . [Visit] the city of Cöpenick [Köpenick]. . . . The Bronze Age finds of Spindlersfeld. Prehistoric settlements from various periods in the surrounding area. . . . The Kietz quarter [of Köpenick]. Across the very sandy and therefore underused Kietz field (prehistoric settlement; St. Mary's grove; beautiful fruit garden; glass factory). . . . Legend of the creation of the Teufelssee. Bismarck monument made of Rüdersdorf limestone. View across the Spree River toward Berlin. . . . Back with the train."[47] On a single day, this itinerary took students into the past via the modern rail. The prehistory, natural beauty, and productivity of Brandenburg mingled with its recent history created by Chancellor Bismarck, and all of this could be seen without losing sight of the imperial capital. A Brandenburg excursion was a lesson in the cohesion of ancient and modern history.

Kiekebusch's instructive displays and pedagogical guide clearly expressed the political potential of local archaeology. In the middle of World War I, he felt that it was important for his compatriots to see their common past and their long-standing relationship with the territory they were called upon to defend. According to Kiekebusch, students—and all Germans—had to learn to love their surrounding area, because "whoever loves the little home must also love the big one. That is the proper goal . . . of local history efforts."[48]

On the eve of World War I, archaeologists were growing stronger in the belief that their discipline could deliver more information about the distant past than linguistics or physical anthropology; and because of this faith, they sought ways to move beyond the collecting and comparing practiced in the nineteenth century and to produce comprehensive interpretations. Gustaf Kossinna developed settlement archaeology to draw conclusions about the relationship between cultures, peoples, and territory, but he used these advances to provide the earliest chapters in a national story that he hoped would inspire pride and solidarity in likeminded Germans. Other versions of prehistory placed the advancement of the field ahead of the advancement of nationalism. Museum directors were certainly nationalists, and they viewed their commitment to domestic archaeology as a patriotic duty. Schuchhardt, Kiekebusch, and Jacob-Friesen were interested in questions of origins and the identification of ancient peoples and cultures, but they did not share Kossinna's exclusive focus on a German prehistory. They recognized that there was a difference between writing the prehistory of what happened on German territory in the prehistoric past and writing an exclusively German prehistory.

They, too, wanted to turn objects into stories, but they highlighted specific issues related to the prehistory of their regions and offered general, not German, introductions to prehistory.

It is fitting that an important example of a new narrative about prehistory grew out of the institution that best represented the endless collecting of the nineteenth century. The Museum für Völkerkunde in Berlin had the largest prehistory collection in the German Empire and was most subject to the museum chaos of this era. The breadth of its holdings, however, also offered an opportunity to produce a sweeping interpretation of European prehistory. In 1919, Carl Schuchhardt published *Alteuropa in seiner Kultur- und Stilentwicklung* (Ancient Europe as seen through the development of its culture and ornamental styles), a survey of European prehistory that was based on Schuchhardt's travels throughout Europe and his work at the Berlin museum. The first lines of the work playfully described how a display case containing early Spanish ceramics caught his attention when he first became director of the prehistory section in 1908. The cups and bowls from Spain "looked longingly over to artifacts from Troy and Mycenae." Schuchhardt could not shake the idea that the two cultures shared some essential characteristics, and his travels suggested further relationships. At Stonehenge in England, Schuchhardt felt he was seeing a model for certain circular structures in Greece, and he sounded even more confident when he asserted that the domed passage graves of Spain and Ireland were not barbaric copies but rather the predecessors to the famous beehive tombs of Mycenae. Observations of ceramics in France, Italy, Malta, Crete, and Athens showed that western Europe presented "a uniform ancient cultural area that in many ways provided the roots for that which would soon thereafter blossom in the eastern Mediterranean."[49]

In *Alteuropa in seiner Kultur- und Stilentwicklung*, Schuchhardt formulated a version of the northern origins of culture, but one that differed from the assertions of *völkisch* thinkers. Whereas Kossinna focused on the consistency and superiority of finds in northern Europe, Schuchhardt's model included a large degree of movement and interaction. A northern style from the Stone Age migrated east and south, interacted with other cultures, and could return to western and northern Europe and flourish in new ways. The most important encounter in this process was the creative contact between what Schuchhardt described as a Nordic cultural current and a cultural current already present in southern and eastern Europe. Schuchhardt described this process as "the *Indogermanisierung* of our continent," and his book set out to demonstrate that this development could be seen by studying the cultural and stylistic evidence that archaeology brought to light.[50]

Schuchhardt explicitly chose the vague word *Strom* (current) to avoid the identification of these formations as peoples. Indeed, he emphasized that the actors in 90 percent of the book remained anonymous. Names of peoples would come only when his coverage reached the historical period and they appeared in sources.[51] On these issues, he differed from pan-Germans, who saw Germans in periods all the way back to the Bronze Age.

Reviews of Schuchhardt's work took aim at the attempt to write such a sweeping book and questioned whether it was possible to arrive at broad conclusions from archaeological evidence. Hugo Mötefindt, writing in the *Historische Zeitschrift*, claimed that Schuchhardt "belonged to those scholars who are less concerned with the intense study of the material and constructing an argument from the smallest building blocks. They look instead at the big picture in order to pursue their own new ideas and theories. Those who approach research in this way run the danger of producing only superficial ideas and, via their own theories, losing contact with the healthy basis in the facts." V. Gordon Childe appreciated the comparisons drawn between the Mediterranean world and northern Europe but noted that Schuchhardt had inverted the direction that most other scholars had given this influence. Here, Childe wrote, "our author seems to have abandoned the method of explaining the known from the known which is the keynote of the best German work and has led to many valuable results."[52]

Culture, *Volk*, and Race after World War I

The mobilization for World War I sent the directors of many museums into the battlefield, including Albert Kiekebusch in Berlin, Karl Hermann Jacob-Friesen in Hannover, Max Ebert in Königsberg, and Peter Goeßler in Stuttgart. This took them away from their museums and put local excavations on hold. The war, however, also brought new opportunities for fieldwork, as the construction of trenches and fortification lines brought prehistoric material to light. Carl Schuchhardt published several notices about wartime finds, including a long report from the autumn of 1914 about Roman sarcophagi and Merovingian pottery along the western front. Schuchhardt then went into great detail about a "quite interesting and valuable recovery" in northern France. After finding a bronze ring and a human skull in a trench near Bucy-le-Long, a Captain Pehlemann supervised an investigation that revealed a shaft tomb containing corpses laid out on gravel but apparently enclosed by timbers. The army unit proceeded to uncover thirty-two La Tène graves that revealed corpses decorated with bronze necklaces and arm rings and surrounded by finely polished ceramic pieces. One grave contained three iron spear tips. Herr Niggemann, an enlisted man from Berlin, delivered much of this material to the Museum für Völkerkunde in Berlin.[53]

World War I also brought deeper intellectual changes that directly affected domestic archaeology and historical narration. *Völkisch* groups reviled the German defeat and the creation of the Weimar Republic. They sought regeneration by venerating a pure and vital Germanic past, and they continued to promote the idea of the northern origins of European culture. They also took aim at the peace negotiations that mandated the loss of German territory in the east. The Treaty of Versailles forced Germany to cede large portions of the provinces of West Prussia, Posen, and Silesia to the newly constituted state of Poland. This transfer, along with the proclamation of Danzig as an international city, gave Poland access to the Baltic Sea, but it also separated the German province of East Prussia from the rest of Germany in the west. Nationalists attacked Poland's new borders with arguments based on history and prehistory. A Pan-German article from 1919, entitled "Deutsch-polnische Vorzeit" (German-Polish prehistory), advanced the idea that the centuries of Slavic inhabitation of eastern Europe merely played "the role of an interlude" and claimed that the region's Indo-Germanic roots dated back to 9000 BCE.[54] Gustaf Kossinna produced a manifesto in 1919 that claimed that Silesia and West Prussia were "natural" German possessions because of Germanic settlements from prehistoric times, and he pushed this text on the German delegates to the Paris Peace Conference.[55] Pride in the "Germanization" of eastern lands had been a common theme in Pan-German circles for decades, but the territorial changes after the war raised the level of vitriol in arguments about the prehistory of the borderland between Germany and Poland.

World War I was also a critical turning point in the development of racial thinking. In the late nineteenth century, physical anthropologists in Germany had largely rejected the idea of a hierarchy of races as unscientific. They continued to measure and record physical characteristics, but they did not draw larger conclusions about their data that would link racial groups to present-day nations. The war years, however, drove a younger generation of scholars to view national combatants in racial terms.[56] This new connection between races and peoples addressed a methodological frustration similar to the museum chaos that stymied archaeologists. Younger physical anthropologists felt that the detachment of the Virchow era had brought their field to a dead end. The field continued to accumulate data, but it lacked a framework for applying this data to broader questions of the day. This generation of scholars therefore placed its faith in genetics as a new field that would allow anthropologists to draw conclusions about the cultural and psychological levels of peoples based on race.[57]

After the war, *Rassenkunde* (racial studies) emerged as a field of academic study that sought to understand racial differences and speculate about the possibility of managing or improving national populations through eugenics. These questions spilled over into prehistory studies, as scholars considered races, peoples, and cultures not only as present-day entities but also as the products of long-term processes that reached back into the distant past. Hans Günther (1891–1968) emerged as the most important popularizer of these ideas. In *Rassenkunde des deutschen Volkes* (An investigation into the racial composition of the German people, 1922), Günther claimed that four main racial groups inhabited central Europe: a Nordic race, a western (often called Mediterranean) race, an eastern (often called Alpine) race, and a Dinaric race (associated with southeastern Europe). Günther then spent the majority of the book detailing the physical and psychological attributes of these groupings.

Although Günther stated that the book was about present-day conditions in Germany, he devoted two chapters to pre- and early history. He treated the origin of races as an unsolvable question (or at least one that natural history, not physical anthropology, would address) and laid a baseline for his discussion in the late Neolithic period. At this point in prehistory, he claimed, "all of today's European races were represented." Günther then explained that the diffusion of culture presented in Carl Schuchhardt's *Alteuropa* had greatly shaped his thinking and that he identified the Nordic race with the cultural current that Schuchhardt saw as moving from north to south and the western race with the current that moved from west to east. Günther did not equate Schuchhardt's concept of culture with his notion of race, but he came very close: "It should be understood that one cannot directly relate Schuchhardt's results to the names of races. . . . The study by the archaeologist Schuchhardt in the first instance examines only *artistic forms and the way they spread.* . . . On the whole, however, Schuchhardt's results line up with those of racial research. The fact emerges that Europe brought forth two creative races: the Mediterranean and the Nordic."[58] Günther's argument brought cultures, peoples, and races closer together. If Schuchhardt was describing ancient cultures and Günther was describing ancient races and both the entities they studied produced European civilization, then in Günther's reasoning the cultures and races had to line up. All that was needed was to identify the people who made up these races and created the cultural artifacts. Nationalist and racist interpreters of the ancient past would perform this triangulation and connect ancient cultures to peoples and, via settlement archaeology, to specific locations in Europe. And since both archaeologists and racial scientists claimed to evaluate the level of civilization (either through the level of cultural production or through the psychological attributes of racial groups), they could then develop a language about the superiority of the Nordic race, which, according to their theories, injected creativity and advancement into the rest of Europe.

During the interwar years, professionals in the field of prehistory took up the question of whether prehistory could elucidate the relationship between races, cultures, and ancient peoples, and they continued to offer a complicated response. In a second edition of *Alteuropa*—now with a different subtitle: *Eine Vorgeschichte unseres Erdteils* (Ancient Europe: A prehistory of our continent, 1926)—Schuchhardt restated his desire to explain the forces of cultural creativity in the ancient past. His book proposed that a "pre-Indo-Germanic" current that covered western and southern Europe and developed from the Paleolithic to the Iron Age and an "Indo-Germanic" one in northern and central Europe had converged and produced the cultures of ancient Greece and Italy and then spread back to the rest of Europe. This process, Schuchhardt argued, revealed "the dualism of ancient Europe, whose back-and-forth interaction produced the sparks that led to new cultural forms and gradually built the basis for Europe in historical times."[59] It was this theme that allowed his work to surpass the earlier efforts to collect more and more material for the museum and reach a point of interpretation. He explained:

> We will not support the humble, helpless view that prehistory is just a "sequence of cultural events." Like history, so too should prehistory endeavor to show the *development* of one thing out of another, seek to identify the driving force at work in the hundreds of "types" found in particular areas, and, in the end, portray the great creative powers and their success. Then it will be clear

that European prehistory is highly relevant, that it offers us the basis for evaluating the entire later distribution of peoples and the characteristics of those
peoples up to today.[60]

Schuchhardt insisted that his work investigated cultures, not races, but his
language about evaluating the "characteristics" of peoples sounded very
similar to the psychological side of *Rassenkunde* that attempted to speak
about creative capacity. A new section of the conclusion to the second edition of *Alteuropa* addressed this issue. Schuchhardt explained that the evidence from prehistory was too sparse to allow physical anthropology to
link skull types or bone measurements to cultures, but the state of affairs
was changing.

> Rudolf Virchow had struggled for several decades of his life to identify the
> conditions, relations, and movements of ancient peoples according to ancient
> skulls and modern peoples. In the end he had many doubts and gave up on
> the enterprise around 1890. [Alfred] Schliz, the archaeologist and anthropol-
> ogist in Heilbronn, took this task up again. . . . And today Hans Günther, a
> young anthropologist, is at work on an explanation of the ethnic composition
> [*ein Völkerbild*] first of Germany and then of all of Europe based on ancient
> and the most recent material that serves as a revelation to a wide audience.[61]

Karl Hermann Jacob-Friesen in Hannover was more skeptical about
the convergence of prehistory and racial studies. His *Grundfragen der
Urgeschichtsforschung* (Basic questions in prehistory research) from 1928
sought to educate prehistorians in the Weimar Republic about the status
and potential of their field. The book evaluated the main scientific path-
ways into the distant past and noted that each field shed light on a specific
concept: physical anthropology studied race, linguistics identified peoples
as language groups, and the cultural sciences (to which archaeology belonged) shed light on the process of cultural development through the
study of artifacts. From the outset, Jacob-Friesen insisted that the concepts
of race, peoples, and culture be kept separate, noting that the methods
and conclusions of one science could not simply be adopted by the other
fields. Thus, the archaeological study of material culture could not identify
races, and the peoples identified by linguistics might or might not line up
with the cultures documented by archaeological finds.[62]

Throughout the text, Jacob-Friesen condemned what he viewed as the
superficial and irresponsible conflation of peoples and cultures, and he

singled out the work of Gustaf Kossinna for criticism on three grounds. First, he criticized Kossinna's tendency to describe more and more material as Germanic, including material that came from before the time when one could speak of Germanic peoples. Second, he charged that Kossinna drew sweeping conclusions from very small amounts of evidence. And, third, Jacob-Friesen rejected the rhetorical style that asserted unfounded claims as confident truths. Evaluating a passage from Kossinna's *Ursprung und Verbreitung der Germanen in vor- und frühgeschichtlicher Zeit* (The origin and expansion of the Germanic peoples in pre- and early history, 1927) where Kossinna extrapolated from the decorative style of one fibula to claim that much of what was Germany's northern coastline was "Germanic" in the later La Tène period, Jacob-Friesen wrote: "A single broken fibula proves therefore 'with complete certainty' [Kossinna's words] an ethnic relationship! According to our view, that could be only a first hint of such a relationship. One can never ever speak with 'complete certainty.'"[63]

Schuchhardt and Jacob-Friesen addressed pressing questions of their day about the origins of peoples and the process of cultural development. Both scholars trusted that the scientific development of their field would yield more complete answers, and they were critical of others who leaped to conclusions about the connections between race, culture, and peoples. The desire to clarify these issues was very strong in the early twentieth century, but scholars cautioned that it was still too early for prehistory to make these claims. Jacob-Friesen especially expressed the need for greater hesitation in the intellectual environment of the late Weimar Republic by introducing his *Grundfragen* with the following statement: "The precondition for science is doubt, not belief."[64]

CHAPTER EIGHT

Prehistory as a National Socialist Narrative

B etween 1900 and 1932, German archaeologists engaged in an open-
ended discussion about the potential of archaeological interpretation. After January 1933, the Nazi regime dramatically shifted this discussion, especially in publications created by party organizations and in
the realm of education. The program of coordination (*Gleichschaltung*),
which meant bringing all aspects of political, social, and cultural life into
line with Nazi ideology, brought a clear emphasis on the nation and the
Volk to the cultural sciences. The field of pre- and early history was to
downplay regional traditions and broader stories of human development
and focus exclusively on a national story. Cautious formulations and honest statements about the limits of archaeological knowledge gave way to
confident declarations about the prehistoric past. In the Nazi worldview,
Germany had to have a single ancient past, and it was to be presented
with the "complete certainty" that Gustaf Kossinna asserted and Karl
Hermann Jacob-Friesen rejected.

National Socialist cultural agencies embraced the idea of prehistory as
"a preeminently national science" in 1933, and Kossinna's ideas became
dogma. Prehistory research was to clarify the origins of the *Volk*, celebrate the "advanced culture" of the Indo-Germanic race, and promote
the idea that European culture, including the civilizations of Greece and
Rome, had its origins in northern Europe. The Nazi vision of the ancient
past portrayed the German people as the "purest" descendants of this

original culture. Additionally, pre- and early history was used to document Germanic settlements throughout central and eastern Europe and thereby teach Germans to view territories in Poland, Czechoslovakia, and the Soviet Union as ancient German possessions.

Coordination also changed the purpose of prehistory research. Nazi leaders viewed pre- and early history as an *Erziehungswissenschaft*, a field of knowledge that would help raise the *Volk* to intellectual maturity. The primary purpose of the field was no longer the investigation of unsettled scientific questions. Under the new regime, it was used to indoctrinate Germans with a version of the ancient past that would inspire pride and racial thinking. Prehistory was to overturn the negative images of the Germanic tribes as barbarians and present them instead as superior to other peoples because of their creativity and willingness to fight for their territory. The regime aggressively pushed this version of prehistory onto the German public through museum exhibits, publications, school curricula, press reports, official speeches, and even in newsreels and films. This historical education contributed to the acceptance of ideological slogans like *Blut und Boden*, the phrase Nazis used to stress Germany's deep temporal roots in "blood and soil," and it formed the affirmative counterpart to the regime's negative rhetoric about Jews, Slavs, and other groups considered to be historic and racial enemies of the state.

Because of the war and genocide waged by Nazi ideologues, historians have rightfully searched for the origins of radical nationalism and racism in Germany. In dealing with archaeology, most of this scholarship emphasizes the connections between *völkisch* thought, Kossinna's writings, and Nazi ideas. Other works trace a longer history of exclusionary nationalism back to the era of the Wars of Liberation.[1] There is certainly great merit in these studies, and nationalist and Nazi authors viewed themselves as the defenders of a nationalist tradition. The problem with these presentations, though, is that they do not account for other ways of narrating the distant past that were lost because of the increasing emphasis on a national story. Previous scholarship concentrates on the ideas that Nazism embraced, but not on the alternatives that Nazism silenced. A focus on narration shows how dramatically different the Nazi version of the ancient past was from earlier, open-ended research and debate.

Under the Nazi regime, archaeologists presented themselves as part of a national tradition that had to fight for recognition and prominence. This line of thinking portrayed scholars from the early nineteenth century like Jacob Grimm and Johann Gustav Büsching as great patriots who researched the fatherland's ancient past. According to Nazi interpreters, their studies did not achieve the attention or respect that they merited because other forces hindered their research. State resources had been dedicated to the ancient history of "foreign" lands (through classical archaeology and philology). Historical associations had brought greater attention to domestic antiquities, but their localism or regionalism was deemed to be narrow-minded. Rudolf Virchow's empiricism (decried as "internationalism") had diverted the focus of archaeology away from the national story. The Nazi history of the field then presented Gustaf Kossinna as a tireless campaigner on behalf of German prehistory whose brilliant research created a basis for indisputable conclusions about the ethnic and racial understanding of archaeological finds. And it was the Third Reich that would finally create the political and intellectual environment that accorded proper attention to this vital line of research and convey this knowledge to the German people.[2]

This story is partly true, and a more critical version of it is accepted. The engagement with archaeology in the nineteenth century partly fostered nationalist thinking, but this focus covers up the vigor of other traditions, the pursuit of nonnational research, and the actual status of Gustaf Kossinna in the field of prehistory during the first decades of the twentieth century. Kossinna did not just help to prepare the way for Nazi ideas. The regime elevated and canonized Kossinna's ideas. To do this, it had to cut off the discussion of research questions that were not settled, impose a political relevance on prehistory in a way that ran counter to the vision of science practiced by most archaeologists in the German Empire and Weimar Republic, and implement their views with fanaticism and the power of the state. With political utility as the primary goal, those who advanced the Nazi vision of prehistory oversimplified narratives about the northern origins of culture, the racial purity of Germans, the heroism of Germanic warriors, and the extent of territory that could be claimed as Germanic or German. This chauvinist national story ran counter to the hesitation that marked pre- and early history research in the previous decades. The Nazi regime was most able to implement its version of the ancient past in realms where it exercised greatest control, including national propaganda, school curricula, and the cultural imperialism exercised in occupied territories in eastern Europe during World War II. Established professionals in museums and at universities partially embraced the ideological mission that the state had for prehistory. They welcomed the recognition given to their field, and they moved toward more nationalist and racist narratives about the ancient past. But many also displayed a reluctance to use their expertise to tell a story of Germanic greatness exclusively. They continued to discuss and present other topics, and they sought to protect the autonomy of their associations and museums from full control by Nazi organizations. In this way, many individuals avoided the sweeping narratives of hypernationalism and racism by concentrating on themes developed before 1933, including studies that focused on local or regional settings and that drew limited conclusions based on empirical evidence.

Coordinating Personnel

In contrast to the German Empire or Weimar Republic, the Third Reich had several high-ranking leaders who were devoted to the study of the Germanic past. Their worldviews drew mystical, cultic, and racist ideas from the *völkisch* milieu, and they turned to archaeology to substantiate their irrational understanding of the past. Alfred Rosenberg (1893–1946), who joined the NSDAP in the early 1920s, expounded outlandish ideas that became central to Nazi ideology in *Der Mythus des 20. Jahrhunderts* (The myth of the twentieth century, 1930), including the idea of a "Jewish world conspiracy" and notions of Germanic greatness. Rosenberg viewed pre- and early history, racial studies, and folklore as ideological disciplines that should be used to raise the historical and racial awareness of the German people. He promoted these ideas during the late Weimar Republic as the leader of the Kampfbund für deutsche Kultur, the Nazi organization for cultural regeneration, and after 1934 in his own cultural office within the Interior Ministry called the Amt Rosenberg (Rosenberg office).[3] The Amt Rosenberg included the Reichsbund für deutsche Vorgeschichte (National association for German prehistory), the umbrella organization that was charged with the coordination of historical associations and regional museums.

Heinrich Himmler, as *Reichsführer* of the Schutzstaffel, believed in the racial superiority of his SS troops and created the SS-Ahnenerbe in 1936 as a subunit that would investigate the Germanic past. The Ahnenerbe funded several archaeological excavations in Germany and abroad, and it attempted to inculcate the ideas of Germanic greatness and Germany's "civilizing mission" in eastern Europe within the army and SS.

The Nazi hierarchy left the relationship between the Amt Rosenberg and SS-Ahnenerbe unclear. Neither Rosenberg nor Himmler could fully implement his vision of prehistory, and overlapping claims of authority and personal rivalries limited the effectiveness of the coordination of prehistory.[4]

Adolf Hitler wrote of a glorious German past and the racial unity of the *Volk*, but he derided *völkisch* fanatics who were obsessed with Germanic heritage. In *Mein Kampf*, he wrote that he had "to warn again and again against those *deutschvölkisch* wandering scholars . . . [who] rave about old Germanic heroism, about dim prehistory, stone axes, spear and shield." And as chancellor, Hitler complained to Albert Speer that an emphasis on the Germanic past was embarrassing: "Why do we call the whole world's attention to the fact that we have no past? It's bad enough that the Romans were erecting great buildings when our forefathers were still living in mud huts: now Himmler is starting to dig up these villages of mud huts and enthusing over every potsherd and stone ax he finds. . . . We should really do our best to keep quiet about this past. . . . The present-day Romans must be having a laugh at these revelations."[5] Despite these disparaging comments, archaeologists in the Third Reich still quoted Hitler's racial ideas from *Mein Kampf* as a testament to the significance of their work.

Kossinna and his circle had complained that a "preeminently national" archaeology had received too few resources compared with the fields of classical archaeology and anthropology, and many looked to the Nazi state to rectify this situation. These hopes were partly realized between 1933 and 1945. Despite the exigencies of economic depression and later war, lectureships and professorships devoted to "German archaeology" quadrupled from thirteen to fifty-two university positions between 1928 and 1941, and the number of scholars engaged in fieldwork and employed by museums doubled over the same time period. Several positions for prehistory that had been temporary or were housed in other departments received independent standing.[6] Archaeologist Bettina Arnold has described these circumstances as a Faustian bargain: many archaeologists actively participated in the presentation of a Germanic past that supported Nazi propaganda, and in return the regime provided the discipline with resources for excavation and publication, as well as a much larger presence at German universities.[7] In this context, individuals with the proper ideological views could climb the academic ladder quickly. Hans Günther, the race researcher at the University of Jena, received a prestigious professorship in Berlin in 1935. Even more dramatic was the rise of Hans Reinerth. After a decade-long search for a permanent academic position,

Reinerth became head of the Reichsbund für deutsche Vorgeschichte in June 1933 and a full professor in Berlin in 1935.[8]

The Reichsbund für deutsche Vorgeschichte was responsible for coordinating historical associations so that they would pursue "German prehistory" and not local or regional history. Unsurprisingly, the Gesellschaft für deutsche Vorgeschichte, the society created by Kossinna in 1909, joined the Reichsbund without difficulties in 1933. It continued to publish the *Nachrichtenblatt für deutsche Vorzeit*, which, after 1933, presented the image of archaeologists and educators eagerly embracing the Nazi version of prehistory.

But many historical and archaeological associations rejected the dream of streamlining hundreds of regional and local associations in a nationalist project. Established scholars viewed Reinerth as a hack, not a scholar, because he placed ideology above science, and they viewed interventions by the Reichsbund with hostility. Reinerth, with the authority of the state on his side, responded with professional and personal attacks. In the summer of 1933, he publicly denigrated classical archaeologists from the Deutsches archäologisches Institut (German archaeological institute) and the Römisch-Germanische Kommission (Roman-Germanic commission), claiming that their "meddling" in German prehistory "had caused setbacks and propagated falsehoods about German prehistory" and that their "international-pacifist spirit" had brought about their "complete failure . . . in terms of their national-political significance against Poland and France." Reinerth attempted to remove the leaders of both institutions, but Theodor Wiegand remained head of the Deutsches archäologisches Institut until he died in December 1936, and Gerhard Bersu, the director of the Römisch-Germanische Kommission, remained in place until the enforcement of the Nuremberg racial laws in 1936.[9]

Reinerth's pettiness and persistence were clear in his dealings with Karl Hermann Jacob-Friesen, who had been director of the Museum für die Provinz Hannover since 1924. Jacob-Friesen should not have been viewed as an enemy of the Reichsbund. After 1933, Jacob-Friesen described himself as a National Socialist. His writings in the 1930s displayed a growing commitment to racial interpretations of prehistory, and he agreed with the policy of forcing Jews out of leadership roles in archaeological associations. Despite this common ground with Nazi ideology, though, Reinerth and others in the Nazi Party did not view Jacob-Friesen's commitment as fervent enough and they attempted to isolate him.

Fellow archaeologists and museum professionals regarded Jacob-

Friesen highly, and they had elected him as the leader of the Arbeits-
gemeinschaft für die Urgeschichte Nordwestdeutschlands (Association
for the prehistory of northwestern Germany) in 1932 and the Nordwestdeutscher Verband für Altertumsforschung (Northwest German association for antiquities studies) in 1934. When Reinerth pursued groups in
western Germany with particular vigor because he felt that their attention
to Roman history and Roman finds came at the expense of research into
the region's Germanic past, he prevented Jacob-Friesen from becoming
the leader of a national union of German archaeologists, the Berufsver-
einigung deutscher Prähistoriker (Professional union of German prehistorians) in 1933, and he demanded that the Nordwestdeutscher Verband
join the Reichsbund in 1935. Jacob-Friesen refused, citing Reinerth's dis-
honorable actions toward the respected archaeologists who had worked
on the Roman sites at Haltern and Xanten. Jacob-Friesen then tried to
avoid Nazi coordination by having the Nordwestdeutscher Verband join a
national association that was not under Reinerth's control. In 1936, Jacob-
Friesen became leader of the Deutsche anthropologische Gesellschaft
and called for a vote to dissolve the organization so that it too would
avoid coordination under Reinerth.[10]

Another charge against Jacob-Friesen related to his critical comments
about Gustaf Kossinna's method. The 1936 edition of Kossinna's *Die
deutsche Vorgeschichte* included an expansive appendix by Werner Hülle
that discussed the relationship between Kossinna's ideas and National Socialism and reviewed the literature on pre- and early history that had appeared since Kossinna's death in 1931. Hülle, who played a leading role in
the organizational and publishing activities of the Reichsbund, explained
that no general handbook for prehistory studies currently existed and
that Jacob-Friesen's *Grundfragen* "must be fiercely rejected because of
its negative position on Kossinna, its mistaken definition of prehistory research, and its explicit refusal to acknowledge the centrality of racial conditions."[11] Reinerth pushed this issue in an open letter to Jacob-Friesen
in 1937, claiming that no real Nazi would have a problem abandoning his
criticisms of Kossinna. Jacob-Friesen held firm, however, claiming that he
never rejected Kossinna as a person or his life's work, but that the most
pointed version of Kossinna's main ideas was not scientifically acceptable.
Furthermore, he cited a positive review of his *Grundfragen* from 1928
by none other than Werner Hülle.[12] A clear thread running through the
treatment of Jacob-Friesen related to his silence on the ethnic conclusions
that the Nazi regime expected. In July 1939, the education supervisor for

the Hannover district noted that it was impossible to work with someone whose "scientific talent takes on such an expressly analytical, negative-critical quality such that he himself refuses again and again to come to a final, clear position based on the evidence that is present." The district therefore worked around Jacob-Friesen. Heinrich Himmler approved this arrangement, noting that "Professor Friesen stands much too far from the political thinking of National Socialism."[13]

In addition to their ideological basis, these kinds of disputes arose because of Reinerth's abuse of power. Organizations accustomed to regional autonomy balked at his combativeness, even when their leaders were sympathetic to nationalist and National Socialist interpretations of prehistory. These soured relations with the Reichsbund provide good evidence of historian Reinhard Bollmus's claim that Nazism destroyed more than it created. By 1936, regional associations were choosing to disband rather than cooperate, and the Ahnenerbe eclipsed the Reichsbund as the organization that received the most resources for excavations. Reinerth, unable to gain recognition from professional archaeologists, dedicated himself to the propaganda uses of prehistory in publications and education.[14]

Teaching Nazi Values through Prehistory

The Nazi state made a consistent effort to relay messages about the glorious Germanic past and the prominence of race in German schools. The Interior Ministry guidelines for history instruction, published in the spring of 1933, placed the highest priority on pre- and early history "not only because it locates the beginning of our continent's historical development in the original central European homeland of our *Volk*, but also because, better than any other field, it serves as 'a preeminently national science' (Kossinna) that counteracts the low opinion of the cultural level of our Germanic ancestors that has been common up to now." According to the guidelines, the presentation of prehistory must emphasize race "as the basis [*Urboden*] for all rooted characteristics of the individual personality and of peoples." Furthermore, it must emphasize the *völkisch* idea as the opposite of "the internationalist idea, whose poison has threatened to devour the German soul over the past century." Finally, prehistory must employ an expansive notion of the *Volk* because a third of Germans lived outside the borders of the *Reich*, and they were to be treated as tribal brothers (*Stammesbrüder*), not as the citizens of other countries.[15] This last point especially shows that prehistory's ability to address contemporary political concerns was more important than the field's actual content.

The guidelines went on to pervert the pedagogical principles that museum directors had implemented in the 1910s and 1920s. Kiekebusch and Jacob-Friesen had reorganized chaotic displays so that visitors could focus on individual objects and receive a clearer understanding of cultural development in prehistory. The Nazi guidelines also stressed clarity, but they wanted clarity about politics, not science. In the classroom, teachers were to ensure that the student would not "sink in the confusing overabundance of individual events but instead recognize the deeper connections and thereby fortify his education in political judgment and political will." Lesson plans, for example, were to connect the supposedly ancient and specifically Germanic form of the heroic idea to the idea of the *Führer* in the present day. "Together these ideas awaken excitement through the power they possess and the force they have to pull at the heart. They will prevent history from becoming the boring accumulation of knowledge."[16] According to the Interior Ministry, students did not study prehistory primarily to learn about the ancient past; they studied prehistory to deepen the roots of their present-day political convictions.

The guidelines closed with brief comments about the content of prehistory, but these too were subordinated to ideological considerations. Schoolbooks were to start with a depiction of the European Ice Age that showed "how specific races (Neanderthal, Aurignac, Cro-Magnon) were the carriers of a native culture. It is to be demonstrated that even in this most ancient period culture is created by race." When covering post–Ice Age Europe, teachers must convey the basic lesson that "the history of Europe is the work of the peoples of the Nordic race, whose cultural level we see not only in the remains of their stone and bronze tools but in the intellectual achievements that science has documented, not the least of which is the highly developed Nordic (Indo-Germanic) foundational language that superseded the languages of Europe's other races." The text then repeated a popularized version of the northern origins of all cultures: it claimed that enough of the Nordic peoples migrated throughout North Africa, the Middle East, Greece, and northern Italy to produce the cultural achievements of ancient civilizations in these areas. Coverage of these cultures was never to lose sight of the racial connection to Nordic peoples. And lessons stressed Houston Stuart Chamberlain's core idea that cultural achievement derived only from "productive" races: "The fundamental significance of the Germanic migrations lies in the fact that they furnished fresh Nordic blood to a Roman Empire that was degenerated by its racial mishmash. This infusion also sparked the new cultural flowering of the Middle Ages, which occurred only in lands where Germanic peoples had settled for longer periods of time."[17] The details of Greek art or Germanic burial sites that had fascinated scholars for decades were now subordinated to the need to show that cultural achievement relied on contact with the Nordic race.

The state wanted these lessons taught immediately, but the development and publication of new textbooks would take up to two years. Archaeologists associated with the Reichsbund therefore organized other measures that could be implemented more quickly. They hosted prehistory lectures and museum tours for teachers, and the *Nachrichtenblatt* listed crash courses on the *Volk* and race in prehistory and archaeology in eastern Europe.[18] (The lessons about eastern Europe were to provide the justification for Nazi claims on Polish territory.) These events occurred often and throughout all of Germany, and dozens of archaeologists and museum employees contributed their time and expertise.[19] The Nationalsozialistischer Lehrerbund (National Socialist teachers league) supported these training sessions, making them quasi-obligatory for teachers who wanted to hold on to their positions. It is true that museum directors had written about public outreach and the use of artifacts in German schools since 1900, but this topic took on a much more strident tone under National Socialism. Already in the first year of the dictatorship, the Reichsbund proclaimed that museums were introducing tens of thousands of schoolchildren to German prehistory. The state museum for prehistory in Dresden reported that over eighteen thousand pupils visited their museum. The natural history museum in Leipzig hosted lectures and tours for school groups, prepared teaching materials for school use, and organized regular field trips so that people could touch ancient objects and experience an excavation. The museum emphasized that prehistory came to life only when people could see the objects themselves and experience a hands-on demonstration of how an archaeologist excavated. The Leipzig report continued that *Heimatmuseen* should play their part in the education of the *Volk* by providing further exposure to archaeological objects in local settings.[20] This is one area where experts played an important role in disseminating the Nazi version of the ancient past.

It is difficult to know exactly what was conveyed in the classroom in the first years of the Third Reich, but evidence from local and national sources suggests that ideas from the guidelines made it into the class-room quickly. In Northeim, teachers introduced "racial theory and Teu-tonic prehistory" into the curriculum in 1933, and students active in the Hitler Youth policed the teachers to ensure that they remained overtly nationalist. Membership in the Nationalsozialistischer Lehrerbund ex-ploded between 1932 and 1933 to include over 70 percent of all teachers. Bernhard Rust, who became *Reichsminister für Wissenschaft, Erziehung und Volksbildung* (minister of education) in 1934, and Hans Schemm, who oversaw the Nationalsozialistischer Lehrerbund, used the organization to promote an "ethnic fundamentalism" in Germany's schools early in the Third Reich. They ensured that history lessons, excursions, and songs taught egalitarianism, obedience to authority, and solidarity as values of the *Volk* so that German students would learn to view themselves and others in racial terms.[21]

By 1935, teaching aids and textbooks with the required ideological outlook were available. Walter Frenzel, a teacher from Bautzen, urged teachers to quit using older maps and posters "that lacked a high artis-tic standard, the latest scientific knowledge, or the proper dedication to the *Volk*" and to acquire new materials produced by himself, Hans Rei-nerth, and other ideologically reliable interpreters of the ancient past.[22] Surveys of prehistory opened with the idea that one could know oneself only through knowing the "great deeds" of earlier Germans. According to Hans Philipp's text, "All of us enter the future as members of a long line of ancestors. We know nothing greater than our ethnicity [*unser Volks-tum*]. We carry its heritage within ourselves; it is our duty to pass it on."[23]

Museum Professionals and Nazi Interpretations

As the primary institutions for presenting the ancient past to a broader public, museums were, of course, expected to stress Germany's racial heri-tage and cultural superiority. The founders of cultural history museums had focused on regional traditions and hoped that intimate knowledge of the local landscape would foster a sense of belonging that was primarily regional but could also be transferred to the nation. Museum directors in the early twentieth century hoped to maintain these emotional connec-tions to the landscape, but they also wanted to raise the scientific status of archaeology and present new research about prehistory and cultural development. Nazi coordination claimed the emotional bonds to territory and the scientific status of knowledge about the distant past, but both of these aspects of archaeology were to be directed toward the German *Volk*. Regional artifacts were to become national ones, and unsettled research questions were to become indisputable scientific facts. This mission was implemented in new displays sponsored by the Reichsbund, but a more complicated picture emerged in regional museums. Scholars like Karl Hermann Jacob-Friesen and Albert Kiekebusch were certainly German nationalists and saw their field as a patriotic endeavor, but they did not fully embrace the subordination of their field to Nazism's ideological goals. By measuring the presentation of pre- and early history at Germany's museums against the Reichsbund's claims about the ancient past—Nordic origins, the primacy of racial thinking, Germanic greatness, the eternal bond between Germans and their territory, and the political utilization of prehistory—we see that established scholars and their institutions partially embraced the ideological mission set for prehistory but also maintained their commitment to the themes of their earlier research.

New displays created by the Reichsbund provided clear examples of presentations driven by ideological considerations. Following the educational goals pursued in Germany's schools, they sought to engage the public with plotlines of Germanic greatness. The exhibit *Lebendige Vorzeit* (Living prehistory) traveled to six German cities in 1936 and 1937 and brought in between forty thousand and a hundred thousand visitors at each location. The display promoted the idea of German cultural superiority through its interpretation of artifacts. The display halls connected this information to the present day with giant Nazi flags hanging from the ceiling.[24]

Other Reichsbund venues were even more dramatic in bringing prehistory to life. The regime planned to have "living prehistory" exhibits in every German district (*Gau*). Hans Reinerth had one of these living prehistory museums constructed at Oerlinghausen in North Rhine-Westphalia. It portrayed life among the hunters and gatherers of the Stone Age as well as Viking festivals.[25] And a "Germanic procession" at the Grunewald Stadium in Berlin brought a parade of warriors to the capital city.[26]

Established museums were to participate in this propaganda also, but many only partially implemented the Nazi interpretation of the ancient past. Nuremberg possessed historical meanings that seemed to be very compatible to Nazi ideology. The city had hosted the Imperial Diet of the Holy Roman Empire throughout the Middle Ages. As a tribute to this historical role, the GNM was founded there to foster an expansive (großdeutsch) idea of unification that would include all German-speaking lands. And as the site of the Nazi Party's rallies from 1927 through 1938, Nuremberg became central to the propaganda efforts of the regime. One might expect that the GNM presented a bombastic version of German prehistory to the participants and visitors who came for the propaganda rallies, but this was only partially true.

In 1934, Louis Adalbert Springer came to the GNM as an intern in the department for pre- and early history. He was finishing a dissertation under the direction of Kurt Tackenberg in Leipzig, and he was tasked with reorganizing the archaeological display. In the *Nachrichtenblatt*, Springer reviewed the critical developments that had taken place at the museum in the late nineteenth and early twentieth centuries. The objects contributed by Alexander Rosenberg in the 1880s allowed for a more complete overview of prehistoric periods in more of Germany's regions, and the museum had integrated the more scientific approach to prehistory that stressed comparative study and chronology. Yet Springer criticized the earlier display, which was organized by "the wide-ranging concepts of the Stone Age, Bronze Age, Early Iron Age, Roman Imperial period, and the Migration Age" as too static. According to Springer, such an exhibit placed no value on the broader meanings of the objects, which for Springer meant conclusions about racial studies and long continuities in German history. The new exhibit, which opened in 1935, reflected "the most current research and the basic principle that objects are presented and interpreted in the nationalist ways that have become a necessity and duty in light of Gustaf Kossinna's contributions to German prehistory research." Visitors followed the "racial and cultural history of Germany" through a series of rooms that displayed stages of development in northern, eastern, and southern Germany, as well as Scandinavia and the Mediterranean region. A final room encouraged visitors to compare impressive Germanic cultural production with supposedly inferior Roman ceramics from the Rhineland, as well as "the decadent art of the late Roman world and Slavic finds from the period of [Germanic] colonization."[27]

Springer's article described a coordinated museum display in the pages of the Reichsbund's journal, but other sources suggest that although the museum conformed to the main ideological message of the regime, it also preserved more moderate positions on several research questions. The treatment of everyday objects in the prehistory display, for example, allowed room for a discussion of daily life that relativized the bombastic language of Germanic cultural superiority. And the guidebook for the exhibit spoke of the Roman Empire's positive influences on central Europe, noting that the Germanic peoples learned stonework, city planning, toolmaking techniques, and other technologies from the empire. And when external agencies asked the museum for assistance with interpretive questions, the museum recommended works by scholars who were not the most nationalistic in their writings, including Jacob-Friesen. Historian Christian Kohler sees this as something of a middle way between the full coordination of the museum's efforts with Nazi ideology and the continuation of pre-Nazi practices.[28]

Coordination was also less than complete in Hannover. The provincial museum, which had stressed its regional cultural heritage during the German Empire and even into the Weimar Republic, was renamed the Landesmuseum Hannover in 1933. This change indicated the regime's desire to erase the nineteenth-century notion of historical landscapes and stress the coherence of the Reich instead. Jacob-Friesen remained the director of the museum, and as mentioned above, although he identified as a National Socialist after 1933, he had a turbulent relationship with Reinerth and Nazi leaders. His revised guide to the prehistory section of the museum, which appeared in 1938, stressed some tenets of the Nazi view of the ancient past but also included many of the less nationalistic ideas that he had presented during the Weimar Republic.

The guidebook, in line with the scientific orientation of the 1920s, defined prehistory as the "study of human development, including the causal connections that explain both intellectual and physical creativity, specifically for time periods for which we have no written sources." This plotline was about what happened on German territory in ancient times, not Kossinna's formulation of an exclusively German prehistory. At the same time, however, the introductory paragraphs linked the three concepts that Jacob-Friesen had insisted in his *Grundfragen* be kept apart (races, peoples, and cultures) in a way that validated National Socialist ideas about racial descent and the critical role of struggle in producing cultural development. According to the 1938 text, prehistory showed "how our ancestors, on the basis of the process of selection and genetic inheritance, became consistently more highly developed (as races), how they grew and strengthened themselves not as individuals but as communities (peoples), and how their achievements (cultures) became ever greater by responding to the challenges of their environment." The study of prehistory therefore aims "to inspire reverence toward past generations of our ancestors and to cultivate pride in our people. This knowledge shows how our own identity strengthened itself only by asserting itself in this hard fight. And finally prehistory promotes the love of homeland [*Heimatliebe*], because the ground that our ancestors were tied to, that they cultivated, and in which they buried their dead has been hallowed land for thousands of years."[29] This rhetoric certainly surpassed the nineteenth-century notion of regional patriotism and Jacob-Friesen's earlier words of caution about conflating races, cultures, and peoples.

The 1938 guidebook also contained speculative comments about racial studies that were not present in the earlier edition. Jacob-Friesen, employing the racial categories developed by Hans Günther, suggested that "the Germanic peoples go back to the merging [in the later Stone Age] of the people who constructed burial mounds with those who interred their dead in megalith chambers, of which one was probably of the Nordic race and the other belonged to the 'Phalian' [*fälische*] race." And later he referred to Lower Saxony as part of a northern cultural area "that had served as the core lands of the original Germanic peoples [*Urgermanen*] since the Bronze Age and from which the 'Germanization' of all central Europe had proceeded."[30]

Yet despite the emphasis on race and struggle, much of the 1938 guidebook looked quite similar to Jacob-Friesen's coverage from 1920. The discussion of the Stone Age maintained the anthropological orientation's broad view of evidence from throughout Europe. From the Bronze Age forward, the exhibit focused on Lower Saxony and therefore did not draw conclusions about all of Germany. The main themes were not Germanic greatness or the exclusive creativity of Nordic peoples. Instead, the guidebook explained how one could see economic and technological developments, as well as cultural change, through the precise observation of artifacts, metalworking techniques, and burial practices. The narrative praised Germanic art but not in a way that suggested there was something exclusively creative or superior about the peoples who produced it. Jacob-Friesen included nothing derogatory about other cultures.

At the Märkisches Museum in Berlin, Kiekebusch had worked since 1909 to increase the public's understanding of archaeology and to foster a stronger sense of local and national patriotism through an appreciation of Brandenburg's past. His 1929 contribution to Ebert's *Reallexikon* on teaching prehistory repeated many of the points he had made in *Bilder aus der märkischen Vorzeit* (Scenes from Brandenburg's prehistory, 1916). He described how prehistory fostered a sense of belonging and how the physical nature of archaeological objects made them ideal pedagogical tools for children. His essay recommended works that were well respected and not overtly nationalist.[31] After 1933, though, Kiekebusch wrote several articles for National Socialist pedagogical journals that exhibited two changes in his views about public outreach and prehistory. First, his evaluation of the period between 1909 and 1932 switched from a positive appraisal of a period when the public interest in prehistory was growing to a condemnation of the Weimar Republic. This shift set the stage for his view of the Nazi era as a new day for prehistory research. Second, he replaced his earlier focus on the regional history of Brandenburg with sweeping statements about the *Volk* and German prehistory. Writing in 1934, he noted: "That initial efforts to teach prehistory in schools mostly took place in the time of distressful dissension is but proof that our German *Volk* was at its core still thinking and feeling in healthy ways and that the love of one's own *Volkstum* did not die during that most horrible period of internationalism." Under the new conditions in 1933, education must make "the common man and simple woman familiar with the uniqueness of their *Volk*." And he encouraged schools to emphasize the three Vs—*Volkskunde, Vererbungslehre*, and *Vorgeschichte* (ethnography, genetics, and prehistory)—teaching students to be proud of the cultural achievements of their ancient ancestors and the decisive role of race in determining the fate of their people.[32]

Amid this embrace of Nazi values, Kiekebusch still held on to his desire to share his love of archaeology with students. He called on teachers to visit museums, organize hikes, and participate in excavations. During these experiences, "one stands there among devout listeners in or near a historic place, then all will sense the silent shiver of history. Such a presentation has the effect of a heroic saga from the ancient past, and it is astonishing how such an impression on the heart of each individual can never be erased."[33] For Kiekebusch, these moments were emotional, almost magical, but they were also based on science. Students should feel connected to the ancient past because "the science of prehistory can document the existence of a people far back into prehistory." Repeating a milder version of Kossinna's conclusions from settlement archaeology, Kiekebusch explained that "German researchers can demonstrate cultural uniformity in a somewhat clearly defined area on the southwestern coast of the Baltic Sea (Schleswig-Holstein, Mecklenburg, Western Pomerania, northwest Brandenburg), tracing a Germanic culture back to at least the middle of the second millennium BCE, perhaps even so far back as 1800 BCE. This area definitely has been occupied by Germanic peoples since this time. Today, there are no differences of opinion on this topic."[34] Kiekebusch then combined the confidence he had in the research of his day with the excitement that novelists had generated in the nineteenth century. The plotlines that allowed readers to imagine ancient customs and battles could now be viewed as true: "The heroic character of the Germanic tribes is not a new topic, but it certainly was not emphasized enough, and it is increasingly illuminated through new observations and finds. What brought us such excitement back in our earlier youth in [Felix] Dahn's depictions of the Goths, we experience today in stronger ways, as we trace the events and migrations and the massive expansion of the Germanic tribes, not only in the Migration Age, but now also all the way back to the Bastarnae and further back to the migrations of the *Urgermanen* across the Rhine."[35]

The Nazi rise to power shaped Kiekebusch's thoughts about prehistory and pedagogy, but he apparently did not fully implement the racial ideas of the regime when it came to personal relations. As part of the seminar at the Märkisches Museum, Kiekebusch had come to know many people throughout Brandenburg who shared his enthusiasm for archaeology. He recommended one of these participants, Alfred Bab, to the state government as a conservator for the antiquities of the Weißensee district (a northern suburb of Berlin). In a letter from October 1934, Bab thanked Kiekebusch for this recognition, but he had to inform his mentor that he "was not fully Aryan as defined by the Civil Service Law" and that is why he had not joined the Nationalsozialistischer Lehrerbund or other organizations. "I am forced to allow most of today's experiences to pass me by," Bab wrote. He found this most unfortunate because "my entire life I have only thought and felt German—and I don't know if perhaps someone somewhere would have reservations about the possibility of a non-Aryan man teaching other people. This could give rise to complications that would be disastrous for me and embarrassing for you, my honored professor." Bab had tried to avoid the issue, explaining that "as conditions became more extreme, I wanted to just disappear from the seminar silently, but I was kept in place by [my love for] prehistory and by your 'personality,' which has served as a pillar to me in recent times."[36] Kiekebusch apparently did not withdraw his recommendation, because Bab became the conservator in Weißensee, and he wrote to Kiekebusch a month later to inform him that the outcome he had feared had occurred. A colleague reported Bab, and he left the position and the seminar so that he would no longer be in the way.[37] This correspondence does not present Kieke-busch as resisting Nazi policies, but it does indicate that the atmosphere in the seminar was not fully polluted by racial themes and the exclusions created by Nazi ideology.

Kiekebusch conveyed his knowledge of the prehistoric past in two surveys directed toward a general audience. Both works described Ger-man prehistory, in contrast to Kiekebusch's earlier works that focused on Brandenburg. The first, *Deutsche Vor- und Frühgeschichte in Einzel-bildern* (Scenes from German pre- and early history), was published by Reclam, a press known for its reliable and affordable little paperbacks. It was relatively mild in its national tone and racial claims. The book opened with the idea that prehistory must be used to teach patriotism and a sense of national belonging, but it did not stress Nordic supremacy or focus ex-clusively on Germanic culture. And Kiekebusch admitted that some ques-tions, like the origins of the earliest inhabitants of Europe and the lineage of Indo-Germanic peoples, were still unsettled.[38] The most nationalis-tic passages described how the Germanic peoples had become stronger through their battles against the Roman Empire. In Kiekebusch's telling, the *limes* stood "as a beautiful testament to German strength" and as "the grandest monument to the glorious deeds of our forefathers."[39]

Kiekebusch's treatment of Slavic culture, however, contained none of the derision found in many Nazi-era writings. After reporting on the settlement of lands east of the Elbe River by Slavs in the sixth century, Kiekebusch described the significance of several sites. He marveled at the excavations near Mittenwalde, carried out by members of his seminar at the Märkisches Museum, for providing the best example of wood pre-served from the foundation of a Slavic redoubt, and he enthusiastically recommended that his readers visit a display near Oppeln that provided "a nearly complete picture of the Slavic culture of the early Middle Ages." These comments presented Slavic sites as part of the prehistory of central and eastern Germany even if they represented another culture. Yet this coverage was a sideshow to the main plotline, which presented the "re-Germanization" (*Wiedereindeutschung*) of this area by colonists, who ex-panded into sparsely populated territories and reestablished, according to Kiekebusch's subtitle, the "German East."[40]

Kiekebusch's descriptions offered a stark contrast to works that deni-grated or even denied the existence of a Slavic cultural presence in Ger-many. A book on Germany and Poland prepared for the 1933 Interna-tional Historical Congress in Warsaw, for example, included Wilhelm

Unverzagt's conclusions about the prehistory of the German-Polish borderland. Unverzagt, the successor to Carl Schuchhardt at the Museum für
Vor- und Frühgeschichte in Berlin, disputed the existence of an original
Slavic homeland in territory that had belonged to the German Empire,
and he claimed that Slavic sites (including the very ones that Kiekebusch
admired) exhibited a low level of cultural development. Unverzagt asserted, "The excavations in Oppeln and Zantoch give a clear impression
of how it looked in the East before the German colonization. The domestic and defensive buildings were constructed in the most primitive block
technique. . . . When one recalls that such houses appear as the residences
of Slav princes, and at a time when the imposing Romanesque churches
were being built on the Rhine and in central Germany, . . . one can understand what the culturally superior Germanic west offered the primitive
Slavic east."[41] According to this view, the Slavic east could be raised up
only by the coming of German settlers.

In preparation for Berlin's seven hundredth anniversary in 1937, Ernst
Vollert, a high-ranking official in the Interior Ministry, passed along Joseph Goebbels's request to the city's mayor and cultural institutions that
celebrations should not draw attention to Berlin's Slavic past. Vollert
wrote that the *Völkischer Beobachter*, the national newspaper published
by the Nazi Party, had been completely silent on this subject, and this example should be followed in exhibits, parades, speeches, press reports, and
publications. He explained that "while the temporary settlement of eastern Germany by the Slavs should not be denied in historical publications,
it would be untoward to underscore this history at events intended for
a general audience or for international guests."[42] If this advice was not
enough, press coverage of the celebrations addressed the issue "scientifically." A local Berlin gazette reported that recent excavations had dis-
proved "the old fairy tale of the Sorbic 'fishing village' as the nucleus of
Berlin."[43] The Slavic roots had no place in the regime's simplified version
of the capital city's prehistory.

A work from 1935, the year of Kiekebusch's death, presented a much
more martial version of Germanic prehistory. In the preface, Kiekebusch
called attention to the racial lesson readers should learn from the different fates of the Celts and Germanic tribes: after all, one people had virtually disappeared, while the other flourished. He claimed that the Celts
had spread themselves thin and intermarried with other racial groups.
"The consequent mixing with foreign peoples weakened that *Volk* and
watered down its race. These Celts, so weakened and frazzled, then suc-
cumbed to the two peoples on the rise from the south and from the north,

the Romans and the Germanic tribes."[44] The majority of the book then detailed what Kiekebusch saw as the Germanic peoples' valiant fight against the Roman Empire. Typical of his telling is his description of events in 12 BCE that pitted Drusus, "a significant general, who had all the power of the Roman Empire in his hands ... and was driven by an unquenchable desire for fame and conquest," against the *Germanen*, "a people that was intimately tied to the land and in possession of its undiminished inherent power."[45] The book curiously omitted coverage of the German east, despite the increased nationalism of its anti-Roman bravado.

Describing the middle way of several established institutions and scholars is not an attempt to exonerate archaeologists or minimize the significance of Nazi ideas about prehistory. Jacob-Friesen and Kiekebusch, both World War I veterans and ardent nationalists, viewed their outreach and pedagogy efforts as a patriotic duty, and this commitment became more fervent after 1933. Yet themes within domestic archaeology from earlier decades persisted despite the opportunities and threats that coordination presented. Museum administrators and scholars retained some of their devotion to regionalism and to prehistory as a general science (not a national one), and they held on to some of the hesitation they had exhibited about questions of race and ethnic continuities that stretched into the deep past.

Nazi Prehistory in the East

With the outbreak of World War II, the Nazi regime used archaeology to justify the annexation of large swaths of territory in eastern Europe. Occupation agencies promoted the idea that the entire space between the Oder, Vistula, and Bug Rivers was an ancestral home to Germanic tribes and that the "resettlement" of this area by Germans was a return to earlier conditions. This territorial claim went well beyond the losses mandated by the Treaty of Versailles to include land in the interwar states of Poland, Czechoslovakia, Romania, and the Soviet Union. A firm narrative about the Germanic nature of this land replaced the interwar debates between Polish and German scholars. The Slavic past was denigrated or denied, and German settlers in the area were taught to view the land as eternally German. There was no middle way in this setting. Prehistory became propaganda, and it served as part of the intellectual preparation for a new and brutal racial hierarchy in occupied lands in the east.

In Poznań, the most important center for prehistory research in Poland during the interwar years, scholars went into hiding or exile after the German invasion in the fall of 1939. Józef Kostrzewski, the leading figure in Polish archaeology, spent the entire war living under an alias and received financial support and scholarly material from friends.[46] SS-Hauptsturmführer Hans Schleif planned to transform Poznań from a center for Polish archaeology into a major research center for Germanic prehistory. University and museum resources were combined into a Landesamt für Vorgeschichte (State office for prehistory) that employed between twenty-seven and thirty-nine people throughout the war. The office was quick to place Nazi archaeologists into key roles and create museum exhibits that taught the new dogma. The Landesamt's journal made clear that scholarship would serve the imperial aims of the regime. In its introductory article, SS-Brigadeführer Robert Schulz explained his vision of archaeology in occupied Poland:

> I view the advancement of prehistory research as one of the most important cultural-political tasks in the administration of the Wartheland. . . . Poles misconstrued and falsified their results in chauvinistic ways. They attempted to deny the Germanic settlement of this land that lasted over a thousand years. In its place they falsely attributed [the evidence of] other cultures to the Slavs with the clear goal of scientifically justifying their claim on this territory, which had been illegally annexed by them with the help of the Western powers in 1919. German scholarship has long proven these Polish claims to be incorrect. . . .
>
> [The publication of this research] should familiarize all the ethnic Germans residing in the area with the ancient history of their homeland, which is so closely tied to the history of the Germanic peoples. May the knowledge of the streams of Nordic, Germanic, and finally German blood that have poured out over the eastern German territories since the first agricultural settlements awaken in us a feeling of commitment to make this territory, which has been hotly contested over the past millennium, once and for all a purely German area of settlement.[47]

The Landesamt planned to silence Polish scholarship and teach settlers to view the land as eternally German. Despite these grandiose plans, it soon encountered personnel shortages because of the continuation and escalation of the war. The office produced only a single year of its journal.

The Nazis infamously rebranded Łódź as Litzmannstadt, naming it after the general who had led German forces into the city during World War I. The occupying authorities during World War II imposed other historical meanings on the city that stretched much further back into the past. Archaeologists associated with the ethnography museum in Łódź had excavated a graveyard near Biała in 1936 and found an ornate urn that contained ashes, a sewing needle, two iron blades, a fibula, and a crescent-shaped comb. Significantly, a series of etchings adorned the base of the urn that included symbols that looked like two intertwined swastikas. When the Nazi occupation began, authorities used artifacts in the museum to claim that the area's Germanic history extended back two thousand years. The urn from Biała became a symbol of the city in 1941, and it was showcased on Łódź's new emblem against a gold and dark-blue background, the colors associated with General Litzmann's unit in World War I.[48] The urn helped to cement the supposedly ancient relationship between Germans and the Wartheland when it circulated as a propaganda image. In addition to appearing as a city symbol, it was featured on a commemorative medallion for the Winter Relief Program under the heading "Ewiger deutscher Osten" (An eternal German east), a slogan that stressed both the imperialist ambition to possess the east forever and the claim that this possession was the natural order of things because of the asserted German presence in ancient times.[49]

Prehistory research was also revamped along ideological lines farther east in the Generalgouvernement (General government). In 1939, over 180 academics were removed from the university in Kraców, and the Institut für deutsche Ostarbeit (Institute for German work in the east) was created to promote the Nazi regime's vision of the occupied territory. This institute included a section for prehistory that oversaw excavations and managed museums. In 1941, it produced an exhibit about "Germanenerbe im Weichselraum" (Germanic heritage along the Vistula River) that stressed the dynamism of Nordic peoples and their contributions to the cultural development of the area.[50] In this occupied territory, the Slavic past was erased.

World War II was to make the east "German" again, but it actually brought enormous damage to the physical infrastructure that displayed Germany's ancient past. Museums were promoted in the Third Reich as the primary vehicles for presenting the Germanic past to a wider public, but by 1939, they were making preparations for the destruction that the war would bring. Major collections throughout Germany were placed in bunkers and other storage facilities for the duration of the war. During half of the Third Reich, their ability to contribute to the regime's propaganda program was severely limited.

The Third Reich used domestic archaeology to tell stories of ethnic continuity and racial superiority with extreme zeal. This practice channeled the familiarity with the ancient past that nineteenth-century traditions had fostered toward the nation and *Volk* in radically new ways. It certainly grew out of *völkisch* thought and Kossinna's conception of "a preeminently national science." By emphasizing the great confidence with which Nazi archaeologists spoke about the ancient past, though, we also see that Nazism's subordination of prehistory to ideology represented a break with earlier practices. Nazi narratives trampled the professional standards established by German and European archaeologists in the early twentieth century. Ideologues refused to admit that some questions were unsettled or to acknowledge that archaeological material often allowed only limited conclusions. Instead, they saturated German society with bombastic claims about the strength and glory of ancient ancestors. Nazi readings of the Germanic past were not the end of the story, however. The postwar period brought both a retreat from exclusively nationalist interpretations and an eventual return to the multiple forces that were at work in the nineteenth century. Once again, archaeologists and museum personnel brought more open-ended approaches to scientific research and public outreach.

Epilogue

The war waged by the Nazi regime destroyed the intellectual climate that promoted domestic archaeology as "a preeminently national science." The racial dreams and desire for European domination that led to the brutality of World War II and the Holocaust exposed the belief in a primordial and superior *Volksgemeinschaft*, or people's community, as destructive and inhuman. In postwar Europe, nationalism was condemned as a root cause of Nazi aggression, and identities were to be based on more humane values: liberal democracy and international human rights in the West and the idea of the people's democracy in Eastern Europe. In these contexts, the most recent and powerful justification for domestic archaeology lost its relevance. Some cities even abandoned their archaeological collections. The town of Oerlinghausen sold off artifacts from the museum founded by Hans Reinerth for a pittance. In the 1950s, the new director of the Märkisches Museum believed Nazi ideology had ruined domestic archaeology, and as part of the antifascist position of the East German state, the museum initially devoted itself to the history of Berlin and did not reestablish its prehistory collection.[1]

Yet the story of domestic archaeology after 1945 was one of eventual recovery. Even in an environment of broad skepticism toward nationalism, most museums restored their collections, and university institutes and offices of historical preservation maintained the gains in personnel that came during the Third Reich. These institutions continued to carry

out excavations and produce new scholarship. Today, pre- and early history remains a part of school curricula, and museums engage a broad
public with educational programs and exhibits. This recovery poses two
important questions: How did the field of pre- and early history address
nationalist and Nazi versions of the ancient past after 1945? And what
narratives about the distant past resonated in the political and intellectual climate of the postwar years? Answers to the first question followed
broader patterns of dealing with the Nazi past in Germany. After decades
of relative silence, a critical engagement with the Nazi past emerged in the
1960s and increased dramatically after German reunification in 1990. Answers to the second question reveal the relevance of nineteenth-century
debates, as museums again stressed the local and regional context of archaeological finds, and scholars developed scientific interpretations that
were not primarily driven by nationalism.

In the immediate postwar period, Allied occupation authorities attempted to root out Nazi ideology. This process focused on the narrow
issue of leadership or direct participation in the actions of the Nazi regime and not the broader but more complicated question of propagating nationalistic and racist views. Denazification review boards evaluated
thousands of state employees, but these proceedings were mild, and many
individuals who had been Nazi Party members were allowed to continue
in their civil service posts after 1945.[2] Decades passed before historians
investigated the prevalence of racism in German society and the roles
played by cultural institutions in supporting the ideology of the regime.

Denazification in the field of pre- and early history followed this pattern. After the war, the most blatantly racist texts disappeared, and Nazi
institutions for pre- and early history were disbanded. Most of the blame
for the politicization of the field was placed on the regime and specifically Hans Reinerth. Reinerth was arrested and incarcerated in 1946.
After a lengthy proceeding, the state commission for denazification in
Baden found Reinerth to be guilty in 1949, citing his leadership role in
the Reichsbund für deutsche Vorgeschichte, his manipulation of the field
of prehistory for political reasons, and his actions as a university professor. Reinerth unsuccessfully appealed this judgment in 1952, but then the
Ministry of Justice in Baden-Württemberg voided his conviction in September 1953, claiming that Reinerth had fought against "the illusory Germanic teachings of researchers under Reichsführer-SS Himmler." Despite
this legal clearance, professional colleagues universally blackballed Reinerth. Archaeologists testified against Reinerth in his denazification hear-

ings, and in 1949, the main association for archaeology research in southwestern Germany issued a formal declaration that archaeologists must
protect the objectivity of their field and guard against the practices associated with Reinerth and the Reichsbund. Scholars later advised the German government's board for research funding to reject Reinerth's appli-
cations for support.[3]

Other scholars who supported the shift toward nationalist and racist
thinking during the Third Reich were able to continue in their prominent
positions. Museum directors and university professors, as state employees, were required to fill out denazification questionnaires. Despite having benefited from the elevated position of prehistory during the Third
Reich, these scholars argued that they had resisted efforts to coordinate
their field, that their scientific works had remained objective, and that
they should be viewed primarily as victims of the regime. In many cases,
review boards ordered a brief detention and investigation but ultimately
classified these individuals as not guilty.

Karl Hermann Jacob-Friesen, for example, was detained by occupation
authorities and initially removed from his position at the Niedersächsisches Landesmuseum, but he was back in his university and museum
positions by 1949. Jacob-Friesen had been a Nazi Party member, and his
scholarship was nationalist and racist. Yet he argued that he had guarded
against the pseudoscientific Germanic fantasies of Wilhelm Teudt and
that he had been a victim of Reinerth's campaigns. In his denazification
questionnaire, he listed the numerous setbacks he had experienced during
the Third Reich, noting that he had been removed as the head of multiple
professional organizations and as the editor of a journal for the prehistory
of Lower Saxony. He wrote that he had been called upon to recant his
Grundfragen three times and that his treatment was most recently cited at
the Nuremberg Trials as evidence of the "inquisitorial intolerance of the
Rosenberg papacy." Fellow museum employees testified on his behalf.[4]
Jacob-Friesen resumed his position at the Landesmuseum Hannover, and
in 1950, restarted his journal *Die Kunde* (The message) with an introductory article that stressed the need to break from the past but also to preserve the links between Germans and prehistory. He wrote, "We Germans
have a right to be proud of the culture of our ancestors, but we must guard
against the possibility of this interest degenerating into wild enthusiasm
or losing its scientific character."[5]

The American military government removed Ernst Wahle from his
university position in Heidelberg in November 1945, but he was rehired a

year later and remained in this position until his retirement in 1957. In a series of essays about the history of the field, Wahle described the abuse of pre- and early history as a general European problem that arose from nationalism. He noted that the Nazi state did not tolerate those who did not go along with its vision for the field, and he cited his criticisms of Kossinna as proof of his objectivity.[6]

In the postwar period, scholars like Jacob-Friesen and Wahle warned of the dangers of the politicization of their field and called for a new era when pre- and early history would be led by science and not ideology. Yet the ethnic and racial content of their Nazi-era work was not critically examined, and scholarship preserved the ethnic paradigm that Kossinna had developed.[7] Researchers no longer transmitted the indoctrinating message of the Nazi regime, but neither did they go out of their way to discuss the recent abuse of the field or evaluate the influence of ethnic and nationalist concepts in pre- and early history.

The postwar career and writings of Ernst Wahle illustrate these conflicting messages, as archaeologist Dietrich Hakelberg has shown. Wahle warned against the political engagement of prehistory and the influence of nationalism on science, but he had directly benefited from the nationalist enthusiasm that advanced the study of prehistory after 1933. He viewed race fanatics as harmful, but he believed that a racialized version of anthropology offered a legitimate set of scholarly questions. Hakelberg documents how nationalist perspectives from the first half of the twentieth century persisted in Wahle's postwar work. Comparing the 1950 edition of *Deutsche Vorzeit*, Wahle's major synthetic work, with the 1932 version, Hakelberg notes that the revised edition omitted the most blatant Germanic language and racial formulations, but it still contained terminology that suggested the values of the earlier time. Wahle, for example, continued to write of a ruling elite (*Herrenschichten*), heroic deeds, and the civilizing mission of certain prehistoric peoples. And he continued to refer to a long continuity of "Nordic productivity" from the Stone Age to the Bronze Age as a feature of European prehistory that had "the greatest historical impact." All of this, Hakelberg notes, maintained the *völkisch* concept of carriers of culture who brought advancement because of their indomitable creativity and skills in war and conquest.[8]

When archaeologists did question the ethnic paradigm, they treated it as an archaeological problem, not as a broader question about the propaganda value of narratives about national origins. In *Einführung in die Vorgeschichte* (Introduction to prehistory, 1959), the most important in-

troductory textbook for archaeology in postwar West Germany, Hans Jürgen Eggers (1906–75) offered a thoughtful review of the development
of key methods and concepts in prehistoric archaeology, and he devoted
two chapters to a thorough criticism of Kossinna's ethnic interpretations.
Eggers had served as an assistant to Wilhelm Unverzagt in Berlin before
receiving a position in the provincial museum in Stettin in 1933. He was
drafted into the army during World War II, and after the war, he was employed by Hamburg's museum for ethnology and prehistory and also held
a professorship in Hamburg.[9]

In his discussion of ethnic interpretations, Eggers drew a clear contrast
between Oscar Montelius and Gustaf Kossinna. Montelius had advanced
archaeological chronologies by developing the study of typologies. His innovations brought new possibilities for interpretation, but Eggers stressed
that they always came with a measure of openness and humility. "While
Montelius may have gone down some wrong paths initially, he still possessed enough character and ability for self-criticism to recognize his own
mistakes and correct them."[10] It was otherwise with Gustaf Kossinna. "For
Kossinna, there were certain basic truths that required no further proof."
The most important of these was the idea that an archaeological culture
clearly indicated a people and a race and specifically that the Germanic
peoples had remained in place from the later Stone Age through the Age
of Migrations.[11] Eggers identified several fundamental problems that
arose from Kossinna's gross overconfidence. He explained how Kossinna
treated the archaeological record as if it were complete and its meaning
crystal clear. Eggers insisted that one must first acknowledge the fragmentary nature of archaeological evidence. He admitted that "not only
is the material available to us incomplete, . . . but every set of artifacts is
to some degree even more fragmentary." Some material deteriorates and
other material does not. Gravesites, hoards, and settlements, as different
kinds of archaeological sites with different purposes for the people who
created them, preserved some kinds of objects but not others. And conditions in the area surrounding an excavation site—whether the area was
a forest or farmland, for example—would affect what remained. Finally,
Eggers acknowledged, an archaeologist's interpretation would certainly
be shaped by earlier research and the interpretation of similar sites. "All
these factors must be considered when we use prehistoric archaeological material critically and when we want to arrive at a proper evaluation
of imperfect sources."[12] For Eggers, archaeology was about the process of
interpretation, not the drive toward ethnic conclusions. This perspective

undercut the certainty with which one could speak about long continuities and connections to modern peoples.

Eggers's guidance sounded like a detailed version of Leopold von Ledebur's earlier words of caution. Writing in the 1830s, Ledebur had noted that it was too soon to attach ethnographic classifications to objects. More than a century later, Eggers believed that archaeologists still needed to acknowledge the limits of archaeological interpretation. He emphasized that when constructing a map that shows the distribution of a specific type of sword or a certain ceramic form, for example, one is not locating the borders of a people or even a culture. Rather, one is seeing only the practice of burying or preserving a certain kind of artifact, and that fact should open up questions about trade relations, political conditions, or religious practices. This led Eggers to a sober conclusion: "One should never view the ethnic interpretation as the only or even as the dominant interpretation, as Kossinna did. When we interrogate the distribution map ourselves and keep its many possible meanings in mind, the map will then speak and certainly deliver historical information to us, even if it is not always information about ethnicity."[13]

Eggers offered a thorough critique of Kossinna's methods, and he urged readers to bring multiple questions to the study of archaeology. Yet his book addressed ethnic interpretations only as an archaeological problem, not as a political or historical problem. He did not discuss the individuals or regime that elevated Kossinna's ideas. Nor did he comment on the propaganda value of ethnic thinking in German society. This kind of evaluation of the most recent past would have been very difficult because of Eggers's proximity to events and to the continued careers of so many individuals who were active during the Nazi period.

For archaeology museums, the first two decades after World War II represented a long period of recovery. Many institutions experienced partial damage or total destruction from the bombing of German cities. The Landesmuseum Hannover, for example, closed in 1943, sustained damage during the war, and was subjected to theft and plunder in 1945. Half of the museum's exhibit rooms were opened to the public in 1950, but the prehistory section was still in disarray in the late 1950s. War damage to the museum was not completely repaired until the 1960s.[14] Conditions were similar in Munich. The Bavarian Vor- und Frühgeschichtliche Staatssammlung (the successor to Johannes Ranke's Prähistorische Sammlung) was closed to the public during the war and artifacts were stored in multiple locations for safekeeping. The building that housed the collection, the

Alte Akademie, was heavily damaged by an air raid in April 1944. After
the war, the museum collection was reassembled, and Friedrich Wagner
organized a series of temporary exhibits beginning in 1948 that displayed
parts of the collection in spaces administered by the BNM. The museum,
existing as an institution without a building, published a series of catalogs
about its holdings to ensure that "the museum did not disappear completely from the public's mind" during this period of provisional exhibits.[15]
In Berlin, the Staatliches Museum für Vor- und Frühgeschichte, located in
the Martin-Gropius-Bau since 1922, closed in August 1939, and the collection was moved to various secure locations throughout Germany during the war. The Gropius building, which was just a few blocks away from
the main headquarters of the Gestapo and Hitler's Reich Chancellery,
suffered damage from multiple air raids between 1943 and 1945.[16] After
the war, the collection, like Berlin and Germany, was divided. Items that
had been stored during the war in what became West Germany formed
the nucleus of a new Museum für Vor- und Frühgeschichte, which opened
in one of the wings of a former summer palace of the Hohenzollerns in
Charlottenburg, West Berlin, in 1960.[17] It again became an important display that presented the prehistory of Europe, not just Germany. A much
smaller number of objects remained in East Berlin and were exhibited
infrequently. The Charlottenburg museum took custody of many of the
items from the Märkisches Museum, including the Bronze Age artifacts
from the royal grave of Seddin that Ernst Friedel and his group had excavated in 1899. The Märkisches Museum eventually built a new archaeological collection with excavations carried out after 1945.[18]

The economic recovery of the Federal Republic of Germany during
the 1950s and 1960s allowed for resources for excavations and the renovation and expansion of archaeology museums. With improvements in 1964
and 1971, the pre- and early history section of the Landesmuseum in Hannover again became an exemplary archaeology museum. Harking back to
Jacob-Friesen's original goals, the museum provided a scientific overview
of all prehistoric epochs, and it increased its public-outreach efforts with
hands-on exhibits designed for children and school groups. In Munich, the
major Bavarian collection was renamed the Prähistorische Staatssammlung (State collection for prehistory), and it received a large, new building dedicated solely to archaeology in 1976. The institution was renamed
the Archäologische Staatssammlung in 2000, and the building, currently
undergoing a major renovation, still houses the exhibits and research
work associated with this collection.

During this phase of museum expansion, domestic archaeology in Germany remained cautious, even antitheoretical. As archaeologist Heinrich Härke has observed, the field adopted a strongly positivist stance that emphasized meticulous excavation and documentation but rarely ventured into larger questions of interpretation. Specifically, in the archaeology of the Bronze and Iron Ages, the eras that had been so closely tied to questions of ethnic identity in the first half of the twentieth century, "the pursuit of typology and chronology was an end in itself."[19] Härke explains this skepticism toward theory in part as a reaction to the abuse of ethnic interpretations during the Nazi period, and he notes that this shortcoming cut German scholarship off from theoretical debates in postwar Anglo-American academic circles. This focus on specific excavations avoided ethnic interpretations, but it left the Nazi past largely untouched.

With more resources and modern buildings, regional museums again presented the prehistoric past to a wide audience. Making a clear break with the Nazi era, museums acknowledged the distinction between prehistory and early history and abandoned the task of tracing peoples and cultures into the Bronze Age and beyond, and all talk of Nordic racial origins disappeared. Prehistory focused on new insights about human evolution and deep time, the use of carbon dating in archaeology, and new excavations carried out by regional offices of historical preservation. Much of this knowledge was conveyed to the public in successful magazines like *Archäologie in Deutschland* and publications that focused on specific regions like *Das archäologische Jahr in Bayern*.

Museums have continued to present Germany's regions as coherent historical landscapes shaped by geography and settlement patterns in early history. This partly has to do with the federal nature of archaeology and historical preservation work in Germany and the wider phenomenon of Europe being a place of regions in a postwar era of supranational institutions, but it was also a continuation of the connections to the local landscape developed in the nineteenth century. In his 1880 request to the Prussian Cultural Ministry, Johannes Ranke had referred to the "all-German exhibit" as a patriotic endeavor that would shed light on the origins of nineteenth-century peoples. This desire to investigate the identity of ancient peoples was alluring, but it stood in contrast to the DAG's scholarly approach that refused to make connections between the exhibit and modern Germans. Similarly, recent exhibits about "Germanic tribes" have appealed to a wide audience by suggesting a relationship between ancient and contemporary peoples or between ancient cultures and nearby landscapes, and they often use possessive language about insights into "our history."[20] These exhibits coincided with the vigorous debate in archaeological and historical circles about ethnic identity in early history. The catalogs that accompanied these exhibits presented the latest archaeological knowledge and information about the constructed nature of early medieval identities, but they still suggested close ties between the land and ancient peoples. Some scholars have rejected presentations that focus primarily on ancient peoples because they lend a coherence to these identities that is problematic. A productive approach suggested by archaeologist Peter Wells would be to move beyond identities as definitive labels placed on excavated material and instead develop frameworks that explain the processes that create identities. In this line of inquiry, museums would explore the meanings that ancient peoples themselves expressed in the cultural and social patterns revealed by an archaeological site.[21]

In Berlin, two archaeological collections that have featured prominently throughout this book have returned to their old settings, suggesting other ways that nineteenth-century traditions interact with contemporary cultural politics. The urns from the royal grave of Seddin were on display in the Museum für Vor- und Frühgeschichte in Charlottenburg for much of the postwar period as wonderful examples of Bronze Age decorative art. In October 1999, though, these items returned to the Märkisches Museum for a new exhibit about the history of Berlin. Here, the Seddin finds were still portrayed as exceptional products of early craftsmanship, but they also offered a tribute to the gentlemen who made Berlin a cultural capital in the late nineteenth century. The urns stood in a display case flanked by photographs of Ernst Friedel and other members of the Brandenburgia society. This cheerful installation celebrated civic pride as much as it displayed important archaeological finds.[22]

In 2009, another change in the Berlin museum landscape occurred. The Neues Museum, completed in the 1850s as the centerpiece of a historicist display of world cultures, had been badly damaged during World War II and remained in a state of disrepair into the 1990s. Its grand opening in 2009 was the crowning achievement of the ambitious refurbishment of the entire complex of cultural institutions on Berlin's Museum Island. The press and architectural circles praised the way David Chipperfield's restoration of the Neues Museum not only showcases the grandeur of Friedrich August Stüler's original entry hall and staircases but also preserves wartime damage. Some walls have been left in their faded state, and bright marble marks areas of the museum that have been reconstructed. This architectural masterpiece stands as a powerful symbol of the new Berlin because it simultaneously celebrates one of the city's world-class museums and acts as a witness to Germany's aggressive and war-torn past. Chancellor Angela Merkel captured the significance of the Neues Museum for contemporary cultural politics by describing it as the property of the *Kulturnation*—a state that appreciates the diversity of human cultures and that has committed the resources needed to reconstruct the wonderful ensemble of institutions on the Museum Island.[23]

The Museum für Vor- und Frühgeschichte received a very prominent place inside the Neues Museum. Part of the collection has returned to the gallery that housed the Sammlung der nordischen Alterthümer from 1854 to the mid-1880s. The room is still decorated with the Scandinavian sagas and fantastic depictions of the Stone, Bronze, and Irons Ages, though many of the murals are faded, marred, or lost because of wartime damage. It is now devoted to a safe journey through the development of domestic archaeology in Germany. A series of display cases presents the forward-thinking work of Leopold von Ledebur in the 1830s, the enthusiasm of historical associations in the nineteenth century, Rudolf Virchow's scientific perspective during the German Empire, and the professionalization of archaeology under Carl Schuchhardt and Wilhelm Unverzagt. The display makes no mention of the meanings these objects had for viewers in the nineteenth century or during World War I, the Weimar Republic, or the Nazi dictatorship. The primary reference to World War II presents the museum as the victim of Soviet looting in a panel about Schliemann's Trojan gold pieces and other precious artifacts that were designated as wartime losses.[24] The third floor of the Neues Museum offers a chronological tour through European prehistory. This central position for the prehistory collection has little to do with an ethnic identification with a Germanic past. Rather, it is a dynamic version of Carl Schuchhardt's 1909 vision of the museum as a premiere research institution for European archaeology and a compelling exhibit for the public.

A critical reading of these two Berlin displays could argue that the Märkisches Museum and the Museum für Vor- und Frühgeschichte are hiding behind nineteenth-century traditions without confronting the museum visitor with the nationalist and racist meanings that the field advanced during the first half of the twentieth century. The exhibits invite viewers to celebrate traditions of civic engagement and scientific development and learn about the exciting insights delivered by current research. A more generous interpretation would recognize these traditions as alternatives within archaeological thought that existed before the early twentieth century. Germany has had many ancient pasts, and the Nazi version of pre- and early history was not the end of the story.

Since the eighteenth century, domestic archaeology in Germany has facilitated many kinds of narratives. The engagement with the land under the Hohenzollern kings led to the sense of wonder captured in Johann Bekmann's historical description and Gotthilf Treuer's questions about religious practices among "the ancient Germans." In the wake of the Napoleonic Wars, voluntary associations set out to find and preserve the past, and they approached archaeology as a new way to shed light on the "darkness of prehistory." Antiquarians excavated and collected artifacts with the hope of verifying classical sources and explaining local conditions in ancient times. They recognized that they were at the beginning of this process, however, and they eschewed narratives in favor of inventories.

In the middle of the nineteenth century, scientific discoveries upset the finite nature of history and posed new questions about cultural development. The field of pre- and early history was shifted from a search for the boundaries of static peoples to research into cultural transfers, migrations, and racial groups. A new European framework was added to local practices, and the amount of evidence one needed to create narratives increased dramatically. Rudolf Virchow and others who pursued this research in the last third of the nineteenth century could not provide satisfactory answers. The documentation of specific sites and the growth of collections became ends in themselves, while reading audiences and museum visitors expected fuller explanations of the ancient past and clearer connections to their present day. Nationalist archaeologists followed the trail blazed by novelists and placed narrative coherence before scientific caution. This new priority culminated in Kossinna's sweeping statements and Nazi dogma about a Germanic past that promised a great German future.

Archaeology continues, as Heiko Steuer has argued, to grapple with issues that were already present in the 1920s. A strong desire to answer questions about ethnic identity persists, and scholars caution that any answer must be scientifically grounded and respect the limits of what archaeology can reveal.[25] At this moment, it is helpful to recall the multiple messages about pre- and early history that archaeologists and museums have delivered to their varied audiences. Archaeology stands as an important area of research and cultural heritage, and it continues to spark the curiosity of children on school field trips and provide enrichment for people interested in local and regional history. Archaeology will avoid narrow history and exclusionary thinking when it steers the curiosity about "ancestors" in multiple directions and engages broader questions about European prehistory and humanity. Like Chipperfield's restoration of the Neues Museum, this approach can acknowledge nineteenth-century traditions, the scars of the twentieth century, and present-day cultural needs simultaneously.

Acknowledgments

While writing this book, I found it quite easy to get excited about the details of a nineteenth-century excavation report or the rich descriptions in a historical novel. It was more difficult to fit these small, curious pieces of the past into a larger framework that examined the tension between archaeological interpretation and historical narration. I am very grateful for the mentors and friends who helped me develop this project into a more wide-ranging work. I am especially thankful for the encouragement and advice I received from Harry Liebersohn and Peter Fritzsche during the development of this project. Suzanne Marchand, Glenn Penny, and George Williamson read the entire manuscript at various points in its development. Their wonderful insights helped me expand both the chronological and the intellectual scope of the book. My appreciation for museums and their history grew immensely through many conversations with Sven Kuhrau in Berlin. Christine Ruane and Joseph Bradley were not only careful readers of my work but important supporters of my professional development as a historian. Conversations with Bettina Arnold and Rupert Gebhard helped me see my project from the archaeologist's perspective. My interest in German studies has benefited from friendships with so many smart and supportive people, including David Bielanski, Sace Elder, Andrew Evans, David Johnson, Molly Wilkinson Johnson, Jennifer Jordan, and Joe Perry. Toward the end of the process, E. J. Carter, Derek Hoff, and Tracie Matysik read large portions of my work on short notice and kept telling me to go for it.

This project has also received generous financial and institutional support. A summer research trip was funded by the Council for European Studies, and I benefited greatly from a fellowship and the fruitful intellectual environment provided by the Berlin Program for Advanced German and European Studies at the Freie Universität Berlin. Writing and follow-up research were also supported by grants from the University of Illinois and Kansas State University. I appreciate the help I received while arranging the illustrations for the book from Silke Geiring at the Bayerische Staatsbibliothek in Munich, Rupert Gebhard at the Archäologische Staatssammlung in Munich, Michaela Hussein-Wiedemann at the Zentralarchiv of the Staatliche Museen zu Berlin—Preußischer Kulturbesitz, Robert Wein at the Stiftung Stadtmuseum in Berlin, and Fred Smith from the Zach S. Henderson Library at Georgia Southern University.

I owe a very large debt to the anonymous reviewers arranged by the University of Chicago Press. Their comments were thorough and challenging, and the project matured greatly because of their input. I am also very thankful for the kindness and never-ending patience that Karen Merikangas Darling, executive editor at the University of Chicago Press, extended during the review and publication processes and for the responsiveness and clarity that Susannah Engstrom offered through her editorial assistance. The manuscript benefited greatly from an in-depth editorial review by Rose Rittenhouse, also at the University of Chicago Press. And I express my thanks to Clinton Otte-Ford for his review of many of my German translations and to Pamela Bruton for her meticulous copyediting work. All these conversation partners and careful readers have helped me develop as a scholar and improved this book immensely. All flaws in the book are my own.

I also wish to express my appreciation for the love and encouragement I have received from family and friends. Visits to Andrea Hartmann and Jochen Müller were welcome diversions during several research trips to Germany. I am deeply thankful for the many ways that my parents, Ron and Mary Maner, have supported my education and indulged my fascination with European history. Janine Stines has read these chapters almost as many times as I have, and she offered great insights at every stage of the process. Most of all, she has been a constant source of love and support. I dedicate this book to her.

Notes

Introduction

1. Thomas, *Skull Wars*, 29–70.

2. Arnold, "The Past as Propaganda"; and Haßmann, "Archaeology in the 'Third Reich.'"

3. Hobsbawm, "Introduction: Inventing Traditions," 14.

4. National case studies are presented in Kohl, Kozelsky, and Ben-Yehuda, *Selective Remembrances*; Kohl and Fawcett, *Nationalism, Politics, and the Practice of Archaeology*; and Diaz-Andreu and Champion, *Nationalism and Archaeology*.

5. Pohl, *Die Germanen*, ix, 2–4.

6. Gillett, "Ethnogenesis," 244.

7. Curta, introduction, 5. Walter Goffart has criticized the ways that the field of early German history (*germanische Altertumskunde*) has adopted assumptions about the contrast between the Roman Empire and a coherent "Germanic world" in *Barbarian Tides*, 1–12, 40–55.

8. Daniel, *Idea of Prehistory*, 109.

9. Geary, *Myth of Nations*, 15.

10. Here I have drawn on historian Rudy Koshar's work on historical preservation. Koshar treats the value and significance of specific buildings and monuments as open-ended. Their meaning is shaped by the stock of existing architecture, the local settings of specific sites, regional and national political developments, wartime destruction, and the outlook of the preservationists themselves. The interaction between these variables "produced not the national continuity of ideologues' fantasies but a malleable yet consistent discourse about continuity." Koshar, *Germany's Transient Pasts*, 11.

11. Hayden White calls attention to the unquestioned way that we view the past as a story to be told. Modern readers view nonnarrative historical approaches with a skeptical eye. Annals provide arbitrary dates without interpretation, and chronicles list achievements without drawing conclusions, but these forms seem incom-

plete because they do not explain all that we want to know. Historians therefore include the elements of a story (explaining motives, uncovering precursors, and providing explanatory conclusions) in order to give past events meaning and coherence. White, *Content of the Form*, 1–25.

12. My study draws inspiration from Celia Applegate's *A Nation of Provincials*, an outstanding analysis of the interaction between political circumstances and cultural activities that contributed to local, regional, and national loyalties.

13. Brather, *Ethnische Interpretationen*, 79–84.

14. This appreciation of other possibilities in the development of pre- and early history follows a general trend that moves beyond an understanding of the nineteenth century primarily as the age of nationalism. For a review of recent works that recover traditions of cosmopolitanism, global connections, and vigorous civic debate in nineteenth-century Germany, see Penny, "Nineteenth Century."

Chapter One

1. Vallibus, "Versuch eines Entwurfs," 1375. The author discusses the richness of place-names and rural legends on 1376–79.

2. Ibid., 1375.

3. Ibid.

4. Bahn, *History of Archaeology*, 54. Similar comments appear in Sklenar, *Archaeology*, 33.

5. Momigliano, "Ancient History and the Antiquarian," 18.

6. The number of these publications increased dramatically from fifty-eight to over seven hundred during the eighteenth century. Sheehan, *German History*, 152–60.

7. Kelley, *Faces of History*, 162–87.

8. I have used the translation of Tacitus's *Germania* by J. B. Rives for all quotations from this work. Subsequent references will be given in parentheses in the text.

9. Philologists did not begin to discuss the *Germania* as a literary and rhetorical work, as opposed to a straightforward report on ancient peoples, until the twentieth century. See See, *Barbar, Germane, Arier*, 31–32; and Rives, introduction, 48–56.

10. Several authors have recounted the story of the discovery and reception of Tacitus's *Germania*: Strauss, *Sixteenth-Century Germany*; and Sklenar, *Archaeology*, 24–25.

11. Schama, *Landscape and Memory*, 93; Strauss, *Sixteenth-Century Germany*, 22–28, 42–43; and Sklenar, *Archaeology*, 22–25.

12. Schama, *Landscape and Memory*, 96.

13. For a wonderful introduction to the history and archaeology of this battle, as well as a review of the main Roman sources, see Wells, *Battle That Stopped Rome*, 15–55, 80–124.

14. Tacitus, *Annals*, 32. This information comes not from a report of the ambush itself but from a description of a military maneuver six years after the event. Germanicus and his soldiers made a detour to the site of the defeat to mourn the dead.

15. Wells, *Battle That Stopped Rome*, 35.

16. Tacitus, *Annals*, 82.

17. Pomian, "Franks and Gauls," 45–51. On the ancient Slavs, see Sklenar, *Archaeology*, 14.

18. Kelley, *Faces of History*, 106–18.

19. Goffart, *Narrators of Barbarian History*.

20. Schnapp, *Discovery of the Past*, 139–42.

21. Klindt-Jensen, *Scandinavian Archaeology*, 49–53.

22. Sklenar, *Archaeology*, 16. The legend about healthy chickens appears in Rhode and Rhode, *Cimbrisch-Hollsteinische Antiquitaeten-Remarques*, February 28, 1719, 66–68.

23. This explanation is related in Pickel, *Beschreibung*, 10–14.

24. Details about these royal collections appear in Sklenar, *Archaeology*, 28–38; Gummel, *Forschungsgeschichte*, 11–12; Goeßler, *Führer durch die K. Staatssammlung*, v; and Vocelka, *Rudolf II.*, 142–43. On the cabinets of curiosities, see Schlosser, *Die Kunst- und Wunderkammer*, 33; and Kenseth, "Age of the Marvelous."

25. Bergmann, "Kurfürst August"; and Gummel, *Forschungsgeschichte*, 11–12.

26. For Rudolf's court, see Sklenar, *Archaeology*, 28–32; and Vocelka, *Rudolf II.*, 142–43. Gummel discusses conditions in Württemberg in *Forschungsgeschichte*, 62.

27. Blackbourn, *Conquest of Nature*, 21–62.

28. Treuer, *Kurtze Beschreibung*, 1–4.

29. Ibid., dedication.

30. Ibid., 11–13.

31. Ibid., 11.

32. Ibid., 25.

33. Quoted in Gummel, *Forschungsgeschichte*, 34.

34. Quoted in Seger, "Maslographia," 2.

35. Rhode and Rhode, *Cimbrisch-Hollsteinische Antiquitaeten-Remarques*, January 31, 1719, 33.

36. Ibid., May 23, 1719, 168; and June 20, 1719, 200.

37. Gummel, *Forschungsgeschichte*, 45.

38. On Stukeley, see Piggott, *Ruins in a Landscape*, 101–32. Archaeologists today interpret Stonehenge as a ceremonial site, but they associate it with anonymous prehistoric peoples, not Stukeley's fanciful Druids. They note that the site was modified several times between 3000 and 1500 BCE. Hill, *Stonehenge*.

39. Franz Xaver von Wegele, "Becmann, Johann Christoph," in *Allgemeine deutsche Biographie*, 2:240–41.

40. The full title of this beautiful book indicates its wide-ranging nature: Bekmann, *Historische Beschreibung der Chur und Mark Brandenburg nach ihrem Ur-*

sprung, Einwohnern, Natürlichen Beschaffenheit, Gewässer, Landschaften, Stäten, Geistlichen Stiftern &c. Regenten, deren Staats- und Religions-Handlungen, Wapen, Siegel und Münzen, Wohlverdienten Geschlechtern Adelichen und Bürgerlichen Standes, Aufnehmen der Wissenschafften und Künste in derselben.

41. Ibid., 1:345.

42. Ibid., 347–55. These chambered tombs, scattered throughout northern Europe and the British Isles, are from the Neolithic age, dating perhaps from before 2500 BCE.

43. The Prussian academy's charter from July 11, 1700, is quoted in Ecker, "Recht und Rechtsgeschichte," 10.

44. Braun, *Die Anfänge*, 16.

45. "Gesetze der churbayerischen Akademie [1759]," 21–22.

46. Dannheimer, "90 Jahre Prähistorische Staatssammlung," 2–3.

47. Aleida Assmann describes a similar shift from history as legacy to history as remembrance in her study of English literature during the seventeenth century. As legacy, history sang the praises of the ruler in order to preserve his glory. Remembrance, on the other hand, was more inclusive. It went beyond the achievements of the ruling house to explain a territory's origins. Assmann, *Erinnerungsräume*, 78.

48. Vallibus, "Versuch eines Entwurfs," 1379.

49. Mosse, *Crisis of German Ideology*; and Krebs, *A Most Dangerous Book*, 182–213.

50. Thucydides, *Peloponnesian War*, 2–13.

51. Smail, "Sacred History."

52. Todd, *Early Germans*, 44–46.

53. Caesar, *Gallic War*, 53, 63.

54. Rhode and Rhode, *Cimbrisch-Hollsteinische Antiquitaeten-Remarques*, January 24, 1719, 26.

55. Ibid., November 14, 1719, 364–67; and November 21, 1719, 370. Patriotic expressions were not limited to descriptions of Cimbri by German scholars. In France, the comte de Caylus was proud that his large comparative work included domestic antiquities, saying, "I think that I can boast of having provided more details on the antiquity of the Gauls than anyone else. I have done my country this small service." Caylus is quoted in Pomian, "Franks and Gauls," 36.

56. Pickel, *Beschreibung*, 6.

57. Tacitus, *Germania*, 81; and Pickel, *Beschreibung*, 3–5.

58. Pickel, *Beschreibung*, 14–16.

59. "Ueber die Gräber der alten Teutschen," 122.

60. This is similar to the phenomenon that Daniel Segal describes as making history "surveyable." It creates narratives that unify ancient and contemporary peoples. Segal, "'Western Civ,'" 777–78.

61. Rhode and Rhode, *Cimbrisch-Hollsteinische Antiquitaeten-Remarques*, July 18, 1719, 227.

62. Kortüm, *Beschreibung*, 3–7.

63. Schuchhardt, *Vorgeschichte von Deutschland*, 313–17.

64. Kortüm, *Beschreibung*, 13.

65. Comte de Caylus, *Recueil d'antiquités égyptiennes, étrusques, grecques et romaines*, vol. 1 (Paris, 1752), iii–iv; quoted in Schnapp, *Discovery of the Past*, 36–37.

66. Ledebur, *Das Königliche Museum*, v.

67. For the activities in Bavaria, see Dannheimer, "90 Jahre Prähistorische Staatssammlung," 2–3. On Württemberg, see Goeßler, *Führer durch die K. Staatssammlung*, v–vi.

68. Gummel, *Forschungsgeschichte*, 57–58.

69. Ledebur, *Das Königliche Museum*, v–vi.

70. Pickel, *Beschreibung*, 28.

71. Kortüm had read Bekmann and was especially encouraged by works on Slavic history.

72. "Ueber die Gräber der alten Teutschen."

73. Hummel, *Bibliothek der deutschen Alterthümer*, iii. Kortüm expressed a similar frustration with his inability to examine many objects from around Neubrandenburg. He was certain that his conclusions about Rhetra would be confirmed if scholars could see all the evidence. Kortüm, *Beschreibung*, 44.

74. Hummel, *Beschreibung entdeckter Alterthümer*, viii–ix.

Chapter Two

1. Ledebur, *Blicke auf die Literatur*, 37.

2. Kruse, Vorrede, v.

3. Gooch, *History and Historians*, 40–55.

4. For examples of these activities, see Agnew, *Origins*; and Hitchins, *Rumanian National Movement*. For archaeology specifically, see Smith, "Authenticity, Antiquity and Archaeology."

5. Iggers, *German Conception*, 34–43; and Leerssen, *National Thought*, 107–14.

6. Heimpel, "Geschichtsvereine," 48–49.

7. Green, *Fatherlands*, 18. I will translate *vaterländisch* as "regional" in the names of nineteenth-century associations and museums to emphasize this important point.

8. Bann, *Romanticism*, 63. On the revolutionary upheaval, see Fritzsche, *Stranded in the Present*, 1–18, 92–131.

9. Schnabel, "Ursprung," 24–25.

10. Marchand, *Down from Olympus*, 152–87.

11. The quoted phrase appears in Crane, "Collecting and Historical Consciousness," 3. On the creation of *Denkmäler*, see Crane, *Collecting*, 29–44.

12. Schnabel, "Ursprung," 12–15.

13. Buchner, *Reise*, 1:39.

14. Huscher, "Beschreibung," 6.

15. Ibid.

16. Ibid., 53.

17. Jensen, *Friedrich*, 29–54. Plates of these paintings are included in Börsch-Supan, *Friedrich*.

18. Levezow, *Andeutungen*, 3.

19. Vorwort to *Baltische Studien* 1 (1832): iii–iv.

20. Hofmann used a small knife, similar to one used to harvest asparagus, for this delicate task. Hofmann also cautioned his workers not to throw out anything that appeared unusual. These items were placed in a basket so that Hofmann could later review them. Only the expert could determine what was valuable. Dorow, *Die Kunst Alterthümer aufzugraben*, 9–12.

21. Levezow, *Andeutungen*, 5–7.

22. Brühl, "Instruction," 12.

23. Nipperdey, "Verein als soziale Struktur," 3. A wide array of voluntary associations were founded in the nineteenth century. The groups that are the focus of my study carried names that indicated their devotion to history and antiquities: *historischer Verein, Altertumsverein, Verein für Altertumskunde, Geschichtsverein*, etc.

24. Ranke, *Die akademische Kommission*, 14. The orders that established the district associations are reprinted in Stetter, "Entwicklung," 78–90.

25. Vorwort to *Oberbayerisches Archiv für vaterländische Geschichte* 1 (1839): iii–iv. Hereafter abbreviated as *OA*.

26. Stetter, "Entwicklung," 27, 33.

27. Eichler, "Museum vaterländischer Alterthümer."

28. The association's financial report for 1844–45 appears in Dürrich and Menzel, *Die Heidengräber am Lupfen*, 27.

29. Gerson, *Pride of Place*, 34. On England, see Piggott, *Ruins in a Landscape*, 171–95.

30. Gerson, *Pride of Place*, 29–30.

31. Levine, *Amateur*, 51, 71.

32. *Die schlesischen Provinzial-Blätter*, Bd. 1, "1827–1869," Geheimes Staatsarchiv, Preußischer Kulturbesitz (hereafter GStA PK), I. HA Rep. 76Vc Kultusministerium, Sekt. 14, Tit. 23, Nr. 4.

33. On the difficult relationships between Dorow, the Prussian administration, and the university in Bonn, see Fuchs, "Zur Geschichte der Sammlungen," 40–78.

34. Applegate, *Nation of Provincials*, 23–27. On Bamberg and Bayreuth, see Kunz, *Verortete Geschichte*.

35. Van Riper, *Men among the Mammoths*, 29.

36. Dorow, *Die Kunst Alterthümer aufzugraben*, 10.

37. Marchand, *Down from Olympus*, 152–87.

38. Ranke, "Character of Historical Science," 45. The distinction between anti-

quarian efforts and scientific history was not limited to Berlin. Oxford historians derided the efforts of English county associations, and novelists turned the antiquarian obsessed with narrow details into a laughable stock figure in British literature. Levine, *Amateur*, 28–29, 71–74.

39. Ranke, *Universal History*, ix.

40. Smith, *Gender of History*, 119. Smith (70) shows how the distinction between "amateur" and "scientific" practices also marginalized history written and read by women.

41. Ziolkowski, *Clio*, 26–32 (quotation on 29).

42. Humboldt, "Historian's Task," 5–6.

43. Novick, *Noble Dream*, 26–30.

44. Iggers and Moltke, introduction, xix–xx.

45. Quoted in White, *Metahistory*, 165.

46. Iggers explains that Ranke found "a larger context" in his belief in a divine order and that his histories explored the modern state as the manifestation of this order in the world. Iggers, *German Conception*, 76–77. See also Hayden White's comments on the impact of these assumptions on narrative forms in *Metahistory*, 7–11.

47. Ledebur, *Blicke auf die Literatur*, 37.

48. Levezow, *Andeutungen*, 6–7.

49. Dorow, *Die Kunst Alterthümer aufzugraben*, 10.

50. Levezow, *Andeutungen*, 6–7.

51. Crane, *Collecting*, 38–44.

52. "Der Thüringisch-Sächsische Verein für Erforschung des vaterländischen Alterthums: Erklärung über den Zweck und Umfang seines Strebens," GStA PK, I. HA Rep. 76Ve Kultusministerium, Sekt. 9, Abt. VI, Nr. 8, Bd. 1, "Der Thüringisch-Sächsische Alterthumsverein in Halle, 1819–1828."

53. Lisch, *Friderico-Francisceum*.

54. Vorwort to *Baltische Studien* 1 (1832): iii–iv.

55. Examples of this practice include excavations near Ulm in 1848. See Thrän, "Bericht über die Ausgrabung"; and Wehrberger, "Ausgrabungen," 62–63.

56. Williamson, *Longing*, 107.

57. In German, these examples are *Burgwall, Brautstein, Heidenkeller, Teufelskeller, Hünengrab, Opferstein*, and *Wendenkirchhof*. Brühl, "Instruction," 12.

58. Similarly, Romantic-era philologists hoped to determine the historical provenance of Nordic sources like the Edda saga and the *Nibelungenlied* but still preserve the divine power of these sources. Williamson, *Longing*, 72–120.

59. "Verzeichniß der bisher bekannt gewordenen Grabhügel in Oberbayern," *OA* 1 (1839): 119–28; "Verzeichniß antiquarischer Funde aus den königlichen Landgerichtsbezirken Burghausen, Laufen und Titmanning," *OA* 1 (1839): 176–205; and "Erster Nachtrag zu dem Verzeichnisse der bisher bekannt gewordenen Grabhügel in Oberbayern," *OA* 1 (1839): 279–80.

60. Joseph von Hefner, "Ueber die Eröffnung germanischer Grabhügel bei St. Andrä," *OA* 1 (1839): 170–75.

61. Ibid., 172.

62. In the 1760s, the Scottish poet James Macpherson "invented" the bard Ossian and popularized this supposed source of ancient Gaelic culture. Ibid., 170–75.

63. Crane, *Collecting*, 49–53; and Krins, "Gründung."

64. The results, which covered fourteen districts around Potsdam, were presented in 1852 in Ledebur, *Die heidnischen Alterthümer*.

65. Wächter, *Statistik*; and "Ausschreiben des Königlichen Ministeriums des Innern [Hannover], die Schonung der Denkmale der Vorzeit betrf," November 4, 1844, and "An die Kgl. Eisenbahn-Direktion," October 8, 1851, GStA PK, I. HA Rep. 76Ve Kultusministerium, Sekt. 11, Abt. VI, Nr. 3, Bd. I, "Die Vorkehrungen zur Ehrhaltung der Denkmäler und Ueberreste der vaterländischen Kunst."

66. Ledebur, Vorwort to *Die heidnischen Alterthümer*, n.p.; and Wächter, *Statistik*, 6–8.

67. *Archiv des historischen Vereins für den Untermainkreis* 1 (1833): 65.

68. *Dritter Jahresbericht des historischen Vereins von und für Oberbayern*, 1840, 79–80.

69. Ledebur, *Das Königliche Museum*, v–vi. The Prussian Cultural Ministry also referred to this collection as the Sammlung Slavo-Germanischer Alterthümer and the Nordische Sammlung. Most common before the 1850s, though, was the name Königliches Museum vaterländischer Alterthümer, which Ledebur used in the title of his catalog.

70. Sheehan, *Museums*, 51.

71. Klindt-Jensen, *Scandinavian Archaeology*, 49–53.

72. Thomsen's guidebook appeared in German in 1837 and in English in 1848. Trigger, *Archaeological Thought*, 73–80.

73. Lisch, *Friderico-Francisceum*. On the correspondence with Danish scholars, see Bertram, "Vom 'Museum vaterländischer Alterthümer,'" 49–52.

74. Nehls, "Einbruch," 216.

75. GStA PK, I. HA Rep. 76Ve Kultusministerium, Sekt. 15, Abt. XI, Nr. 3, "Die Sammlung Slavo-Germanischer Alterthümer im Schloß Monbijou, 1831–1893." See also Bertram, "Vom 'Museum vaterländischer Alterthümer,'" 37–43.

76. Nehls, "Einbruch," 220; and Bernau, "Chronologie," 9–10.

77. The 1842 museum guide was Weyl, *Sammlung*. On the robbery, see Nehls, "Einbruch," 222–26.

78. Ledebur, *Das Königliche Museum*, 3.

79. Ibid., vii–viii.

80. Ibid., ix.

81. Ibid., 1–3.

82. Ibid., 3, 60.

83. Ibid., 162. Ledebur did note that new associations had recently been

founded in Paderborn and Wetzlar to complement the work of a group in Münster and that this offered a start to addressing this situation.

84. Hoppe, "Gesamtverein," 1–8.

85. Braun, "Limesforschung in Bayern," 60.

86. On Aufseß's historical interests, see the wonderful coverage in Crane, *Collecting*, 69–72.

87. Hans von Aufseß, "Sendschreiben an die erste allgemeine Versammlung deutscher Rechtsgelehrten, Geschichts- und Sprachforscher zu Frankfurt am Main"; quoted in Burian, "Das Germanische Nationalmuseum," 148–49.

88. Burian, "Das Germanische Nationalmuseum," 132–35.

89. Menghin, "Sammlungs- und Forschungsgeschichte."

90. Böhner, "Das Römisch-Germanische Zentralmuseum," 21–22.

91. Ibid., 29–31.

Chapter Three

1. Menzel, *Geschichte der Deutschen*, 1068.

2. Ibid., 43–44.

3. Ibid., 7–8.

4. Ibid., 1068.

5. Dürrich and Menzel, *Die Heidengräber am Lupfen*. My thanks to George Williamson for directing me to Menzel's archaeological activity.

6. Pohl, *Die Germanen*, 86–87.

7. Kaufhold, "Friderizianische Agrar-, Siedlungs- und Bauernpolitik," 191–99.

8. Schlette, "Büsching"; and Gerber, *Gesellschaft*, 59–63.

9. Kruse, *Budorgis*, v.

10. Büsching, *Abriß*, 10–11. His comments about the distribution of various types of artifacts appear on 21–36.

11. Ibid., 5–11.

12. Kruse, *Budorgis*, 34.

13. Ibid., 56–163.

14. Ibid., 145–48.

15. Ibid., 147–48.

16. Tacitus described the Lugii in *Germania*, 137. Kruse's ethnographic speculations appear in *Budorgis*, 34, 56, 102–3.

17. Kruse, *Budorgis*, xi.

18. References to this local correspondence appear in ibid., 59, 60, 64, 69, 73, 94, 98, 100, 103, 105.

19. Kruse, Einleitung, 5.

20. *Meyers Konversations-Lexikon*, 4th ed. (1885–92), s.v. "Schlesien"; and *Brockhaus: Handbuch des Wissens in vier Bänden*, 1928 ed., s.v. "Schlesien."

21. Hagen, *Germans*, 118–36.

22. Rączkowski, "'Drang nach Westen'?"

23. Wahle, "Vorgeschichtsforschung," 705.

24. Koch-Sternfeld, "Beinfeld," 181.

25. Bayerisches Hauptstaatsarchiv, Munich, MK 14527, "Bey Fridolfing (Lg. Titmaning) ausgegrabene alterthümliche Gegenstände betr.," p. 3.

26. Koch-Sternfeld, *Fürsten-, Volks- und Culturgeschichte.*

27. Koch, "Aufklärung."

28. Wiesend, "Archäologische Funde," 28.

29. Weber, "Mitteilungen."

30. Bayerisches Hauptstaatsarchiv, Munich, MK 14527, "Bey Fridolfing (Lg. Titmaning) ausgegrabene alterthümliche Gegenstände betr."

31. Putzer, "Staatlichkeit."

32. "Neue Kreiseinteilung und Benennung der Kreise," reprinted in Stetter, "Entwicklung," 89–90.

33. Schreiber, *Taschenbuch*, 1:135–36.

34. Sklenar, *Archaeology*, 93–95; and Böhner, "Das Römisch-Germanische Zentralmuseum," 11.

35. Koch, "Aufklärung," 85 (emphasis in the original).

36. Lindenschmit, *Räthsel*, 13 (emphasis in the original).

37. Lindenschmit and Lindenschmit, *Todtenlager*, 30.

38. Ibid.

39. Ibid., 32.

40. Wiesend, "Archäologische Funde," 27.

41. Ibid., 14–28.

42. Koch, "Aufklärung," 112.

43. Föringer, "Matthias Koch," 163–64.

44. Dürrich and Menzel, *Die Heidengräber am Lupfen*, 3–7.

45. Ibid., 4.

46. Ibid., 3–7.

47. Ibid., 23.

48. Ibid., 5.

49. Ibid., 23.

50. Ibid., 23–26 (quotation on 23).

51. Construction had stopped when the area converted to Protestantism in 1531. Today, the 530-foot steeple (the tallest of its kind in Germany) extends above rows of red, tiled roofs, making a beautiful impression as one approaches the city. Greiner, "Verein," 116–52.

52. Thrän, "Bericht über die Ausgrabung." See also Wehrberger, "Ausgrabungen," 62–63.

53. Schenk, "Haßler"; and Karl Gustav Veesenmeyer, "Haßler, Konrad Dietrich," in *Allgemeine deutsche Biographie*, 11:15–20.

54. Haßler, *Todtenfeld*, 3–4; and Wehrberger, "Ausgrabungen," 68.

55. Haßler, *Todtenfeld*, 4.

56. Ibid., 32.

57. Ibid., 10.

58. Ibid., 1.

59. Ibid., 11–23, 36.

60. Ibid., 1.

61. Sir. 46:23 (NAB). The full story of Samuel's return from the dead appears in 1 Sam. 29:8–20. Haßler, *Todtenfeld*, 1.

62. Weyl, *Sammlung*, 3–4.

63. Sheehan, *Museums*, 84–85, 132–34.

64. Bayerisches Hauptstaatsarchiv, Munich, MInn 45405, "Die Verschmelzung der ältern und neuern Gebietstheile des Königreiches."

65. Smith, *Politics*, 36–39.

66. Stein, "Riehl," 490. See also Smith, *Politics*, 40–44.

67. Henry Simonsfeld, "Riehl, Wilhelm Heinrich," in *Allgemeine deutsche Biographie*, 53:362–83.

68. Riehl, "Volkskunde," 213–14.

69. Aretin, Vorwort, iv. On the founding of the museum, see Bauer, "Maximilian II." The building housed the BNM from 1867 to 1894, when a new building for it was planned on Prinzregentenstraße. Today, the original BNM building houses the Five Continents Museum, Munich's ethnology museum.

70. Aretin, Vorwort, vi.

71. Bauer, "Maximilian II.," 22.

72. *Das Bayerische Nationalmuseum*, 9–10.

73. Koeglmayr, *Mosaikboden*, 12. Several nineteenth-century sources, including Koeglmayr's, refer to the site as Westenhofen. The village is known as Westerhofen today. I thank Dr. Rupert Gebhard (Archäologische Staatssammlung, Munich) for supplying me with Koeglmayr's text.

74. *Das Bayerische Nationalmuseum*, 3–9.

75. Ibid., 11–12.

76. Ibid., 9–14.

77. Spruner, *Wandbilder*, iv.

78. Ibid., 1–2.

79. Ibid., 7.

80. Toews, *Becoming Historical*, 200–204.

81. *Führer durch die Königlichen Museen*, 6–7.

82. Stüler, *Das Neue Museum*, 4.

83. Bertram, "Vom 'Museum vaterländischer Alterthümer,'" 54.

84. Stüler, *Das Neue Museum*, 4.

85. Ibid., 2. This connection is also emphasized in *Führer durch die Königlichen Museen*, 128–29.

86. Schasler, *Die Königlichen Museen*, 119–22; and Löwe, *Das Neue Museum*, 23–29.

87. *Führer durch die Königlichen Museen*, 127–35.

Part Two

1. Arminius was often referred to by the "Germanized" name Hermann. Benario, "Arminius." The Hermannsdenkmal was begun in the 1840s as an expression of German nationalism, but it was not completed until after the establishment of the German Empire. Dörner, "Mythos."

2. Nipperdey, *Deutsche Geschichte*, 2:262.

Chapter Four

1. "Die sechste Allgemeine Versammlung," 77 (emphasis in the original). Subsequent references to the proceedings of the 1875 meeting will be given in parentheses in the text.

2. Lubbock, *Pre-historic Times*, viii.

3. James Ussher (1581–1656), an archbishop from Armagh, Ireland, placed the beginning of the world at 4004 BCE. This chronology was generally accepted in European learned circles. Fagan, *Eyewitness*, 12.

4. Smail, "Sacred History," 1338.

5. Donald Grayson provides examples from across Europe of attempts to maintain the distinction between an ancient natural world and a more recent human history. Grayson, *Establishment of Human Antiquity*, 27–42, 97–98.

6. Schnapp, *Discovery of the Past*, 311–12.

7. Grayson, *Establishment of Human Antiquity*, 195.

8. *Verhandlungen des naturhistorischen Vereines der preussischen Rheinlande und Westphalens* 16 (1859): 153; quoted in Leverkus, "Geschichte." See also Grayson, *Establishment of Human Antiquity*, 212; and Van Riper, *Men among the Mammoths*, 134.

9. Keller, "Vorbemerkung" to *Pfahlbauten*, n.p. See also Speck, "Pfahlbauten."

10. Lisch, *Pfahlbauten*; and Virchow, "Pfahlbauten." *Pfahlbaufieber* is the title of an exhibit in the archaeology museum at the Eberhard-Karls University in Tübingen, Germany. The room details the enthusiastic search for pile dwellings in Württemberg during the 1860s.

11. Coles and Coles, *People of the Wetlands*, 51–54.

12. Ecker, "Berechtigung," 4 (emphasis in the original).

13. Of Virchow's 1,180 anthropological publications, 518 were related to archaeology. Ackerknecht, *Virchow*, 219–20.

14. Quotation from Andree, *Virchow*, 49–50. See also Virchow, "Zur Geschichte von Schivelbein"; and Virchow, "Schivelbeiner Alterthümer." On the fake idols, see Virchow, "Die sogenannten Idole von Prillwitz."

15. Virchow, *Briefe*, 47.

16. The conversation with the American colleague is quoted in Ackerknecht, *Virchow*, 207–8.

17. Ibid., 89–98.

18. McNeely, *"Medicine on a Grand Scale."*

19. Virchow, "Hünengräber und Pfahlbauten," 5–6. Virchow's lectures were published as the first volume of the *Sammlung gemeinverständlicher wissenschaftlicher Vorträge* (Collection of scientific lectures for a general audience). Virchow was very attached to this project as a way to spread scientific knowledge, and he served as the coeditor of this series until 1901, the year before his death.

20. Virchow, "Hünengräber und Pfahlbauten," 32.

21. Ibid., 8–15 (quotations on 8 and 12).

22. Ibid., 12.

23. Ibid., 17.

24. Andree, "Geschichte."

25. Ecker, "Berechtigung," 4. For excellent coverage of the DAG's pursuit of anthropology and ethnology, see Zimmerman, *Anthropology and Antihumanism*; and Penny, *Objects of Culture*.

26. "Protocoll über die erste Sitzung."

27. On these terms, see Fetten, "Archaeology and Anthropology," 162–66. Members of the DAG mostly used the terms *Urgeschichte* or *Prähistorie* to refer to their interest in the early development of humankind.

28. The association shortened its name to the Historischer Verein von Oberbayern around 1873.

29. GStA PK, I. HA Rep. 76Vc, Kultusministerium, Sekt. 1, Tit. 11, Teil I, Nr. 4, Bd. I, "Die Gesellschaft für Anthropologie, Ethnologie und Urgeschichte," p. 230.

30. GStA PK, I. HA Rep. 76Vc, Kultusministerium, Sekt. 1, Tit. 11, Teil I, Nr. 4, Bd. II, "Die Gesellschaft für Anthropologie, Ethnologie und Urgeschichte," pp. 14–18. The DAG anticipated additional income from entrance fees charged for the exhibit, set at three marks for conference participants and fifty pfennig for the public.

31. GStA PK, I. HA Rep. 76Vc, Kultusministerium, Sekt. 1, Tit. 11, Teil I, Nr. 4, Bd. I, "Die Gesellschaft für Anthropologie, Ethnologie und Urgeschichte," p. 243.

32. "Verhandlungen der XI. allgemeinen Versammlung," 3. Subsequent references to the proceedings of the 1880 meeting will be given in parentheses in the text.

33. "Einladung zur Beschickung der Ausstellung anthropologischer und vorgeschichtlicher Funde Deutschlands, welche in Verbindung mit der XI. allgemeinen Versammlung der deutschen anthropologischen Gesellschaft im August 1880 in

Berlin stattfinden wird," GStA PK, I. HA Rep. 76Vc, Kultusministerium, Sekt. 1,
Tit. 11, Teil I, Nr. 4, Bd. I, "Die Gesellschaft für Anthropologie, Ethnologie und
Urgeschichte." The commission also appealed to associations by explaining that
other countries were far ahead in creating national collections.

34. Ibid.

35. Ibid.

36. Zimmerman, *Anthropology and Antihumanism*, 131–34.

37. "A. Voss' neues Prachtwerk."

38. Ibid.

39. *Neue preußische Zeitung*, August 7, 1880. This paper is also known as the
Kreuzzeitung.

40. *Berliner Tageblatt*, July 25, 1880 (emphasis in the original).

41. *National-Zeitung* (Berlin), Abend-Ausgabe, August 5, 1880.

42. Massin, "From Virchow to Fischer," 86–94; and Evans, *Anthropology at War*,
57–80.

43. Zimmerman, *Anthropology and Antihumanism*, 38–44 (quotations on 38
and 39).

44. Ibid., 48.

45. Ibid., 146.

46. Hannaford, *Race*, 203–4.

47. Rudolf Virchow famously resisted the interpretation of the Neanderthal
skull as protohuman, arguing instead that its shape appeared to be the result of
a deformity caused by disease. Historians have viewed Virchow's position as evidence of his hostility toward Darwinian evolution. Andree argues, though, that
Virchow's comments follow a pattern of cautious judgments. Virchow refrained
from comment until he had seen the skull, and he did not rule out the possibility
that the skull was ancient. He simply stated that there was not enough evidence to
prove such a claim. Andree, *Virchow*, 51–52, 151–64.

48. Rowley-Conwy, *From Genesis to Prehistory*, 60–65.

49. Zimmerman, *Anthropology and Antihumanism*, 86–107.

50. Quoted in Snyder, *Race*, 117.

51. Ackerknecht, *Virchow*, 208–11.

52. The results of the *Schulstatistik* appeared in *Archiv für Anthropologie* 16
(1886): 275–476.

53. Zimmerman, "Anti-Semitism as Skill," 415, 428 (emphasis in the original),
409.

Chapter Five

1. "Eine hervorragend nationale Wissenschaft" was the subtitle to Kossinna's
best-known book, *Die deutsche Vorgeschichte*. He elaborated prehistory's national
significance on p. v.

2. Applegate, *Nation of Provincials*, 65–68. For a listing of the historical associations created during this era, see Hoppe and Lüdtke, *Die deutschen Kommissionen*, 239–54.

3. Penny, *Objects of Culture*, 39–49.

4. Pinder, *Die Aufgaben der Provinzialmuseen*, 9–10, 19–20.

5. The kingdom was first arranged into smaller districts (*Kreise*), but this gave way to the provincial system. Hahn, *Provinzial-Ordnung*, 195.

6. Gärtner, "Begründer," 80–83.

7. Voß, "Die ethnologische und nordische Sammlung," 160.

8. GStA PK, I. HA Rep. 76Ve Kultusministerium, Sekt. 15, Abt. XI, Nr. 3, "Die Sammlung Slavo-Germanischer Alterthümer im Schloß Monbijou, 1831–1893."

9. The activity in Lower Saxony is documented in GStA PK, I. HA Rep. 76Ve Kultusministerium, Sekt. 11, Abt. VI, Nr. 3, Bd. VII, "Die Vorkehrungen zur Erhaltung der Denkmäler und Überreste der vaterländischen Kunst, 1887–1890 (Hannover)." For Münster, see GStA PK, I. HA Rep. 76Ve Kultusministerium, Sekt. 15, Abt. XI, Nr. 3, "Die Sammlung Slavo-Germanischer Alterthümer im Schloß Monbijou, 1831–1893."

10. *Führer durch das Museum für Völkerkunde*, 10–47.

11. Penny, *Objects of Culture*, 187–97.

12. *Vossische Zeitung*, July 6, 1900; quoted in Gärtner, "Begründer," 85.

13. The inability to do more with the collection marked the scholarly career of Albert Voß. An otherwise-kindhearted obituary noted that Voß published very little despite his position as the overseer of Germany's largest collection for pre- and early history. This criticism is perhaps indicative of what was possible. Voß was constantly trying to keep up with the growing collection, and it would be more accurate to say that he published a lot in the way of inventories, guidebooks, and manuals. But all these texts were attempts to organize artifacts, not to interpret them. A narrative about European or German prehistory was not yet possible. "Zur Erinnerung an Albert Voß."

14. "Ausgrabungen der Denkmäler der Vorzeit."

15. "Instruction der Commission für die Rheinischen Provinzial-Museen zu Bonn und Trier" (1875); quoted in *Rheinisches Landesmuseum Bonn*, iii–iv.

16. *Bonner Zeitung*, August 2, 1886; quoted in Fuchs, "Zur Geschichte der Sammlungen," 119.

17. Lehner, *Führer durch das Provinzialmuseum zu Bonn*, 18–20.

18. Ibid., 4.

19. Ibid., 35–36.

20. Hettner, *Illustrierter Führer*, 115–35. In addition to the museums in Bonn and Trier, the city government in Düsseldorf created a historical museum that included an archaeological collection, and the Cologne branch of the DAG founded a museum specifically for pre- and early history in 1907.

21. Müller included a notice each year in the association's report about items acquired for the Museum für Kunst und Wissenschaft. The association's outreach

efforts are discussed in the 37. *Nachricht über den historischen Verein für Niedersachsen*, 1875, 11–12. Müller also worked on a major inventory of domestic antiquities that was published after his death as Müller, *Vor- und frühgeschichtliche Alterthümer.*

22. Katenhusen, "150 Jahre Niedersächsisches Landesmuseum Hannover," 72.

23. Kossinna, "Eine archäologische Reise," 25.

24. *Jahrbuch des Provinzial-Museums zu Hannover*, 1901–4, 3–4. The Historischer Verein für Niedersachsen maintained a list of sites they could not excavate and hoped future researchers would be able to explore them. This annual report also included a floor plan of the new museum.

25. *Jahrbuch des Provinzial-Museums zu Hannover*, 1905–6, 1. Reimers suggested the creation of a separate museum for natural history as the best solution to the space issues.

26. Buchholz and Pniower, *Das Märkische Provinzial-Museum*, 6–15.

27. In these years, four-fifths of the museum's acquisitions were gifts and only one-fifth were purchases. Ibid., 6–7.

28. Ibid., 19.

29. As historian David Blackbourn has explained, such administrative titles "had nothing to do with pseudoaristocratic status, rather [they] placed the state seal of approval on their business activities and their position in the bourgeois social order." Blackbourn, *Long Nineteenth Century*, 367–68.

30. Mielke, "Das Märkische Museum." The meetings and outings are described in the society's journal. See, e.g., "Besichtigung der Brauerei."

31. Friedel, *Führer durch das Köllnische Rathaus*. Buchholz and Pniower's *Das Märkische Provinzial-Museum* includes several photographs of the exhibit. The earlier museum on Klosterstraße (1876–80) had less display space but a similar arrangement of the material. One main difference was that in the earlier museum the order of the rooms began with the prehistory section and moved forward in time. Friedel, *Eintheilungs-Plan* (1879).

32. Friedel, "Archäologische Streifzüge."

33. "Die Exkursion nach dem Spreewalde," *Berliner Tageblatt*, August 9, 1880. The outing to the Römerschanze was described in "Congreß deutscher Anthropologen zu Berlin," *Neue preußische Zeitung*, August 14, 1880.

34. Mielke, "Das Märkische Museum," 7–9. Mielke was a regular participant in these outings from 1895 to 1914.

35. Friedel, "Der Bronzefund," 37.

36. The "Urfidele Buddellied" lyrics are reprinted in Michel, "'In Berlin höchst wunderbar,'" 157. Brandenburgia members also enjoyed the "Historisches Buddler-Lied," reprinted in "Bericht über die Wanderfahrt nach Brandenburg, 1892/93," *Brandenburgia* 1 (1892–93): 74.

37. Michel, "'Und nun kommen Sie auch gleich noch mit 'ner Urne.'"

38. Friedel, "Die Funde aus dem Königsgrab von Seddin," 33–38; and "Hünengrab von Perleberg," *Die Gartenlaube*, no. 42 (1899).

39. Hare, *Excavating Nations*, 72–82.

40. Eichler, "Museum vaterländischer Alterthümer," 91–113.

41. Rüster, "Geschichte des Museums," 72–77.

42. Goeßler, "Die K. Altertümersammlung," 3–10 (quotation on 3).

43. Strobel, "Paulus," 127–28.

44. Gummel, *Forschungsgeschichte*, 234–41.

45. Goeßler, *Führer durch die K. Staatssammlung*, 1–63.

46. *Führer durch das königlich Bayerische Nationalmuseum*, 46. To reconstruct the changes in the pre- and early history section of the BNM described below, I consulted editions of the museum guidebook from 1868, 1881, 1882, 1887, 1892, and 1894. See also Graf, "Aus der Geschichte des Museums," 7–28.

47. Ranke, *Die akademische Kommission*, 26.

48. Ranke, "Unsere Ziele," iv–v.

49. Ranke, *Die akademische Kommission*, 38–41.

50. Hager and Mayer, *Die vorgeschichtlichen, römischen und merowingischen Alterthümer*.

51. *Führer durch das Bayerische Nationalmuseum* (1894), 14.

52. *Führer durch das Bayerische Nationalmuseum* (1908), 22. The building on Prinzregentenstraße is still the home of the BNM today.

53. On the map, see Beltz, "Erläuterung der Karten zur Vorgeschichte"; and Stopfel, "Geschichte der badischen Denkmalpflege."

54. Krause's notice appeared in *Verhandlungen der Berliner Gesellschaft für Anthropologie, Ethnologie und Urgeschichte*, November 18, 1899, 656–57, and was reprinted in *Brandenburgia* 9 (1900–1901): 76–77.

55. "Erhaltung der Funde an Münzen.

56. Runde, "Nachrichten über vor- und frühgeschichtliche Altertumsfunde."

57. Mestorf, *Die vaterländischen Alterthümer*, 4–5, 27–29. Other museums also offered guidance to those who embarked on excavations. The guidebook for the Märkisches Provinzial-Museum included an appendix with Karl von Brühl's excavation instructions from 1835.

58. Voß, *Merkbuch, Alterthümer aufzugraben und aufzubewahren*.

59. Voß, "Die ethnologische und nordische Sammlung," 160.

60. Review of *Anleitung zu anthropologisch-vorgeschichtlichen Beobachtungen im Gebiete der deutschen und österreichischen Alpen*, by Johannes Ranke, *Literarisches Centralblatt für Deutschland*, December 3, 1881, 1673–74.

61. "Mittheilungen aus den Lokalvereinen," 7.

62. On the images from the province of Saxony, see Rüster, "Geschichte des Museums," 79–80. Plate 17 of the article depicts the poster. On the West Prussian images, see "Vorgeschichtliche Wandtafeln für Westpreussen," *Naturwissenschaftliche Wochenschrift*, August 6, 1899, 376–79.

63. Examples include *Illustrirte Zeitung* 57 (1871), issues 1461 and 1463.

64. References to Der Sammler in the *Augsburger Abendzeitung* appear throughout Hager and Mayer, *Die vorgeschichtlichen, römischen und merowing-*

ischen Alterthümer. On the *Pfälzisches Museum*, see Applegate, *Nation of Provincials*, 92.

65. Hoppe and Lüdtke, *Die deutschen Kommissionen*, 203–4.

66. Mestorf, *Die vaterländischen Alterthümer*, 4–5.

67. Fritz, *Museum für Kunst und Kulturgeschichte*, 7–15. On school collections in Germany more broadly, see Mötefindt, "Verzeichnis der Sammlungen."

68. Conwentz, *Die Heimatkunde in der Schule*, 55–56.

69. Drück, "Die vaterländische Altertumskunde," 5–6.

70. Königliches Generalconservatorium der Kunstdenkmale und Alterthümer Bayerns (Royal conservatorium for Bavaria's art and antiquities) to Königliches Staatsministerium des Innern f. Kirchen- und Schulangelegenheiten (Royal state interior ministry for religious and school affairs), January 11, 1890, in Bayerisches Hauptstaatsarchiv, Munich, MK 11746, "Archäologische und prähistorische Karten und Tafeln."

71. Bayerisches Hauptstaatsarchiv, Munich, MK 11746, *Amtsblatt des königlichen Staatsministeriums des Innern. Königreich Bayern*, February 7, 1890.

72. Pinder, *Die Aufgaben der Provinzialmuseen*, 18–23.

73. The Prussian Cultural Ministry received lists from each of the Prussian provinces. Inventories from the province of Westphalia, for example, came from historical associations in Bielefeld, Essen, and Münster and from museums in Herford, Dortmund, and the Sauerland. Inventories from the province of Hesse-Nassau came from Ems, Hanau, Lahnstein, Wiesbaden, Kassel, Frankfurt am Main, and Herborn. GStA PK, I. HA Rep. 76Ve Kultusministerium, Sekt. 12, Abt. III, Nr. 5, "Altertums- und frühgeschichtliche Sammlungen in Westfalen, 1890–1912"; and GStA PK, I. HA Rep. 76Ve Kultusministerium, Sekt. 13, Abt. III, Nr. 22, "Altertums- und vorgeschichtliche Sammlungen in Hessen-Nassau, 1888–1912."

74. Telegrams from Dr. Heinemann to Ernst Friedel, September 16 and 19, 1899, Stiftung Stadtmuseum (Berlin), Märkisches Museum, Abteilung Ur- und Frühgeschichte (hereafter SSM MM), "Seddin. Zeddin. West-Prignitz."

75. Traugott von Jagow to the Märkisches Provinzial-Museum, 1912, and Albert Kiekebusch to Max Viereck, May 19, 1927, SSM MM, "Seddin. Zeddin. West-Prignitz."

76. *Der Ausschuss des histor. Vereins*, 1–10. Plans for Ranke's Museums-Verein were discussed in *Mittheilungen des Museums-Vereins für vorgeschichtliche Alterthümer Baierns* 2 (1885): 1–7.

77. Dannheimer, "90 Jahre Prähistorische Staatssammlung."

78. 131. Sitzung vom 7. Mai 1900, *Verhandlungen der Kammer der Abgeordneten des bayerischen Landtages* (1899–1900), Stenographischer Bericht, vol. 4, p. 377.

79. GStA PK, I. HA Rep. 76Va Kultusministerium, Sekt. 9, Tit. X, Nr. 10, Bd. III, "Das Schleswig-Holsteinische Museum vorgeschichtlicher Altertümer bei der Uni Kiel, 1892–1905."

80. *Jahrbuch des Provinzial-Museums zu Hannover*, 1901–4, 4.

81. Hahne, "Zur Ausgestaltung der vorgeschichtlichen Sammlung."

82. Wells, *Battle That Stopped Rome*, 43–55.

83. Mommsen, *Die Örtlichkeit*, vii.

84. GStA PK, I. HA Rep. 76Vc Kultusministerium, Sekt. 1, Tit. 11, Teil V, Bd. 41, "Forschungen über die Örtlichkeit der Varusschlacht, April 1901–Juli 1916."

85. Ibid.

86. Ibid.

87. Hartmann, "Das Lager bei Erle," *Westfälischer Merkur*, June 23, 1907.

88. Friedrich Koepp in the *Osnabrücker Zeitung*, October 12, 1907.

Chapter Six

1. Applegate, "Mediated Nation."

2. Stein, "Riehl," 491.

3. Quoted in Tatlock, "Realist Historiography," 61.

4. Stein, "Riehl," 502.

5. Kipper, *Germanenmythos*, 84.

6. In the 1950s, Theodor Heuss, the first president of the Federal Republic of Germany, cited *Rulaman* as the greatest German work in juvenile literature, and *Die Zeit* featured *Rulaman* as a German favorite as recently as 1999. See Günter Herburger, "Rulaman: Mein Jahrhundertbuch," *Die Zeit*, January 14, 1999; and Schäfer, "Rulaman und Höhlenkinder." Information on past editions of these novels appears in the *National Union Catalog*.

7. Binder, "David Friedrich Weinland."

8. Blackbourn, *Long Nineteenth Century*, 274.

9. Walter, *Mit Kindern unterwegs*.

10. Weinland, Vorwort zur ersten Auflage, *Kuning Hartfest*, 3.

11. Weinland, *Kuning Hartfest*, 106.

12. Weinland, Vorwort zur ersten Auflage, *Rulaman*, 5.

13. Weinland, Vorwort zur ersten Auflage, *Kuning Hartfest*, 4 (emphasis in the original).

14. Weinland, *Kuning Hartfest*, 239.

15. Rudolf Marggraff, "Nordendorfer Alterthümer," *Beilage zur Allgemeinen Zeitung*, November 20, 1844.

16. Weinland, *Kuning Hartfest*, footnotes to chapters 2, 21, and 31.

17. Weinland, *Rulaman*, 23–38.

18. Weinland, Vorwort zur ersten Auflage, *Kuning Hartfest*, 3.

19. In Hartfest's dream, the Suebi encounter the "Römling, dieses weiche Volk schalhaariger Zwerge." Weinland, *Kuning Hartfest*, 23–25, 271–73.

20. Tacitus, *Germania*, 78.

21. Weinland, *Kuning Hartfest*, 272.

22. Some of the most sentimental lines in the novel come when Weinland describes Ulf, the young Germanic warrior-to-be, and Berchta, the object of his affection. Berchta embodies the Suebi's harmony with nature and eternal beauty: "Is this noble virgin, ringed with golden hair, not as beautiful as the May morning? Innocence and joy shine from her blue eyes, and on her forehead resides the proud virtue of the ancestors" (*Kuning Hartfest*, 23–25). These descriptions recall not only Tacitus's images of the Germanic peoples as more in tune with nature than the Romans but also Weinland's sympathetic portrait of the innocent Aimats in *Rulaman*.

23. The membership list for the society appears in *Brandenburgia* 1 (1892–93): 12.

24. Fontane, *Wanderungen*, 178–81.

25. Ibid., 181. Fontane refers specifically to the Elsengund, a name that does not appear anywhere else in the *Wanderungen*. Several sources, including the *Topographische Uebersicht des Appellationsgerichts-Departements Frankfurt an der Oder* edited by Güthlein (Frankfurt an der Oder: Harnecker, 1856), list Elsengrund as another name for the Büssower Försterei, which was located in the Prussian district of Friedeberg in the northeastern corner of Brandenburg.

26. Kipper, *Germanenmythos*, 85–90.

27. Tatlock, "Realist Historiography," 59. Tatlock's essay is devoted to *Bilder aus der deutschen Vergangenheit*, but a similar interpretation of Freytag based on *Die Ahnen* appears in Holz, *Flucht aus der Wirklichkeit*, 80.

28. Freytag, *Bilder aus der deutschen Vergangenheit*, 134.

29. Ibid., 30.

30. Ibid., 30–31.

31. Ibid., 28–29.

32. Ibid., 29.

33. Tacitus, *Germania*, 87. The modern archaeological record shows this to be incorrect: there were clearly demarcated fields throughout northern Europe that were used year after year, and scholars believe that the ancient authors applied assumptions about nomadic peoples to the Germanic peoples because they lacked information about them. Ibid., 221–22.

34. Freytag, *Bilder aus der deutschen Vergangenheit*, 71–76.

35. Smith, *Politics*, 130–39.

36. Freytag, *Bilder aus der deutschen Vergangenheit*, 57–60.

37. Ibid., 36–37.

38. Brather, "Slawenbilder."

39. Gustav Freytag, dedication to *Die Ahnen*, n.p.

40. Freytag, *Die Ahnen*, 113.

41. Holz, *Flucht aus der Wirklichkeit*, 95.

42. Gregorovius, *Die Verwendung historischer Stoffe*, 5, 14.

43. Ibid., 11–14.

44. Ibid., 19.

45. Frech, "Felix Dahn," 686.

46. Ibid., 690; and Hull, *Absolute Destruction*, 179.

47. Hodgkin, review of *Geschichte der deutschen Urzeit*, 152–54.

48. The article referred to *Urgeschichte der germanischen und romanischen Völker* and *Geschichte der deutschen Urzeit*. *Encyclopedia Britannica*, 1911 ed., s.v. "Dahn, Julius Sophus Felix."

49. Dahn, *Urgeschichte*, 1:3–5.

50. Ibid., 17–18.

51. Ibid., 15 (emphasis in the original).

52. Ibid., 106–7.

53. Hildebrand reminds his young listeners that Roman leaders had offered peace to the Goths a generation earlier. This would have led to easier times, but Hildebrand extols the Goths because they rightfully refused and remained true to their distinctive identity. Dahn, *Ein Kampf um Rom*, 9–11.

54. Ibid., 10–14.

55. This concept also meshed nicely with the idea of social stability that Riehl and Freytag favored. The hierarchical order of the monarchy produced success and confidence, not social conflict. Frech, "Felix Dahn," 694.

56. Dahn, *Ein Kampf um Rom*, 398.

57. Schindler, "Geschichte als tragisches Schicksal," 257.

58. Cethegus's first plan of deception comes in a pact with Cardinal Silverius. Dahn, *Ein Kampf um Rom*, 29–36. There are several other examples throughout the novel.

59. Frech, "Felix Dahn," 694–95.

60. Dahn, *Urgeschichte*, 1:83–85.

61. Belgum, *Popularizing the Nation*, 174–76.

62. Rademacher, *Führer durch das Städtische Prähistorische Museum*, 135–36.

63. The biographical details in the following paragraphs are from Grünert, *Kossinna*.

64. Kossinna, "Ueber die vorgeschichtliche Ausbreitung der Germanen," 109.

65. Kossinna specifically mentioned Otto Schrader's well-known *Sprachvergleichung und Urgeschichte* as an example of the method he found so frustrating. Kossinna, "Ueber die vorgeschichtliche Ausbreitung der Germanen," 110.

66. Grünert, *Kossinna*, 171–84; and more broadly, Marchand, *Down from Olympus*, 153–87.

67. Kossinna, "Ueber die vorgeschichtliche Ausbreitung der Germanen," 112.

68. Ibid., 111.

69. Chamberlain, *Foundations of the Nineteenth Century*, 1:263.

70. Ibid., 320, 324.

71. Ibid., 264.

72. Ibid., 257.

73. Ibid., 385. There are twenty-two references to Rudolf Virchow, fifteen to Johannes Ranke, nine to Felix von Luschan, and five to Julius Kollmann in *Foundations of the Nineteenth Century*.

74. Ibid., 265–67, 530.

75. Ibid., 392–93. The criticism of Virchow's speech appears on 260.

76. Ibid., 262, 294.

77. Ibid., 532–33.

78. Ibid., 267–68.

79. Puschner, *Die völkische Bewegung*.

80. Wiwjorra, "Die deutsche Vorgeschichtsforschung," 195; and Puschner, "Germanenideologie und völkische Weltanschauung."

81. Chamberlain, *Foundations of the Nineteenth Century*, 1:297.

82. Kossinna began a twenty-five-year correspondence with Ludwig Schemann, the anti-Semitic propagandist and founder of Germany's Gobineau-Vereinigung, in 1903. Heinrich Class, the president of the Pan-German League from 1908 to 1939, joined Kossinna's society for prehistory in 1912. Grünert, *Kossinna*, 229–39.

83. Kossinna, "Die indogermanische Frage," 208.

Chapter Seven

1. Kossinna [anon.], "Professuren für deutsches Altertum."

2. Sommer, "Teaching of Archaeology," 211.

3. Grünert, *Kossinna*, 140–60.

4. Around 1908, as the museum gained a larger and more professional staff and a new building, publications began to refer to the museum as the Märkisches Museum, dropping the "Provinzial" from its name.

5. Gummel, *Forschungsgeschichte*, 396–471. Kossinna's students included Ernst Wahle, Walter Schulz, Martin Jahn, and Józef Kostrzewski.

6. Bode, *Mein Leben*, 2:221–22.

7. This exhibit was supposed to be temporary, but it stayed in place until 1921. Schuchhardt came to Berlin amid a heated debate about the future of the Museum für Völkerkunde's collections and the possibility of building a new museum in Dahlem. These plans were put on hold by the outbreak of World War I and the economic conditions that followed. Menghin, "Vom Zweiten Kaiserreich in die Weimarer Republik," 126–35.

8. Schmidt, "Vorgeschichtliche Abteilung," 88 (emphasis in the original).

9. Ibid., 89.

10. Schuchhardt, "Römerschanze," in *Reallexikon der Vorgeschichte*, 11:154–55.

11. Schuchhardt, "Die Römerschanze bei Potsdam," 213.

12. Schuchhardt, "Römerschanze," in *Reallexikon der Vorgeschichte*, 11:154.

13. Hoops, *Reallexikon*, 1:v, vi.

14. Ibid., vii.

15. Schliz, "Vorgeschichtliches deutsches Siedlungswesen," 444–47.

16. Grünert, *Kossinna*, 174–84.

17. Kossinna, "Zum Geleit."

18. Kossinna was most interested in the finds themselves, not their context (the *Fund*, not the *Befund*), and he was primarily concerned with settlements as signs of the occupation of a given territory, not with settlements as sites where one could investigate daily life and cultural change among prehistoric peoples (the *Besiedlung* of the land, not individual *Siedlungen*). Grünert, *Kossinna*, 71–94.

19. Kossinna, *Die Herkunft der Germanen*, 3 (emphasis in the original).

20. Kossinna, *Die deutsche Vorgeschichte*, iv.

21. Ibid., 83.

22. Kossinna, *Die Herkunft der Germanen*, 4, 9.

23. Childe, review of *Ursprung und Verbreitung der Germanen*.

24. Trigger, *Archaeological Thought*, 167.

25. Grünert, *Kossinna*, 94, 223–24.

26. Kügler, "Albert Kiekebusch."

27. Kiekebusch, "Die vorgeschichtliche Abteilung," 131–32.

28. Kuhn, "Aufgaben der Museen," 29. See also Hochreiter, *Vom Musentempel zum Lernort*.

29. Kiekebusch, "Die vorgeschichtliche Abteilung," 135–37.

30. Kiekebusch, "Die Vorgeschichte der Mark Brandenburg"; and Kiekebusch, *Bilder aus der märkischen Vorzeit*.

31. Kiekebusch, "Vorgeschichtliche Wohnstätten."

32. Kiekebusch, "Die Vorgeschichte der Mark Brandenburg," 393.

33. Hahne, "Zur Ausgestaltung der vorgeschichtlichen Sammlung."

34. On Jacob-Friesen's efforts in Leipzig, see Penny, *Objects of Culture*, 143. His study in Stockholm is recorded in the *Jahrbuch des Städtischen Museums für Völkerkunde zu Leipzig* 3 (1908–9): vii.

35. Jacob-Friesen, "Die museumstechnische Auswertung," 59–61.

36. Ibid., 68–69.

37. Ibid., 59.

38. Hahne, "Das neue Provinzialmuseum," 125–33. The interior of the museum offered an example of the sometimes fuzzy line between modernist and *völkisch* conceptions of art. The walls of the main entry and staircase were covered by modernist frescoes by Paul Thiersch. One side of the stairwell depicted scenes that valorized the individual in Germanic life: a face-off between two warriors, the moment of death, and a burial. The other side celebrated community events with scenes depicting the dedication of youth and the selection of a ruler. Hahne, an ardent nationalist, bragged about the originality of this décor, noting that "as far as we know, these are the first large-format 'expressionist' frescoes, and it is fitting that they have been executed in a spirit that is related through and through

to the traditional German and Germanic artistic sensibility." Hahne's description of a "traditional German and Germanic artistic sensibility" was quite ironic, for Thiersch was closely associated with the Werkbund and Bauhaus movements. A decade after this judgment, Hahne joined the Kampfbund für deutsche Kultur, the association that sought to purge modernist art from German society. Ibid., 130.

39. Ibid., 125.

40. Ibid., 143–45.

41. Wiwjorra, "'Ex oriente lux.'"

42. Jacob, "Der Schutz der vorgeschichtlichen Denkmäler," 76–80. This would be a new arrangement in Prussia, as Bavaria was the only state in the German Empire that included an archaeologist as part of its state office for the preservation of artifacts and monuments.

43. Dorka, "40 Jahre siedlungsarchäologische Übungen," 74–77.

44. Kiekebusch, "Siedlungsarchäologische Übungen," 118.

45. Kiekebusch, *Die heimische Altertumskunde*, 40–41. Between 1924 and 1928, about 150,000 students saw a traveling exhibit sponsored by the Märkisches Museum. Dorka, "40 Jahre siedlungsarchäologische Übungen," 77.

46. Kiekebusch, *Die heimische Altertumskunde*, 7–8.

47. Ibid., 53.

48. Ibid., 9.

49. Schuchhardt, *Alteuropa* (1919), vii–viii.

50. Ibid., ix.

51. Ibid., viii–ix.

52. Mötefindt, review of *Alteuropa in seiner Kultur- und Stilentwicklung*; and Childe, review of *Alteuropa in seiner Kultur- und Stilentwicklung*.

53. Schuchhardt, "Kriegsarchäologie." The report also appeared in a weekly for a more general reading audience: *Die Umschau: Wochenschrift über die Fortschritte in Wissenschaft und Technik* 19 (1915): 877.

54. "Deutsch-polnische Vorzeit," *Alldeutsche Blätter* 29 (1919): 163; quoted in Wiwjorra, "Die deutsche Vorgeschichtsforschung," 201.

55. Kossinna, *Altgermanische Kulturhöhe*, 33. For the Paris delegates, see Grünert, "Ur- und Frühgeschichtsforschung," 121.

56. Historian Andrew Evans details how German scholars studied prisoners of war from other countries during World War I. In this heated patriotic context, "leading members of the youngest generation of anthropologists began to conflate the concepts of nation and race in their research on the POW population. Working in the camps, they investigated and portrayed the European enemies of the Central Powers as racial 'others,' assigning distinct racial and biological identities to European peoples and nations in the process." Evans, *Anthropology at War*, 132.

57. Ibid., 199–214.

58. Günther, *Rassenkunde*, 225–26 (emphasis in the original).

59. Schuchhardt, *Alteuropa* (1926), x.

60. Ibid., 4.

61. Ibid., 280.

62. Jacob-Friesen, *Grundfragen der Urgeschichtsforschung*, 2.

63. Ibid., 207.

64. Ibid., 1.

Chapter Eight

1. Arnold, "The Past as Propaganda." On the connections to the nineteenth century, see Leerssen, *National Thought*.

2. The most nationalistic version of this history appeared in Stampfuß, *Kossinna*. Milder variations include Wahle, "Vorgeschichtsforschung," 705–10; and Gummel, *Forschungsgeschichte*.

3. Bollmus, *Das Amt Rosenberg*, 178.

4. Bollmus (ibid.) and Kater (*Das "Ahnenerbe" der SS*) document the competition within these institutions that led to opportunism and personal rivalries among prehistorians.

5. The translated *Mein Kampf* quotation appears in Mees, "Hitler and *Germanentum*," 255. Albert Speer's translated recollection is quoted in Arnold, "The Past as Propaganda," 469.

6. Haßmann, "Archaeology in the 'Third Reich,'" 88–92.

7. Arnold, "Dealing with the Devil."

8. Schöbel, "Hans Reinerth," 323–36; Bollmus, *Das Amt Rosenberg*, 160–61.

9. Bollmus, *Das Amt Rosenberg*, 167–68.

10. Wegner, "Auf vielen und zwischen manchen Stühlen," 397–417.

11. Hülle, appendix to *Die deutsche Vorgeschichte*, 272.

12. Wegner, "Auf vielen und zwischen manchen Stühlen," 410–11.

13. Quoted in ibid., 412.

14. Bollmus, *Das Amt Rosenberg*, 249.

15. "Neue Richtlinien," 81–82.

16. Ibid.

17. Ibid., 82–83.

18. "Stand der Vorgeschichte," 185.

19. In his survey of ten representative archaeologists during the National Socialist period, Wolfgang Pape ("Zehn Prähistoriker") found that six were in the Nationalsozialistischer Lehrerbund and eight participated in at least one ideological seminar.

20. "Tätigkeitsbericht des Staatlichen Museums für Vorgeschichte," 52–53; and "Die Vorgeschichte in der Volksbildungsarbeit."

21. For the Northeim curriculum, see Allen, *Nazi Seizure of Power*, 258. On the centrality of "ethnic fundamentalism," see Koonz, *Nazi Conscience*, 136–37.

22. Walter Frenzel's *Grundzüge der Vorgeschichte Deutschlands und der Deutschen* is quoted in Haßmann, "Archäologie und Jugend," 113.

23. Philipp, *Vor- und Frühgeschichte des Nordens*, 1.

24. Haßmann, "Archaeology in the 'Third Reich,'" 88–92.

25. Schmidt, "Die Rolle der musealen Vermittlung," 155.

26. Kiekebusch, "Der Germanenzug."

27. Springer, "Die Neuaufstellung der vorgeschichtlichen Sammlungen," 199–201.

28. Kohler, *Ein ruhiges Fortbestehen?*, 62–64.

29. Jacob-Friesen, *Wegweiser durch die urgeschichtliche Abteilung*, 2.

30. Ibid., 6, 10.

31. Kiekebusch, "Vorgeschichte im öffentlichen Unterricht."

32. Kiekebusch, "Heimische Vorgeschichte," 218–19.

33. Ibid., 223.

34. Kiekebusch, "Die Welt des Germanentums," 307.

35. Ibid., 310–11.

36. Alfred Bab to Albert Kiekebusch, October 19, 1934, SSM MM.

37. Alfred Bab to Albert Kiekebusch, November 3, 1934, SSM MM.

38. Kiekebusch, *Deutsche Vor- und Frühgeschichte*, 47.

39. Ibid., 116–17.

40. Ibid., 142, 151–53.

41. Wilhelm Unverzagt's "Zur Vorgeschichte des ostdeutschen Raumes," in *Deutschland und Polen: Beiträge zu ihren geschichtlichen Beziehungen*, ed. Albert Brackmann (Munich: Oldenbourg, 1933), 3–5, is quoted in Burleigh, *Germany Turns Eastwards*, 66.

42. Copy of letter from July 1, 1937, SSM MM.

43. "Berlin gräbt Germanen-Siedlung aus," *Berliner Lokal-Anzeiger*, August 8, 1937; translation quoted from Weiss, "Panem et Circenses," 244.

44. Kiekebusch, *Germanische Geschichte*, 12.

45. Ibid., 51.

46. Makiewicz, "Archäologische Forschung in Poznań," 522–24.

47. Schulz, "Geleitwort."

48. Blombergowa, "Archäologische Funde im Dienst der Propaganda," 289–90.

49. The medallion is featured in the German Historical Museum's data bank of images: http://www.dhm.de/datenbank/dhm.php?seite=5&fld_0=ZD020625.

50. Burleigh, *Germany Turns Eastwards*, 276–77.

Epilogue

1. Schmidt, "Die Rolle der musealen Vermittlung," 155; Seyer, "Die ur- und frühgeschichtliche Sammlung," 133; and Heinz Seyer (archaeologist at the Märkisches Museum), in discussion with the author, May 5, 1999.

2. The focus on twenty-two top-ranking political and military leaders at the

Nuremberg Trials was the most prominent example of this policy. Judt, *Postwar*, 51–62.

3. Reinerth spent the next thirty-five years isolated from the academic field of prehistory, but he continued to manage a private museum devoted to lake dwellings in Unteruhldingen. The court proceedings are quoted in Schöbel, "Hans Reinerth," 358–61.

4. Quotations from Jacob-Friesen's denazification questionnaire appear in Wegner, "Auf vielen und zwischen manchen Stühlen," 413–14.

5. Quoted in ibid., 414.

6. Hakelberg, "Deutsche Vorgeschichte als Geschichtswissenschaft," 274–84.

7. Steuer, "Deutsche Prähistoriker."

8. Hakelberg, "Deutsche Vorgeschichte als Geschichtswissenschaft," 274–84.

9. Carnap-Bornheim, "Hans Jürgen Eggers," 173–77.

10. Eggers, *Einführung in die Vorgeschichte*, 199.

11. Ibid., 238–39.

12. Ibid., 270.

13. Ibid., 295.

14. Katenhusen, "150 Jahre Niedersächsisches Landesmuseum Hannover," 67–70.

15. Dannheimer, "90 Jahre Prähistorische Staatssammlung," 25–29 (quotation on 28).

16. Bertram, "Wilhelm Unverzagt," 174–76.

17. Nawroth, "Aus Trümmern erstanden," 197–99.

18. Seyer, "Die ur- und frühgeschichtliche Sammlung," 134.

19. Härke, "All Quiet on the Western Front?," 193.

20. These exhibits included *Germanen, Hunnen und Awaren: Schätze der Völkerwanderungszeit* at the GNM in Nuremberg in 1987; *Die Bajuwaren: Von Severin bis Tassilo, 488–788*, in Rosenheim, Germany, and Mattsee, Austria, in 1988; and *Die Alamannen*, in Stuttgart, Zurich, and Augsburg in 1997 and 1998.

21. Wells, *Beyond Celts, Germans, and Scythians*, 13–33.

22. Weinland and Winkler, . . . *schaut auf diese Stadt!*, 24–27.

23. Nicola Kuhn, "Die Schatzkammer ist geöffnet," *Der Tagesspiegel*, October 17, 2009.

24. The Schliemann pieces were displayed in Saint Petersburg and Moscow in the mid-1990s, and negotiations about their return and the broader issue of wartime looting are ongoing. Meyer, "Priam's Treasure."

25. Steuer, "Deutsche Prähistoriker," 19.

Bibliography

Archival Collections

Bayerisches Hauptstaatsarchiv, Munich, Germany.
Geheimes Staatsarchiv, Preußischer Kulturbesitz, Berlin, Germany.
Stadtarchiv, Munich, Germany.
Stiftung Stadtmuseum, Märkisches Museum, Abteilung Ur- und Frühgeschichte, Berlin, Germany.

Printed Material

Ackerknecht, Erwin. *Rudolf Virchow: Doctor, Statesman, Anthropologist*. Madison: University of Wisconsin Press, 1953.
Agnew, Hugh LeCaine. *Origins of the Czech National Renascence*. Pittsburgh, PA: University of Pittsburgh Press, 1993.
Allen, William Sheridan. *The Nazi Seizure of Power: The Experience of a Single German Town, 1922–1945*. Rev. ed. Danbury, CT: Watts, 1984.
Andree, Christian. "Geschichte der Berliner Gesellschaft für Anthropologie, Ethnologie und Urgeschichte, 1869–1969." In *Festschrift zum hundertjährigen Bestehen der Berliner Gesellschaft für Anthropologie, Ethnologie und Urgeschichte, 1869–1969*, edited by Hermann Pohle and Gustav Mahr, 9–139. Berlin: Heßling, 1969.
———. *Rudolf Virchow als Prähistoriker*. Vol. 1, *Virchow als Begründer der neueren deutschen Ur- und Frühgeschichtswissenschaft*. Cologne: Böhlau, 1976.
Applegate, Celia. "The Mediated Nation: Regions, Readers, and the German Past." In *Saxony in German History: Culture, Society, and Politics*, edited by James Retallack, 33–50. Ann Arbor: University of Michigan Press, 2000.
———. *A Nation of Provincials: The German Idea of Heimat*. Berkeley: University of California Press, 1990.

Aretin, Karl Maria von. Vorwort to *Das Bayerische Nationalmuseum*, iii–viii. Munich: Wolf, 1868.

Arnold, Bettina. "Dealing with the Devil: The Faustian Bargain of Archaeology under Dictatorship." In *Archaeology under Dictatorship*, edited by Michael Galaty and Charles Watkinson, 191–212. New York: Kluwer/Plenum, 2004.

———. "The Past as Propaganda: Totalitarian Archaeology in Nazi Germany." *Antiquity* 64 (1990): 464–78.

Arnold, Bettina, and Henning Hassmann. "Archaeology in Nazi Germany: The Legacy of the Faustian Bargain." In *Nationalism, Politics, and the Practice of Archaeology*, edited by Philip L. Kohl and Clare Fawcett, 70–81. Cambridge: Cambridge University Press, 1995.

Assmann, Aleida. *Erinnerungsräume: Formen und Wandlungen des kulturellen Gedächtnisses*. Munich: Beck, 1999.

"Ausgrabungen der Denkmäler der Vorzeit, Erhaltung der Funde von Alterthümern." *Centralblatt für die gesammte Unterrichts-Verwaltung in Preußen*, 1887, 609–10.

Der Ausschuss des histor. Vereins von Oberbayern an die hochgeehrten Mitglieder. Munich: Kgl. Hof- und Universitätsbuchdruckerei, 1885.

"A. Voss' neues Prachtwerk unserer Wissenschaft." *Correspondenz-Blatt der deutschen Gesellschaft für Anthropologie, Ethnologie und Urgeschichte* 12 (1881): 16.

Bahn, Paul, ed. *The Cambridge Illustrated History of Archaeology*. Cambridge: Cambridge University Press, 1996.

Bann, Stephen. *Romanticism and the Rise of History*. New York: Twayne, 1995.

Bauer, Ingolf. "König Maximilian II., sein Volk und die Gründung des Bayerischen Nationalmuseums." *Bayerisches Jahrbuch für Volkskunde*, 1988, 1–38.

Das Bayerische Nationalmuseum. Munich: Wolf, 1868.

Bekmann, Johann Christoph. *Historische Beschreibung der Chur und Mark Brandenburg nach ihrem Ursprung, Einwohnern, Natürlichen Beschaffenheit, Gewässer, Landschaften, Städen, Geistlichen Stiftern &c. Regenten, deren Staats- und Religions-Handlungen, Wapen, Siegel und Münzen, Wohlverdienten Geschlechtern Adelichen und Bürgerlichen Standes, Aufnehmen der Wissenschafften und Künste in derselben*. Edited by Bernhard Ludwig Bekmann. 2 vols. Berlin: Voß, 1751–53.

Belgum, Kirsten. *Popularizing the Nation: Audience, Representation, and the Production of Identity in "Die Gartenlaube," 1853–1900*. Lincoln: University of Nebraska Press, 1998.

Beltz, Robert. "Erläuterung der Karten zur Vorgeschichte von Mecklenburg." Pts. 1–4. *Correspondenz-Blatt der deutschen Gesellschaft für Anthropologie, Ethnologie und Urgeschichte* 32 (1901): 10–16, 20–23, 30–32, 37–39.

Benario, Herbert. "Arminius into Hermann: History into Legend." *Greece and Rome* 51 (2004): 83–94.

Bergmann, A. R. "Kurfürst August und Kurfürstin Anna in ihren Beziehungen zur prähistorischen Forschung." *Sitzungsberichte der naturwissenschaftlichen Gesellschaft Isis in Dresden*, 1894, 9–10.

Bernau, Nikolaus. "Chronologie oder Geographie? Die Ersteinrichtungen des Museums für Vor- und Frühgeschichte in Berlin 1830, 1855 und 1876: Ein Beitrag zur Museumsgeschichte." Diplomarbeit, Hochschule der Künste Berlin, 1996.

Bertram, Marion. "Vom 'Museum vaterländischer Alterthümer' im Schloss Monbijou zur 'Sammlung der Nordischen Alterthümer' im Neuen Museum: Die Ära Ledebur 1829 bis 1873." In Menghin, "Das Berliner Museum für Vor- und Frühgeschichte," 31–79.

———. "Wilhelm Unverzagt und das Staatliche Museum für Vor- und Frühgeschichte." In Menghin, "Das Berliner Museum für Vor- und Frühgeschichte," 162–92.

"Besichtigung der Brauerei unseres Mitgliedes, des Herrn Julius Bötzow." *Brandenburgia* 5 (1896–97): 105–7.

Binder, Hans. "David Friedrich Weinland: Zoologe, Jugendbuchautor, 1829–1915." *Lebensbilder aus Schwaben und Franken* 13 (1977): 314–40.

Blackbourn, David. *The Conquest of Nature: Water, Landscape and the Making of Modern Germany*. London: Cape, 2006.

———. *The Long Nineteenth Century: A History of Germany, 1780–1918*. New York: Oxford University Press, 1998.

Blombergowa, Maria Magdalena. "Archäologische Funde im Dienst der Propaganda am Beispiel der Ereignisse in Łódź in den Jahren 1939–1945." In Leube, *Prähistorie und Nationalsozialismus*, 289–92.

Bode, Wilhelm von. *Mein Leben*. 2 vols. Berlin: Reckendorf, 1930.

Böhner, Kurt. "Das Römisch-Germanische Zentralmuseum—eine vaterländische und gelehrte Gründung des 19. Jahrhunderts." *Jahrbuch des Römisch-Germanischen Zentralmuseums, Mainz* 25 (1978): 1–48.

Bollmus, Reinhard. *Das Amt Rosenberg und seine Gegner: Studien zum Machtkampf im nationalsozialistischen Herrschaftssystem*. Stuttgart: Deutsche Verlags-Anstalt, 1970.

Boockmann, Hartmut, Arnold Esch, Hermann Heimpel, Thomas Nipperdey, and Heinrich Schmidt, eds. *Geschichtswissenschaft und Vereinswesen im 19. Jahrhundert: Beiträge zur Geschichte historischer Forschung in Deutschland*. Göttingen: Vandenhoeck und Ruprecht, 1972.

Börsch-Supan, Helmut. *Caspar David Friedrich*. Translated by Sarah Twohig. New York: Braziller, 1974.

Brather, Sebastian. *Ethnische Interpretationen in der frühgeschichtlichen Archäologie: Geschichte, Grundlagen und Alternativen*. Ergänzungsbände zum Reallexikon der germanischen Altertumskunde, edited by Heinrich Beck, Dieter Geuenich, and Heiko Steuer, vol. 42. Berlin: de Gruyter, 2004.

———. "Slawenbilder: 'Slawische Altertumskunde' im 19. und 20. Jahrhundert."
Archeologické rozhledy 53 (2001): 717–51.

Braun, Rainer. *Die Anfänge der Erforschung des rätischen Limes*. Stuttgart:
Kärcher, 1984.

———. "Die Anfänge der Limesforschung in Bayern." *Jahrbuch für fränkische
Landesforschung* 42 (1982): 1–66.

Brühl, Karl Friedrich Moritz Paul von. "Instruction für die beim Chausseebau
beschäftigten Beamten in Beziehung auf die in der Erde sich findenden Alterthümer heidnischer Vorzeit [1835]." In *Eintheilungs-Plan des Märkischen
Provinzial-Museums der Stadtgemeinde Berlin*, 11–15. 4th ed. Berlin, 1879.

Buchholz, Rudolf, and Otto Pniower, eds. *Das Märkische Provinzial-Museum der
Stadtgemeinde Berlin von 1874 bis 1899: Festschrift zum fünfundzwanzigjähri-
gen Bestehen*. Berlin: Stankiewicz, 1901.

Buchner, Andreas. *Reise auf der Teufels-Mauer: Eine Untersuchung über die Über-
bleibsel der römischen Schutz-Anstalten im jenseits der Donau gelegenen Rhae-
tien*. 3 vols. Regensburg: Montag-Weissische Buchhandlung, 1818.

Burian, Peter. "Das Germanische Nationalmuseum und die deutsche Nation." In
*Das Germanische Nationalmuseum Nürnberg, 1852–1977: Beiträge zu seiner
Geschichte*, edited by Bernward Deneke and Rainer Kahsnitz, 127–262. Mu-
nich: Deutscher Kunstverlag, 1978.

Burleigh, Michael. *Germany Turns Eastwards: A Study of Ostforschung in the
Third Reich*. Cambridge: Cambridge University Press, 1988.

Büsching, Johann Gustav. *Abriß der deutschen Alterthums-Kunde*. Weimar: Verlag
des Landes-Industrie-Comptoirs, 1824.

Caesar, Julius. *The Gallic War*. Translated by H. J. Edwards. Cambridge, MA: Harvard University Press, 1952.

Carnap-Bornheim, Claus von. "Hans Jürgen Eggers und der Weg aus der Sack-
gasse der ethnischen Deutung." In Steuer, *Eine hervorragend nationale Wissenschaft*, 173–97.

Chamberlain, Houston Stewart. *Foundations of the Nineteenth Century*. Translated
by John Lees. 2 vols. New York: Fertig, 1968.

Childe, V. Gordon. Review of *Alteuropa in seiner Kultur- und Stilentwicklung*, by
Carl Schuchhardt. *Journal of Hellenic Studies* 43 (1923): 79–80.

———. Review of *Ursprung und Verbreitung der Germanen in vor- und frühgeschichtlicher Zeit*, by Gustaf Kossinna. *Man* 27 (1927): 54–55.

Coles, Bryony, and John Coles. *People of the Wetlands: Bogs, Bodies, and Lake-
Dwellers*. London: Thames and Hudson, 1989.

Confino, Alon. *The Nation as a Local Metaphor: Württemberg, Imperial Germany,
and National Memory, 1871–1918*. Chapel Hill: University of North Carolina
Press, 1997.

Conwentz, Hugo. *Die Heimatkunde in der Schule: Grundlagen und Vorschläge zur
Förderung der naturgeschichtlichen und geographischen Heimatkunde in der
Schule*. 2nd ed. Berlin: Borntraeger, 1906.

Conze, Werner. "'Deutschland' und 'deutsche Nation' als historische Begriffe." In *Die Rolle der Nation in der deutschen Geschichte und Gegenwart*, edited by Otto Büsch and James J. Sheehan, 21–38. Berlin: Colloquium, 1985.

Crane, Susan A. *Collecting and Historical Consciousness in Early Nineteenth-Century Germany*. Ithaca, NY: Cornell University Press, 2000.

———. "Collecting and Historical Consciousness: New Forms for Collective Memory in Early 19th-Century Germany." PhD diss., University of Chicago, 1992.

Curta, Florin. Introduction to *Borders, Barriers, and Ethnogenesis: Frontiers in Late Antiquity and the Middle Ages*, edited by Florin Curta, 1–9. Turnhout, Belgium: Brepols, 2005.

Dahn, Felix. *Felicitas: Historischer Roman aus der Völkerwanderung*. Vol. 3 of *Gesammelte Werke*. Leipzig: Breitkopf und Härtel, 1921.

———. *Ein Kampf um Rom*. Vol. 1 of *Gesammelte Werke*. Leipzig: Breitkopf und Härtel, 1921.

———. *Urgeschichte der germanischen und romanischen Völker*. 2 vols. Berlin: Grote, 1881.

Daniel, Glyn. *The Idea of Prehistory*. London: Watts, 1962.

Dannheimer, Hermann. "90 Jahre Prähistorische Staatssammlung München." *Bayerische Vorgeschichtsblätter* 40 (1975): 1–33.

Dauber, Albrecht. "Zur Geschichte der archäologischen Denkmalpflege in Baden." *Denkmalpflege in Baden-Württemberg* 3 (1983): 47–51.

Diaz-Andreu, Margarita, and Timothy Champion, eds., *Nationalism and Archaeology in Europe*. Boulder: Westview Press, 1996.

Dorka, Gertrud. "40 Jahre siedlungsarchäologische Übungen und Studien in Berlin." *Berliner Blätter für Vor- und Frühgeschichte* 4 (1955): 73–80.

Dörner, Andreas. "Der Mythos der nationalen Einheit: Symbolpolitik und Deutungskämpfe bei der Einweihung des Hermannsdenkmals im Jahre 1875." *Archiv für Kulturgeschichte* 79 (1997): 389–417.

Dorow, Wilhelm. *Erlebtes aus den Jahren 1813–1820*. Vol. 1. Leipzig: Hinrichs, 1843.

———. *Die Kunst Alterthümer aufzugraben und das Gefundene zu reinigen und zu erhalten*. Hamm: Schulz und Wundermann, 1823.

Drück, T. "Die vaterländische Altertumskunde im Gymnasialunterricht." In *Wissenschaftliche Beilage zum Programm des königlichen Gymnasiums in Ulm*, 1–20. Ulm: Wagner, 1894.

Dürrich, Ferdinand von, and Wolfgang Menzel. *Die Heidengräber am Lupfen (bei Oberflacht)*. Stuttgart: Arnold, 1847.

Ebert, Max, ed. *Reallexikon der Vorgeschichte*. 15 vols. Berlin: de Gruyter, 1924–32.

Ecker, Alexander. "Die Berechtigung und die Bestimmung des Archivs." *Archiv für Anthropologie* 1 (1866): 1–6.

Ecker, Alexander. "Recht und Rechtsgeschichte in der Bayerischen Akademie der Wissenschaften von 1759 bis 1827." PhD diss., University of Regensburg, 2004.

Effros, Bonnie. *Uncovering the Germanic Past: Merovingian Archaeology in France, 1830–1914*. Oxford: Oxford University Press, 2012.

Eggers, Hans Jürgen. *Einführung in die Vorgeschichte.* Munich: Piper, 1959.

———. "Der Wagen Odins: Ein Beitrag zur Frühzeit der Vorgeschichtsforschung in der Mark Brandenburg." In *Gandert-Festschrift zum sechzigsten Geburtstag von Otto-Friedrich Gandert am 8. August 1958*, edited by Adriaan von Müller and Wolfram Nagel, 31–41. Berlin: Lehmann, 1959.

Eichler, Hans. "Das Museum vaterländischer Alterthümer zu Münster." *Westfälische Zeitschrift* 124–25 (1974–75): 91–115.

———. "Zur Geschichte des Landesmuseums für Kunst und Kulturgeschichte." *Westfalen* 36 (1958): 137–43.

"Erhaltung der Funde an Münzen &c. für die Museen oder andere Sammlungen." Pts. 1–2. *Centralblatt für die gesammte Unterrichts-Verwaltung in Preußen*, 1865, 202; 1872, 668.

Esch, Arnold. "Limesforschung und Geschichtsvereine: Romanismus und Germanismus, Dilettantismus und Facharchäologie in der Bodenforschung des 19. Jahrhunderts." In Boockmann et al., *Geschichtswissenschaft und Vereinswesen im 19. Jahrhundert*, 163–91.

Evans, Andrew. *Anthropology at War: World War I and the Science of Race in Germany.* Chicago: University of Chicago Press, 2010.

Fagan, Brian, ed. *Eyewitness to Discovery: First-Person Accounts of More than Fifty of the World's Greatest Archaeological Discoveries.* New York: Oxford University Press, 1996.

Fetten, Frank. "Archaeology and Anthropology in Germany before 1945." In Härke, *Archaeology, Ideology and Society*, 143–82.

Fontane, Theodor. *Vor dem Sturm: Roman aus dem Winter 1812 auf 13.* Munich: dtv, 1994.

———. *Wanderungen durch die Mark Brandenburg: Die Grafschaft Ruppin.* Vol. 9 of *Sämtliche Werke.* Munich: Nymphenburg, 1959.

Föringer, Heinrich Konrad. "Matthias Koch." *Jahres-Bericht des historischen Vereines von Oberbayern* 39–40 (1876–77): 161–70.

Frech, Kurt. "Felix Dahn: Die Verbreitung völkischen Gedankenguts durch den historischen Roman." In Puschner, Schmitz, and Ulbricht, *Handbuch zur "Völkischen Bewegung," 1871–1918*, 685–98.

Freytag, Gustav. *Die Ahnen.* Vol. 8 of *Gesammelte Werke.* 3rd ed. Leipzig: Hirzel, 1910.

———. *Bilder aus der deutschen Vergangenheit.* Vol. 17 of *Gesammelte Werke.* 3rd ed. Leipzig: Hirzel, 1910.

Friedel, Ernst. "Archäologische Streifzüge durch die Mark Brandenburg." Pts. 1–3. *Zeitschrift für Ethnologie* 3 (1871): 175–97; 5 (1873): 245–58; 11 (1879): 370–76.

———. "Der Bronzefund von Spindlersfeld bei Coepenick." *Brandenburgia* 1 (1892–93): 37–38.

———. *Eintheilungs-Plan des Märkischen Provinzial-Museums der Stadtgemeinde Berlin.* 4th ed. Berlin: Städtische Behörde, 1879.

————. *Eintheilungs-Plan des Märkischen Provinzial-Museums der Stadtgemeinde Berlin*. 6th ed. Berlin: Theinhardt, 1882.

————. *Führer durch das Köllnische Rathaus*. Berlin: Theinhardt, 1881.

————. "Die Funde aus dem Königsgrab von Seddin, Kreis West-Prignitz." In *Das Märkische Provinzial-Museum der Stadtgemeinde Berlin von 1874 bis 1899*, edited by Rudolf Buchholz and Otto Pniower, 33–44. Berlin: Stankiewicz, 1901.

————. *Vorgeschichtliche Funde aus Berlin und Umgegend*. Berlin: Mittler, 1880.

Fritz, Rolf. *Museum für Kunst und Kulturgeschichte der Stadt Dortmund*. Hamburg: Cram, de Gruyter, 1964.

Fritzsche, Peter. *Stranded in the Present: Modern Time and the Melancholy of History*. Cambridge, MA: Harvard University Press, 2004.

Fuchs, Reinhard. "Zur Geschichte der Sammlungen des Rheinischen Landesmuseums Bonn." In *Rheinisches Landesmuseum Bonn: 150 Jahre Sammlungen, 1820–1970*, edited by Landschaftsverband Rheinland, 1–158. Düsseldorf: Rheinland, 1971.

Führer durch das Bayerische Nationalmuseum. Munich: Straub, 1894.

Führer durch das Bayerische Nationalmuseum. 8th ed. Munich: Verlag des Bayerischen Nationalmuseums, 1908.

Führer durch das königlich Bayerische Nationalmuseum in München. Munich: Straub, 1882.

Führer durch die Königlichen Museen. Edited by the Generalverwaltung der Königlichen Museen zu Berlin. Berlin: Weidmann, 1880.

Führer durch das Museum für Völkerkunde. Berlin: Spemann, 1898.

Gärtner, Tobias. "Begründer einer international vergleichenden Forschung—Adolf Bastian und Albert Voß (1874–1906)." In Menghin, "Das Berliner Museum für Vor- und Frühgeschichte," 80–102.

Geary, Patrick. *The Myth of Nations: The Medieval Origins of Europe*. Princeton, NJ: Princeton University Press, 2002.

Gerber, Michael Rüdiger. *Die Schlesische Gesellschaft für vaterländische Cultur (1801–1945)*. Beihefte zum Jahrbuch der Schlesischen Friedrich-Wilhelms-Universität zu Breslau, vol. 9. Sigmaringen: Thorbecke, 1988.

Gerson, Stephane. *The Pride of Place: Local Memories and Political Culture in Nineteenth-Century France*. Ithaca, NY: Cornell University Press, 2003.

"Gesetze der churbayerischen Akademie der Wissenschaften [1759]." *Almanach der königlichen bayerischen Akademie der Wissenschaften*, 1843, 14–25.

Gillett, Andrew. "Ethnogenesis: A Contested Model of Early Medieval Europe." *History Compass* 4 (2006): 241–60.

Goeßler, Peter. *Führer durch die K. Staatssammlung vaterländischer Kunst- und Altertumsdenkmäler in Stuttgart*. 2nd ed. Stuttgart: Kohlhammer, 1906.

————. "Die K. Altertümersammlung in Stuttgart und ihr archäologischer Bestand von 1862–1912." In *Festschrift zur Feier des fünfzigjährigen Bestehens der K. Altertümersammlung in Stuttgart*, 3–16. Stuttgart: Deutsche Verlags-Anstalt, 1912.

Goffart, Walter. *Barbarian Tides: The Migration Age and the Later Roman Empire.* Philadelphia: University of Pennsylvania Press, 2006.

———. *The Narrators of Barbarian History (A.D. 550–800): Jordanes, Gregory of Tours, Bede, and Paul the Deacon.* Princeton, NJ: Princeton University Press, 1988.

Gooch, G. P. *History and Historians of the Nineteenth Century.* 2nd ed. Boston: Beacon Hill, 1952.

Graf, Hugo. "Aus der Geschichte des Museums." In *Führer durch das Bayerische Nationalmuseum in München,* 7–28. 6th ed. Munich: Bayerisches Nationalmuseum, 1904.

Grayson, Donald. *The Establishment of Human Antiquity.* New York: Academic Press, 1983.

Green, Abigail. *Fatherlands: State-Building and Nationhood in Nineteenth-Century Germany.* Cambridge: Cambridge University Press, 2001.

Gregorovius, Leo. *Die Verwendung historischer Stoffe in der erzählenden Literatur.* Munich: Buchholz und Werner, 1891.

Greiner, Johannes. "Der Verein für Kunst und Altertum in Ulm und Oberschwaben." *Württembergische Vierteljahrshefte für Landesgeschichte* 30 (1921): 116–55.

Grünert, Heinz. *Gustaf Kossinna (1858–1931): Vom Germanisten zum Prähistoriker: Ein Wissenschaftler im Kaiserreich und in der Weimarer Republik.* Rahden, Germany: Leidorf, 2002.

———. "Ur- und Frühgeschichtsforschung in Berlin." In *Geschichtswissenschaft in Berlin im 19. und 20. Jahrhundert: Persönlichkeiten und Institutionen,* edited by Reimer Hansen and Wolfgang Ribbe, 91–148. Berlin: de Gruyter, 1992.

Gummel, Hans. *Forschungsgeschichte in Deutschland.* Berlin: de Gruyter, 1938.

———. *Lehrerschaft, Ausgrabungsgesetz und Denkmalschutz.* Greifswald: Moninger, 1926.

Günther, Hans. *Rassenkunde des deutschen Volkes.* Munich: Lehmann, 1922.

Hagen, William. *Germans, Poles, and Jews: The Nationality Conflict in the Prussian East, 1772–1914.* Chicago: University of Chicago Press, 1980.

Hager, Georg, and Josef Alois Mayer. *Die vorgeschichtlichen, römischen und merowingischen Alterthümer des Bayerischen Nationalmuseums.* Vol. 1 of *Kataloge des Bayerischen Nationalmuseums.* Munich: Rieger, 1892.

Hahn, Oskar. *Provinzial-Ordnung für die Provinzen Preußen, Brandenburg, Pommern, Schlesien und Sachsen.* Berlin: Guttentag, 1875.

Hahne, Hans. "Das neue Provinzialmuseum für Vorgeschichte zu Halle." *Museumskunde* 14 (1919): 125–46.

———. "Zur Ausgestaltung der vorgeschichtlichen Sammlung des Provinzialmuseums in Hannover als Hauptstelle für vorgeschichtliche Landesforschung in der Provinz Hannover." Pts. 1–2. *Jahrbuch des Provinzial-Museums zu Hannover,* 1908–9, 21–35; 1909–10, 45–47.

Hakelberg, Dietrich. "Deutsche Vorgeschichte als Geschichtswissenschaft—Der Heidelberger Extraordinarius Ernst Wahle im Kontext seiner Zeit." In Steuer, *Eine hervorragend nationale Wissenschaft*, 199–310.

Hanisch, Manfred. *Für Fürst und Vaterland: Legitimitätsstiftung in Bayern zwischen Revolution 1848 und deutscher Einheit.* Munich: Oldenbourg, 1991.

Hannaford, Ivan. *Race: The History of an Idea in the West.* Washington, DC: Woodrow Wilson Center, 1996.

Hare, J. Laurence. *Excavating Nations: Archaeology, Museums, and the German-Danish Borderlands.* Toronto: University of Toronto Press, 2015.

Härke, Heinrich. "All Quiet on the Western Front? Paradigms, Methods and Approaches in West German Archaeology." In *Archaeological Theory in Europe: The Last Three Decades*, edited by Ian Hodder, 187–222. London: Routledge, 1991.

———, ed. *Archaeology, Ideology and Society: The German Experience.* Rev. ed. Frankfurt am Main: Lang, 2002.

Haßler, Konrad Dietrich. *Das alemannische Todtenfeld bei Ulm.* Ulm: Wagner, 1860.

Haßmann, Henning. "Archaeology in the 'Third Reich.'" In Härke, *Archaeology, Ideology and Society*, 67–142.

———. "Archäologie und Jugend im 'Dritten Reich.'" In Leube, *Prähistorie und Nationalsozialismus*, 107–46.

Heimpel, Hermann. "Geschichtsvereine einst und jetzt." In Boockmann et al., *Geschichtswissenschaft und Vereinswesen im 19. Jahrhundert*, 45–73.

Hettner, Felix. *Illustrierter Führer durch das Provinzialmuseum in Trier.* Trier: Lintz, 1903.

Hill, Rosemary. *Stonehenge.* Cambridge, MA: Harvard University Press, 2008.

Hitchins, Keith. *The Rumanian National Movement in Transylvania, 1780–1849.* Cambridge, MA: Harvard University Press, 1969.

Hobsbawm, Eric. "Introduction: Inventing Traditions." In *The Invention of Tradition*, edited by Eric Hobsbawm and Terence Ranger, 1–14. Cambridge: Cambridge University Press, 1983.

Hochreiter, Walter. *Vom Musentempel zum Lernort: Zur Sozialgeschichte deutscher Museen, 1800–1914.* Darmstadt: Wissenschaftliche Buchgesellschaft, 1994.

Hodgkin, Thomas. Review of *Geschichte der deutschen Urzeit*, by Felix Dahn. *English Historical Review* 4 (1889): 152–54.

Holz, Claus. *Flucht aus der Wirklichkeit: "Die Ahnen" von Gustav Freytag; Untersuchungen zum realistischen historischen Roman der Gründerzeit 1872–1880.* Frankfurt am Main: Lang, 1983.

Hoops, Johannes, ed. *Reallexikon der germanischen Altertumskunde.* 4 vols. Strasbourg, France: Trübner, 1911–19.

Hoppe, Willy. "Einhundert Jahre Gesamtverein." *Blätter für deutsche Landesgeschichte* 89 (1952): 1–38.

Hoppe, Willy, and Gerhard Lüdtke, eds. *Die deutschen Kommissionen und Vereine für Geschichte und Altertumskunde*. Berlin: de Gruyter, 1940.

Hull, Isabel. *Absolute Destruction: Military Culture and the Practices of War in Imperial Germany*. Ithaca, NY: Cornell University Press, 2005.

Hülle, Werner. Appendix to *Die deutsche Vorgeschichte: Eine hervorragend nationale Wissenschaft*, by Gustaf Kossinna, 269–92. 7th ed. Leipzig: Kabitzsch, 1936.

Humboldt, Wilhelm von. "On the Historian's Task." In *The Theory and Practice of History*, edited by Georg Iggers and Konrad von Moltke, translated by Wilma Iggers and Konrad von Moltke, 5–23. Indianapolis: Bobbs-Merrill, 1973.

Hummel, Bernhard Friedrich. *Beschreibung entdeckter Alterthümer in Deutschland*. Edited by Christian Friedrich Carl Hummel. Nuremberg: Grattenauer, 1792.

———. *Bibliothek der deutschen Alterthümer, systematisch geordnet und mit Anmerkungen versehen*. Nuremberg: Grattenauer, 1787.

Huscher, Friedrich Wilhelm. "Beschreibung und historische Erläuterung eines teutschen Runensteines, des einzigen, der bisher in Teutschland selbst entdeckt worden ist, und einiger anderen merkwürdigen Alterthümer germanischer Vorzeit, welche auf einem Waldgebirge bey Großhabersdorf, 4 Stunden von Ansbach, sich vorfinden." *Variscia. Mittheilungen aus dem Archive des Voigtländischen Alterthumsforschenden Vereins* 2 (1830): 1–54.

Iggers, Georg. *The German Conception of History: The National Tradition of Historical Thought from Herder to the Present*. Rev. ed. Middletown, CT: Wesleyan University Press, 1983.

Iggers, Georg, and Konrad von Moltke. Introduction to *The Theory and Practice of History*, edited by Georg Iggers and Konrad von Moltke, translated by Wilma Iggers and Konrad von Moltke, xv–lxxi. Indianapolis: Bobbs-Merrill, 1973.

Jacob, Karl Hermann. "Der Schutz der vorgeschichtlichen Denkmäler." *Prähistorische Zeitschrift* 9 (1917): 75–81.

Jacob-Friesen, Karl Hermann. *Grundfragen der Urgeschichtsforschung: Stand und Kritik der Forschung über Rassen, Völker und Kulturen in urgeschichtlicher Zeit*. Hannover: Helwing, 1928.

———. "Die museumstechnische Auswertung vorgeschichtlicher Sammlungen nach dem pädagogischen Prinzip." *Museumskunde* 16 (1920): 56–100.

———. *Wegweiser durch die urgeschichtliche Abteilung des Landesmuseums Hannover*. Hannover: Landesmuseum Hannover, 1938.

Jankuhn, Herbert. *Haithabu: Eine germanische Stadt der Frühzeit*. Neumünster: Wachholtz, 1937.

Jensen, Jens Christian. *Caspar David Friedrich: Life and Work*. Translated by Joachim Neugroschel. Woodbury, NY: Barron's, 1981.

Judt, Tony. *Postwar: A History of Europe since 1945*. New York: Penguin, 2005.

Katalog der Ausstellung Prähistorischer und Anthropologischer Funde Deutschlands. Berlin: Berg und Holten, 1880.

Katenhusen, Ines. "150 Jahre Niedersächsisches Landesmuseum Hannover." In *Das Niedersächsische Landesmuseum Hannover 2002: 150 Jahre Museum in Hannover—100 Jahre Gebäude am Maschpark; Festschrift zum Jahr des Doppeljubiläums*, edited by Heide Grape-Albers, 18–94. Hannover: Niedersächsisches Landesmuseum, 2002.

Kater, Michael H. *Das "Ahnenerbe" der SS, 1935–1945: Ein Beitrag zur Kulturpolitik des Dritten Reiches*. Stuttgart: Deutsche Verlags-Anstalt, 1974.

Kauffmann, Friedrich. *Deutsche Altertumskunde*. Vol. 1, *Von der Urzeit bis zur Völkerwanderung*. Munich: Beck, 1913.

Kaufhold, Karl Heinrich. "Friderizianische Agrar-, Siedlungs- und Bauernpolitik." In *Kontinuität und Wandel: Schlesien zwischen Österreich und Preußen*, edited by Peter Baumgart, 167–201. Sigmaringen: Thorbecke, 1990.

Keller, Ferdinand. *Die keltischen Pfahlbauten in den Schweizerseen*. Zurich: Meyer und Zeller, 1853.

Kelley, Donald. *Faces of History: Historical Inquiry from Herodotus to Herder*. New Haven, CT: Yale University Press, 1998.

Kenseth, Joy. "The Age of the Marvelous: An Introduction." In *The Age of the Marvelous*, edited by Joy Kenseth, 25–59. Hanover, NH: Hood Museum of Art, 1991.

Kiekebusch, Albert. *Bilder aus der märkischen Vorzeit: Für Freunde der heimischen Altertumskunde insbesondere für die Jugend und ihre Lehrer*. Berlin: Reimer, 1916.

———. *Deutsche Vor- und Frühgeschichte in Einzelbildern: Vom ersten Auftreten des Menschen bis zur Wiedergewinnung des deutschen Ostens*. Leipzig: Reclam, 1934.

———. "Der Germanenzug im Berliner Grunewaldstadion." *Nachrichtenblatt für deutsche Vorzeit* 9 (1933): 178–82.

———. *Germanische Geschichte und Kultur der Urzeit: Vom ersten Auftreten der Germanen in der Geschichte bis zum Beginn der Völkerwanderung*. Leipzig: Quelle und Meyer, 1935.

———. *Die heimische Altertumskunde in der Schule: Ein Beitrag zur Um- und Ausgestaltung des heimatkundlichen Unterrichts*. Berlin: Siegismund, 1915.

———. "Heimische Vorgeschichte als Quelle völkischer Bildung." *Die Deutsche Schule: Erziehungswissenschaftliche Monatsschrift für den Bereich der Volksschule* 38 (1934): 217–24.

———. "Siedlungsarchäologische Übungen und Studien im Märkischen Museum." *Brandenburgia* 24 (1915): 117–20.

———. "Die Vorgeschichte der Mark Brandenburg." In *Die Volkskunde*, edited by Ernst Friedel and Robert Mielke, 345–458. Vol. 3 of *Landeskunde der Provinz Brandenburg*. Berlin: Reimer, 1912.

———. "Vorgeschichte im öffentlichen Unterricht." In Ebert, *Reallexikon der Vorgeschichte*, 14: 200–216.

———. "Die vorgeschichtliche Abteilung des Märkischen Museums der Stadt Berlin." *Mannus: Zeitschrift für Vorgeschichte* 1 (1909): 130–37.

———. "Vorgeschichtliche Wohnstätten und die Methode ihrer Untersuchungen." *Korrespondenz-Blatt der Deutschen Gesellschaft für Anthropologie, Ethnologie und Urgeschichte* 43 (1912): 63–68.

———. "Die Welt des Germanentums vor seinem Eintreten in die Geschichte." *Nationale Erziehung: Monatsschrift für Eltern und Erzieher* 12 (1933): 305–11.

Kipper, Rainer. *Der Germanenmythos im Deutschen Kaiserreich: Formen und Funktionen historischer Selbstthematisierung.* Göttingen: Vandenhoeck und Ruprecht, 2002.

Klejn, L. S. "Gustaf Kossinna (1858–1931)." In *Encyclopedia of Archaeology: The Great Archaeologists,* edited by Tim Murray, 1:233–46. Santa Barbara, CA: ABC-CLIO, 1999.

Klemm, Gustav. *Handbuch der germanischen Alterthumskunde.* Dresden: Walther, 1836.

Klindt-Jensen, Ole. *A History of Scandinavian Archaeology.* London: Thames and Hudson, 1975.

Koch, Matthias. "Aufklärung über die Schlacht zu Fridolfing durch die neuesten antiquarischen Funde." *Oberbayerisches Archiv für vaterländische Geschichte* 6 (1845): 77–112.

Koch-Sternfeld, Joseph von. "Das Beinfeld bei Fridolfing." *Bayerische Blätter für Geschichte, Statistik, Literatur und Kunst,* May 19, 1832, 181–82.

———. *Zur bayerischen Fürsten-, Volks- und Culturgeschichte zunächst im Uebergange vom 5. in das 6. Jh. nach Christus.* Munich: Königliche Akademie der Wissenschaften, 1837.

Koeglmayr, J. *Der Mosaikboden in Westenhofen.* Munich: Schurich, 1856.

Kohl, Philip, and Clare Fawcett, eds. *Nationalism, Politics, and the Practice of Archaeology.* Cambridge: Cambridge University Press, 1995.

Kohl, Philip, Mara Kozelsky, and Nachman Ben-Yehuda, eds. *Selective Remembrances: Archaeology in the Construction, Commemoration, and Consecration of National Pasts.* Chicago: University of Chicago Press, 2007.

Kohler, Christian. *Ein ruhiges Fortbestehen? Das Germanische Nationalmuseum im "Dritten Reich."* Münster: LIT, 2011.

Koonz, Claudia. *The Nazi Conscience.* Cambridge, MA: Harvard University Press, 2003.

Kortüm, J. C. P. *Beschreibung eines neulich bey Neubrandenburg gefundenen wendischen Monuments, mit historischen Erläuterungen zur näheren Bestimmung der Lage des alten Rhetra.* Neubrandenburg: Korb, 1798.

Koshar, Rudy. *Germany's Transient Pasts: Preservation and National Memory in the Twentieth Century.* Chapel Hill: University of North Carolina Press, 1998.

Kossinna, Gustaf. *Altgermanische Kulturhöhe: Ein Kriegsvortrag.* Jena: Nornen, 1919.

———. "Eine archäologische Reise durch Teile Norddeutschlands." *Deutsche Geschichtsblätter: Monatsschrift zur Förderung der landesgeschichtlichen Forschung* 2 (1900): 23–26.

———. *Die deutsche Vorgeschichte: Eine hervorragend nationale Wissenschaft.* Würzburg: Kabitzsch, 1912.

———. *Die Herkunft der Germanen: Zur Methode der Siedlungsarchäologie.* Würzburg: Kabitzsch, 1911.

———. "Die indogermanische Frage archäologisch beantwortet." *Zeitschrift für Ethnologie* 34 (1902): 161–222.

——— [anon.]. "Professuren für deutsches Altertum." *Die Grenzboten: Zeitschrift für Politik, Litteratur und Kunst* 55, no. 2 (1896): 600–605.

———. "Ueber die vorgeschichtliche Ausbreitung der Germanen in Deutschland." *Correspondenz-Blatt der deutschen Gesellschaft für Anthropologie, Ethnologie und Urgeschichte* 26 (1895): 109–12.

———. "Zum Geleit." *Mannus: Zeitschrift für Vorgeschichte* 1 (1909): 1–3.

Krebs, Christopher B. *A Most Dangerous Book: Tacitus's "Germania" from the Roman Empire to the Third Reich.* New York: Norton, 2011.

Krins, Hubert. "Die Gründung der staatlichen Denkmalpflege in Baden und Württemberg." *Denkmalpflege in Baden-Württemberg* 12 (1983): 34–42.

Kruse, Friedrich. *Budorgis oder etwas über das alte Schlesien vor Einführung der christlichen Religion besonders zu den Zeiten der Römer nach gefundenen Alterthümern und den Angaben der Alten.* Leipzig: Hartknoch, 1819.

———. Einleitung to *Archiv für alte Geographie, Geschichte und Alterthümer insonderheit der Germanischen Völkerstämme* 1 (1821): 1–14.

———. Vorrede to *Archiv für alte Geographie, Geschichte und Alterthümer insonderheit der Germanischen Völkerstämme* 1 (1821): v–xxxii.

Kügler, Hermann. "Albert Kiekebusch." *Brandenburgia* 39 (1930): 3–12.

Kuhn, Alfred. "Aufgaben der Museen in der Gegenwart." *Museumskunde* 15 (1920): 26–38.

Kunz, Georg. *Verortete Geschichte: Regionales Geschichtsbewußtsein in den deutschen Historischen Vereinen des 19. Jahrhunderts.* Göttingen: Vandenhoeck und Ruprecht, 2000.

Ledebur, Leopold von. *Blicke auf die Literatur des letzten Jahrzehents zur Kenntniß Germaniens zwischen Rhein und Weser, mit besonderer Rücksicht auf: Das Land und Volk der Bructerer.* Berlin: Enslin, 1837.

———. *Die heidnischen Alterthümer des Regierungsbezirks Potsdam: Ein Beitrag zur Alterthümer-Statistik der Mark Brandenburg.* Berlin: Gebauer, 1852.

———. *Das Königliche Museum vaterländischer Alterthümer im Schlosse Monbijou zu Berlin.* Berlin: Königliche Akademie der Wissenschaften, 1838.

Leerssen, Joep. *National Thought in Europe: A Cultural History.* Amsterdam: Amsterdam University Press, 2006.

Lehner, Hans. *Führer durch das Provinzialmuseum zu Bonn.* Bonn: Georgi, 1901.

Leube, Achim, ed. *Prähistorie und Nationalsozialismus: Die mittel- und osteuropä-ische Ur- und Frühgeschichtsforschung in den Jahren 1933–1945.* In collabora-tion with Morten Hegewisch. Heidelberg: Synchron, 2002.

Leverkus, Erich. "Zur Geschichte des Neandertalerfundes." *Archäologie Online,* October 24, 2001. http://www.archaeologie-online.de/magazin/thema/mythos-neandertaler/zur-geschichte-des-neandertalerfundes.

Levezow, Konrad. *Andeutungen über die wissenschaftliche Bedeutung der allmählig zu Tage geförderten Alterthümer germanischen, slavischen und anderweitigen Ursprungs der zwischen der Elbe und Weichsel gelegenen Länder.* Stettin: Ef-fenbart, 1825.

Levine, Philippa. *The Amateur and the Professional: Antiquarians, Historians and Archaeologists in Victorian England, 1838–1886.* Cambridge: Cambridge Uni-versity Press, 1986.

Lindenschmit, Ludwig, ed. *Die Alterthümer unserer heidnischen Vorzeit.* 5 vols. Mainz: Zabern, 1858–1911.

———. *Die vaterländischen Alterthümer der Fürstlich Hohenzoller'schen Samm-lungen zu Sigmaringen.* Mainz: Zabern, 1860.

Lindenschmit, Ludwig, and Wilhelm Lindenschmit. *Das germanische Todtenlager bei Selzen in der Provinz Rheinhessen.* 1848. Facsimile of the first edition, with an introduction by Kurt Böhner. Mainz: Zabern, 1969.

Lindenschmit, Wilhelm. *Die Räthsel der Vorwelt, oder: Sind die Deutschen einge-wandert?* Mainz: Seifert, 1846.

Lisch, Georg Christian Friedrich. *Friderico-Francisceum oder Grossherzogliche Al-terthümersammlung aus der altgermanischen und slavischen Zeit Meklenburgs zu Ludwigslust.* Leipzig: Breitkopf und Härtel, 1837.

———. *Pfahlbauten in Mecklenburg.* Schwerin: Stiller, 1865.

Löwe, Philipp. *Das Neue Museum: Eine ausführliche Beschreibung seiner Kunst-werke und Sehenswürdigkeiten.* Berlin: Logier, 1857.

Lubbock, John. *Pre-historic Times, as Illustrated by Ancient Remains, and the Man-ners and Customs of Modern Savages.* 4th ed. London: Norgate, 1878.

Makiewicz, Tadeusz. "Archäologische Forschung in Poznań während des Zweiten Weltkrieges." In Leube, *Prähistorie und Nationalsozialismus,* 517–33.

Manias, Chris. *Race, Science, and the Nation: Reconstructing the Ancient Past in Britain, France, and Germany.* New York: Routledge, 2013.

Marchand, Suzanne. *Down from Olympus: Archaeology and Philhellenism in Ger-many, 1750–1970.* Princeton, NJ: Princeton University Press, 1996.

Massin, Benoit. "From Virchow to Fischer: Physical Anthropology and 'Modern Race Theories' in Wilhelmine Germany." In *Volksgeist as Method and Ethic: Essays on Boasian Ethnography and the German Anthropological Tradition,* ed-ited by George W. Stocking Jr., 79–154. Madison: University of Wisconsin Press, 1996.

McCann, W. J. "'Volk und Germanentum': The Presentation of the Past in Nazi

Germany." In *The Politics of the Past*, edited by Peter Gathercole and David Lowenthal, 74–88. London: Unwin Hyman, 1990.

McNeely, Ian. *"Medicine on a Grand Scale": Rudolf Virchow, Liberalism, and the Public Health*. London: Wellcome Trust, 2002.

Mees, Bernard. "Hitler and *Germanentum*." *Journal of Contemporary History* 39 (2004): 255–70.

Menghin, Wilfried, ed. "Das Berliner Museum für Vor- und Frühgeschichte: Festschrift zum 175-jährigen Bestehen." Special issue, *Acta Praehistorica et Archaeologica* 36–37 (2004–5).

————. "Sammlungs- und Forschungsgeschichte." In *Die Vor- und Frühgeschichtliche Sammlung des Germanischen Nationalmuseums*, edited by Gerhard Bott, 7–48. Stuttgart: Theiss, 1983.

————. "Vom Zweiten Kaiserreich in die Weimarer Republik: Die Ära Schuchhardt." In Menghin, "Das Berliner Museum für Vor- und Frühgeschichte," 122–61.

Menzel, Wolfgang. *Geschichte der Deutschen bis auf die neuesten Tage*. 3rd ed. Stuttgart: Cotta, 1837.

Mestorf, Johanna. *Die vaterländischen Alterthümer Schleswig-Holsteins: Ansprache an unsere Landsleute*. Hamburg: Meißner, 1877.

Meyer, Karl. "The Hunt for Priam's Treasure." *Archaeology* 46 (1993): 26–32.

Michel, Kai. "Die Geschichte des Märkischen Provinzial-Museums." *Jahrbuch Stiftung Stadtmuseum Berlin* 2 (1996): 180–95.

————. "'In Berlin höchst wunderbar / buddelt man das ganze Jahr': Gedanken zu Trägergruppen und Adressatenkreisen des Märkischen Provinzial-Museums." In *Renaissance der Kulturgeschichte? Die Wiederentdeckung des Märkischen Museums in Berlin aus einer europäischen Perspektive*, edited by Alexis Joachimides and Sven Kuhrau, 151–65. Dresden: Verlag der Kunst, 2001.

————. "'Und nun kommen Sie auch gleich noch mit 'ner Urne. Oder ist es bloß 'ne Terrine?' Das Märkische Provinzial-Museum in Berlin (1874–1908)." In *Mäzenatisches Handeln: Studien zur Kultur des Bürgersinns in der Gesellschaft*, edited by Thomas Gaehtgens and Martin Schieder, 60–81. Berlin: Bostelmann und Siebenhaar, 1998.

Mielke, Robert. "Das Märkische Museum und die Brandenburgia." *Brandenburgia* 33 (1924): 3–9.

"Mittheilungen aus den Lokalvereinen." *Correspondenz-Blatt der deutschen Gesellschaft für Anthropologie, Ethnologie und Urgeschichte* 21 (1890): 6–8.

Momigliano, Arnaldo. "Ancient History and the Antiquarian." In *Studies in Historiography*, edited by Arnaldo Momigliano, 1–39. New York: Harper Torchbooks, 1966.

Mommsen, Theodor. *Die Örtlichkeit der Varusschlacht*. Berlin: Weidmann, 1885.

Mosse, George L. *The Crisis of German Ideology: Intellectual Origins of the Third Reich*. New York: Grosset and Dunlap, 1964.

————. *The Nationalization of the Masses: Political Symbolism and Mass Movements in Germany from the Napoleonic Wars through the Third Reich*. New York: Fertig, 1975.

Mötefindt, Hugo. Review of *Alteuropa in seiner Kultur- und Stilentwicklung*, by Carl Schuchhardt. *Historische Zeitschrift* 123 (1921): 483–85.

————. "Verzeichnis der Sammlungen vor- und frühgeschichtlicher Altertümer Deutschlands." *Korrespondenz-Blatt der Deutschen Gesellschaft für Anthropologie, Ethnologie und Urgeschichte* 48 (1917): 27–50.

Müller, Johannes. *Vor- und frühgeschichtliche Alterthümer der Provinz Hannover*. Edited by Jacobus Reimers. Hannover: Schulze, 1893.

Müller, Uwe. *Infrastrukturpolitik in der Industrialisierung: Der Chausseebau in der preußischen Provinz Sachsen und dem Herzogtum Braunschweig vom Ende des 18. Jahrhunderts bis in die siebziger Jahre des 19. Jahrhunderts*. Berlin: Duncker und Humblot, 2000.

Nawroth, Manfred. "Aus Trümmern erstanden: Der Neuanfang im Westteil der Stadt (1945–1963)." In Menghin, "Das Berliner Museum für Vor- und Frühgeschichte," 193–211.

Nehls, Harry. "Der Einbruch in das königliche Museum vaterländischer Altertümer in Monbijou im Jahre 1841." *Acta Praehistorica et Archaeologica* 26–27 (1994–95): 213–36.

"Neue Richtlinien des Reichsinnenministeriums für den Geschichtsunterricht." *Nachrichtenblatt für deutsche Vorzeit* 9 (1933): 81–84.

Nipperdey, Thomas. *Deutsche Geschichte, 1800–1918*. 3 vols. Munich: Beck, 1998.

————. "Verein als soziale Struktur in Deutschland im späten 18. und frühen 19. Jahrhundert." In Boockmann et al., *Geschichtswissenschaft und Vereinswesen im 19. Jahrhundert*, 1–44.

Novick, Peter. *That Noble Dream: The "Objectivity Question" and the American Historical Profession*. Cambridge: Cambridge University Press, 1988.

Pape, Wolfgang. "Zehn Prähistoriker aus Deutschland." In Steuer, *Eine hervorragend nationale Wissenschaft*, 55–88.

Paret, Oscar. "Die Anfänge der Urgeschichtsforschung in Württemberg." *Württembergische Vierteljahrshefte für Landesgeschichte*, n.s., 35 (1930): 1–37.

Penny, H. Glenn. "The Fate of the Nineteenth Century in German Historiography." *Journal of Modern History* 80 (2008): 81–108.

————. *Objects of Culture: Ethnology and Ethnographic Museums in Imperial Germany*. Chapel Hill: University of North Carolina Press, 2002.

Philipp, Hans. *Vor- und Frühgeschichte des Nordens und des Mittelmeerraumes: Ein Handbuch für Schule und Haus*. Berlin: Mittler, 1937.

Pickel, Ignaz. *Beschreibung verschiedener Alterthümer welche in Grabhügeln alter Deutschen nahe bey Eichstätt sind gefunden worden*. 1789. Reprint, Fürth: VKA, 1990.

Piggott, Stuart. *Ruins in a Landscape: Essays in Antiquarianism*. Edinburgh: Edinburgh University Press, 1976.

Pinder, Eduard. *Die Aufgaben der Provinzialmuseen.* Leipzig: Schloemp, 1881.

Pohl, Walter. *Die Germanen.* Munich: Oldenbourg, 2000.

———. *Die Völkerwanderung: Eroberung und Integration.* Stuttgart: Kohlhammer, 2002.

Pomian, Krzysztof. "Franks and Gauls." In *Realms of Memory: Rethinking the French Past,* vol. 1, *Conflicts and Divisions,* edited by Pierre Nora, translated by Arthur Goldhammer, 27–76. New York: Columbia University Press, 1996.

Preusker, Karl. *Ueber Mittel und Zweck der vaterländischen Alterthumsforschung.* Leipzig: Nauck, 1829.

Proctor, Robert. "From *Anthropologie* to *Rassenkunde* in the German Anthropological Tradition." In *Bones, Bodies, Behavior: Essays on Biological Anthropology,* edited by George W. Stocking Jr., 138–79. Madison: University of Wisconsin Press, 1988.

"Protocoll über die erste Sitzung der zweiten allgemeinen Versammlung der deutschen anthropologischen Gesellschaft zu Schwerin, am 22. September 1871, im Saale des Schauspielhauses." *Correspondenz-Blatt der deutschen Gesellschaft für Anthropologie, Ethnologie und Urgeschichte* 2 (1871): 41–80.

Puschner, Uwe. "Germanenideologie und völkische Weltanschauung." In *Zur Geschichte der Gleichung "germanisch-deutsch." Sprache und Namen, Geschichte und Institutionen,* edited by Heinrich Beck, Dieter Geuenich, Heiko Steuer, and Dietrich Hakelberg, 103–29. Ergänzungsbände zum Reallexikon der germanischen Altertumskunde, edited by Heinrich Beck, Dieter Geuenich, and Heiko Steuer, vol. 34. Berlin: de Gruyter, 2004.

———. *Die völkische Bewegung im wilhelminischen Kaiserreich: Sprache–Rasse–Religion.* Darmstadt: Wissenschaftliche Buchgesellschaft, 2001.

Puschner, Uwe, Walter Schmitz, and Justus Ulbricht, eds. *Handbuch zur "Völkischen Bewegung," 1871–1918.* Munich: Saur, 1996.

Putzer, Peter. "Staatlichkeit und Recht nach der Säkularisation." In *Geschichte Salzburgs: Stadt und Land,* vol. 2, *Neuzeit und Zeitgeschichte,* edited by Heinz Dopsch and Hans Spatzenegger, 620–59. Salzburg, Austria: Pustet, 1988.

Rączkowski, Włodzimierz. "'Drang nach Westen'? Polish Archaeology and National Identity." In Diaz-Andreu and Champion, *Nationalism and Archaeology in Europe,* 189–217.

Rademacher, Carl. *Führer durch das Städtische Prähistorische Museum im Bayenturm zu Cöln.* Cologne: Bachem, 1910.

Ranke, Johannes. *Die akademische Kommission für Erforschung der Urgeschichte und die Organisation der urgeschichtlichen Forschung in Bayern durch König Ludwig I.* Munich: Verlag der k. b. Akademie, 1900.

———. "Unsere Ziele." *Beiträge zur Anthropologie und Urgeschichte Bayerns* 1 (1877): iii–vi.

Ranke, Leopold von. "On the Character of Historical Science." In *The Theory and Practice of History,* edited by Georg Iggers and Konrad von Moltke, translated by Wilma Iggers and Konrad von Moltke, 33–46. Indianapolis: Bobbs-Merrill, 1973.

————. *Universal History: The Oldest Historical Group of Nations and the Greeks.* Translated by D. C. Tovey and G. W. Prothero. New York: Harper, 1885.

Rheinisches Landesmuseum Bonn. Vol. 1, *Archivalien im Archiv des Landschafts-verband Rheinland, 1820–ca. 1954.* http://www.afz.lvr.de/media/de/archive_im _rheinland/archiv_des_lvr/findbuch_rheinisches_landesmuseum_bonn_1820 _ca_1954.pdf.

Rhode, Christian Detlev, and Andreas Albert Rhode. *Cimbrisch-Hollsteinische Antiquitaeten-Remarques, oder: Accurate und umständliche Beschreibung derer in denen Grab-Hügeln derer alten Heydnischen Hollsteiner der Gegend Hamburg gefundenen Reliquien, als Urnen / Wehr und Waffen / Zierrahten / Ringe / Arm–Bänder / &c.&c. welche durch häuffige Untersuchung und Aufgrabung derer Tumulorum aus selbigen hervor geholet worden.* Hamburg: Piscator, 1720.

Riehl, Wilhelm Heinrich. "Die Volkskunde als Wissenschaft: Ein Vortrag." In *Culturstudien aus drei Jahrhunderten,* 205–29. Stuttgart: Cotta, 1859.

Rives, J. B. Introduction to *Germania,* by Cornelius Tacitus, 1–74. Translated and edited by J. B. Rives. Oxford: Clarendon Press, 1999.

Rowley-Conwy, Peter. *From Genesis to Prehistory: The Archaeological Three Age System and Its Contested Reception in Denmark, Britain, and Ireland.* Oxford: Oxford University Press, 2007.

Ruf, Peter. "Aufgeklärter Absolutismus und expansive Machtpolitik: Das friderizianische Preußen, 1740–1786." In *Preußen-Ploetz: Eine historische Bilanz in Daten und Deutungen,* edited by Manfred Schlenke, 163–68. Freiburg: Ploetz, 1983.

Runde, H. "Nachrichten über vor- und frühgeschichtliche Altertumsfunde aus der Provinz Hannover." *Jahrbuch des Provinzial-Museums zu Hannover,* 1906–7, 13–14.

Rüster, Brigitte. "Geschichte des Museums von 1884 bis 1912." *Jahresschrift für mitteldeutsche Vorgeschichte* 67 (1984): 72–86.

Schäfer, Martina. "Rulaman und Höhlenkinder." *Archäologie in Deutschland* 21 (2005): 62–65.

Schama, Simon. *Landscape and Memory.* New York: Knopf, 1995.

Schasler, Max. *Die Königlichen Museen von Berlin: Ein praktisches Handbuch zum Besuch der Galerien, Sammlungen und Kunstschätze derselben.* 6th ed. Berlin: Nicolai, 1867.

Schenk, Georg. "Konrad Dietrich Haßler: Schulmann, Sprach- und Geschichtsforscher, Politiker, Landeskonservator, 1803–1873." *Lebensbilder aus Schwaben und Franken* 10 (1966): 361–74.

Schiek, Siegwalt. "Zur Geschichte der archäologischen Denkmalpflege in Württemberg und Hohenzollern." *Denkmalpflege in Baden-Württemberg* 3 (1983): 52–58.

Schindler, Alfons. "Geschichte als tragisches Schicksal: Dahn." In *Einführung in die deutsche Literatur des 19. Jahrhunderts,* vol. 2, *März-Revolution, Reichs-*

gründung und die Anfänge des Imperialismus, edited by Josef Jansen, 243–61. Opladen: Westdeutscher Verlag, 1984.

Schlette, Friedrich. "Die Anfänge einer Ur- und Frühgeschichtsforschung in Halle bis zur Gründung des Provinzialmuseums." *Jahresschrift für mitteldeutsche Vorgeschichte* 67 (1984): 9–27.

———. "Büsching, ein Pionier der Urgeschichtswissenschaft." *Ethnographisch-Archäologische Zeitschrift* 20 (1979): 523–32.

Schliz, Alfred. "Vorgeschichtliches deutsches Siedlungswesen." In Hoops, *Reallexikon der germanischen Altertumskunde*, 4: 444–67.

Schlosser, Julius von. *Die Kunst- und Wunderkammer der Spätrenaissance: Ein Beitrag zur Geschichte des Sammelwesens*. Leipzig: Klinkhardt und Bierman, 1908.

Schmidt, Hubert. "Vorgeschichtliche Abteilung der Kgl. Museen zu Berlin." *Prähistorische Zeitschrift* 1 (1909): 88–90.

Schmidt, Martin. "Die Rolle der musealen Vermittlung in der nationalsozialistischen Bildungspolitik: Die Freilichtmuseen deutscher Vorzeit am Beispiel von Oerlinghausen." In Leube, *Prähistorie und Nationalsozialismus*, 147–59.

Schnabel, Franz. "Der Ursprung der vaterländischen Studien." *Blätter für deutsche Landesgeschichte* 88 (1951): 4–27.

Schnapp, Alain. *The Discovery of the Past: The Origins of Archaeology*. Translated by Ian Kinnes and Gillian Varndell. London: British Museum Press, 1996.

Schöbel, Gunter. "Hans Reinerth: Forscher—NS-Funktionär—Museumsleiter." In Leube, *Prähistorie und Nationalsozialismus*, 321–96.

Schreiber, Heinrich. *Die neuentdeckten Hünengräber im Breisgau*. Freiburg: Wagner, 1826.

———, ed. *Taschenbuch für Geschichte und Alterthum in Süddeutschland*. 5 vols. Freiburg: Emmerling, 1839–46.

Schuchhardt, Carl. *Alteuropa: Eine Vorgeschichte unseres Erdteils*. 2nd ed. Berlin: de Gruyter, 1926.

———. *Alteuropa in seiner Kultur- und Stilentwicklung*. Berlin: Trübner, 1919.

———. "Kriegsarchäologie." *Prähistorische Zeitschrift* 6 (1914): 359–60.

———. "Römerschanze." In Ebert, *Reallexikon der Vorgeschichte*, 11:154–55.

———. "Die Römerschanze bei Potsdam nach den Ausgrabungen von 1908 und 1909." *Prähistorische Zeitschrift* 1 (1909): 209–38.

———. *Vorgeschichte von Deutschland*. Munich: Oldenbourg, 1928.

Schulz, Robert. "Geleitwort." *Posener Jahrbuch für Vorgeschichte* 1 (1944): 5.

"Die sechste Allgemeine Versammlung der Deutschen Gesellschaft für Anthropologie, Ethnologie und Urgeschichte zu München am 9. bis 11. August 1875." Edited by Julius Kollmann. Supplement, *Correspondenz-Blatt der deutschen Gesellschaft für Anthropologie, Ethnologie und Urgeschichte*. Munich: Oldenbourg, 1875.

See, Klaus von. *Barbar, Germane, Arier: Die Suche nach der Identität der Deutschen*. Heidelberg: Winter, 1994.

Segal, Daniel. "'Western Civ' and the Staging of History in American Higher Education." *American Historical Review* 105 (2000): 770–805.

Seger, Hans. "Maslographia, 1711–1911." *Schlesiens Vorzeit in Bild und Schrift: Zeitschrift des schlesischen Altertumsvereins*, n.s., 6 (1912): 1–16.

Seyer, Heinz. "Die ur- und frühgeschichtliche Sammlung und die Bodendenkmalpflege in Berlin." *Jahrbuch des Märkischen Museums* 9 (1983): 132–35.

Sheehan, James J. *German History, 1770–1866.* Oxford: Oxford University Press, 1989.

————. *Museums in the German Art World: From the End of the Old Regime to the Rise of Modernism.* Oxford: Oxford University Press, 2000.

Sklenar, Karel. *Archaeology in Central Europe: The First 500 Years.* Translated by Iris Lewitova. New York: St. Martin's, 1983.

Smail, Dan. "In the Grip of Sacred History." *American Historical Review* 110 (2005): 1337–61.

Smith, Anthony. "Authenticity, Antiquity and Archaeology." *Nations and Nationalism* 7 (2001): 441–49.

————. *The Ethnic Origins of Nations.* Oxford: Basil Blackwell, 1986.

Smith, Bonnie. *The Gender of History: Men, Women, and Historical Practice.* Cambridge, MA: Harvard University Press, 1998.

Smith, Woodruff. *Politics and the Sciences of Culture in Germany, 1840–1920.* New York: Oxford University Press, 1991.

Smolla, Günter. "Gustaf Kossinna nach 50 Jahren: Kein Nachruf." *Acta Praehistorica et Archaeologica* 16–17 (1984–85): 9–14.

————. "Das Kossinna-Syndrom." *Fundberichte aus Hessen* 19–20 (1979–80): 1–9.

Snyder, Louis. *Race: A History of Modern Ethnic Theories.* New York: Longmans, Green, 1939.

Sommer, Ulrike. "The Teaching of Archaeology in West Germany." In Härke, *Archaeology, Ideology and Society*, 205–43.

Speck, Josef. "Pfahlbauten: Dichtung oder Wahrheit? Ein Querschnitt durch 125 Jahre Forschungsgeschichte." *Helvetia Archaeologica* 12 (1981): 98–138.

Springer, Louis Adalbert. "Die Neuaufstellung der vorgeschichtlichen Sammlungen des Germanischen Nationalmuseums Nürnberg." *Nachrichtenblatt für deutsche Vorzeit* 11 (1935): 199–204.

Spruner, Karl von. *Die Wandbilder des bayerischen National-Museums.* Munich: Albert, 1868.

Stampfuß, Rudolf. *Gustaf Kossinna: Ein Leben für die deutsche Vorgeschichte.* Leipzig: Kabitzsch, 1935.

"Stand der Vorgeschichte in der Lehrerausbildung." *Nachrichtenblatt für deutsche Vorzeit* 9 (1933): 185–86.

Stein, Mary Beth. "Wilhelm Heinrich Riehl and the Scientific-Literary Formation of 'Volkskunde.'" *German Studies Review* 24 (2001): 487–512.

Stengel, Walter. "Chronik des Märkischen Museums der Stadt Berlin." *Jahrbuch für brandenburgische Landesgeschichte* 30 (1979): 7–51.

Stetter, Gertrud. "Die Entwicklung der historischen Vereine in Bayern bis zur Mitte des 19. Jahrhunderts." PhD diss., Ludwig-Maximilians-University (Munich), 1963.

Steuer, Heiko. "Deutsche Prähistoriker zwischen 1900 und 1995—Begründung und Zielsetzung des Arbeitsgesprächs." In Steuer, *Eine hervorragend nationale Wissenschaft*, 1–54.

————, ed. *Eine hervorragend nationale Wissenschaft: Deutsche Prähistoriker zwischen 1900 und 1995*. Ergänzungsbände zum Reallexikon der germanischen Altertumskunde, edited by Heinrich Beck, Dieter Geuenich, and Heiko Steuer, vol. 29. Berlin: de Gruyter, 2001.

Stopfel, Wolfgang. "Geschichte der badischen Denkmalpflege und ihrer Dienststellen Karlsruhe, Straßburg und Freiburg." *Denkmalpflege in Baden-Württemberg* 32 (2003): 202–10.

Strauss, Gerald. *Sixteenth-Century Germany: Its Topography and Topographers*. Madison: University of Wisconsin Press, 1959.

Strobel, Richard. "Eduard Paulus der Jüngere, zweiter Landeskonservator in Württemberg, gestorben vor 100 Jahren am 16. April 1907." *Denkmalpflege in Baden-Württemberg* 36 (2007): 122–30.

Stüler, Friedrich August. *Das Neue Museum in Berlin*. Berlin: Ernst und Korn, 1862.

Tacitus, Cornelius. *The Annals*. Translated by A. J. Woodman. Indianapolis: Hackett, 2004.

————. *Germania*. Translated by J. B. Rives. Oxford: Clarendon, 1999.

Tacke, Charlotte. *Denkmal im sozialen Raum: Nationale Symbole in Deutschland und Frankreich im 19. Jahrhundert*. Göttingen: Vandenhoeck und Ruprecht, 1995.

"Tätigkeitsbericht des Staatlichen Museums für Vorgeschichte und des Archivs urgeschichtlicher Funde Sachsens in Dresden für die Zeit vom 1. April 1932 bis 31. März 1933." *Nachrichtenblatt für deutsche Vorzeit* 9 (1933): 52–55.

Tatlock, Lynne. "Realist Historiography and the Historiography of Realism: Gustav Freytag's *Bilder aus der deutschen Vergangenheit*." *German Quarterly* 63 (1990): 59–74.

Thomas, David Hurst. *Skull Wars: Kennewick Man, Archeology, and the Battle for Native American Identity*. New York: Basic Books, 2000.

Thrän, Ferdinand. "Bericht über die Ausgrabung von deutschen Grabhügeln in den Waldungen Ansang, Frauenhau und vorderer Hühnerberg im Forstrevier Ringingen, Oberamts Blaubeuren." *Verhandlungen des Vereins für Kunst und Alterthum in Ulm und Oberschwaben* 7 (1850): 45–47.

Thucydides. *On Justice, Power, and Human Nature: The Essence of Thucydides' "History of the Peloponnesian War."* Edited and translated by Paul Woodruff. Indianapolis: Hackett, 1993.

Todd, Malcolm. *The Early Germans*. 2nd ed. Malden, MA: Blackwell, 2004.

Toews, John. *Becoming Historical: Cultural Reformation and Public Memory in*

Early Nineteenth-Century Berlin. Cambridge: Cambridge University Press, 2004.

Treuer, Gotthilf. *Kurtze Beschreibung der Heidnischen Todten-Töpffe, In welchen Die Heiden ihrer verbrannten Todten überbliebene Gebeine und Aschen aufgehoben, unter der Erden beygesetzet Und Bey den jetzigen Zeiten in der Chur- und Marck Brandenburg Hauffen-weise ausgegraben werden*. Nuremberg: Hoffmann, 1688.

Trigger, Bruce. *A History of Archaeological Thought*. Cambridge: Cambridge University Press, 1989.

Tröltsch, Eugen von. *Altertümer aus unserer Heimat: Rhein- und deutsches Donau-Gebiet*. Stuttgart: Kohlhammer, 1890.

———. *Die Pfahlbauten des Bodenseegebietes*. Stuttgart: Enke, 1902.

"Ueber die Gräber der alten Teutschen, welche in unsern Gegenden gefunden worden sind." *Fränkisches Archiv* 2 (1790): 112–22.

Vallibus, Desalis. "Versuch eines Entwurfs, nach welchem man die Alterthümer des platten Landes am glücklichsten aufsuchen und am brauchbarsten beschreiben kan." *Hannoverische Gelehrte Anzeigen* 4, no. 98 (1754): 1375–86.

Van Riper, A. Bowdoin. *Men among the Mammoths: Victorian Science and the Discovery of Human Prehistory*. Chicago: University of Chicago Press, 1993.

"Verhandlungen der XI. allgemeinen Versammlung der Deutschen Gesellschaft für Anthropologie, Ethnologie und Urgeschichte zu Berlin im August 1880." Edited by Johannes Ranke. Supplement, *Correspondenz-Blatt der Deutschen Gesellschaft für Anthropologie, Ethnologie und Urgeschichte*. Munich: Straub, 1880.

Virchow, Rudolf. *Briefe an seine Eltern, 1839 bis 1864*. Edited by Marie Rabl. Leipzig: Engelmann, 1907.

———. "Die Pfahlbauten im nördlichen Deutschland: Vortrag, gehalten in der Sitzung der Berliner Anthropologischen Gesellschaft am 11. Dezember 1869." *Zeitschrift für Ethnologie* 1 (1869): 401–16.

———. "Schivelbeiner Alterthümer." *Baltische Studien* 21 (1866): 179–96.

———. "Die sogenannten Idole von Prillwitz." *Verhandlungen der Berliner Gesellschaft für Anthropologie, Ethnologie und Urgeschichte*, 1878, 264–68.

———. "Über die culturgeschichtliche Stellung des Kaukasus, unter besonderer Berücksichtigung der ornamentirten Bronzegürtel aus transkaukasischen Gräbern." *Abhandlungen der Königlichen Akademie der Wissenschaften zu Berlin*, 1895, 1–66.

———. "Ueber Hünengräber und Pfahlbauten." In *Sammlung gemeinverständlicher wissenschaftlicher Vorträge*, edited by Rudolf Virchow and Franz von Holtzendorff, 5–36. Berlin: Lüderitz, 1866.

———. "Zur Geschichte von Schivelbein." *Baltische Studien* 13 (1847): 1–33.

Vocelka, Karl. *Rudolf II. und seine Zeit*. Vienna: Böhlau, 1985.

"Die Vorgeschichte in der Volksbildungsarbeit des Naturkundlichen Museums der Stadt Leipzig." *Nachrichtenblatt für deutsche Vorzeit* 9 (1933): 55–57.

Voß, Albert. "Die ethnologische und nordische Sammlung." In *Zur Geschichte der Königlichen Museen in Berlin: Festschrift zur Feier ihres fünfzigjährigen Bestehens am 3. August 1880*, 154–60. Berlin: Reichsdruckerei, 1880.

———. *Merkbuch, Alterthümer aufzugraben und aufzubewahren: Eine Anleitung für das Verfahren bei Aufgrabungen, sowie zum Konserviren vor- und frühgeschichtlicher Alterthümer*. Berlin: Mittler, 1888.

Wächter, Johann Karl. *Statistik der im Königreiche Hannover vorhandenen heidnischen Denkmäler*. Hannover: Historischer Verein für Niedersachsen, 1841.

Wahle, Ernst. "Die deutsche Vorgeschichtsforschung in der Gegenwart." *Deutsches Bildungswesen* 3 (1935): 705–15.

———. *Kurze Übersicht der wichtigsten Literatur der Vorgeschichte Mitteleuropas auf Grund des vorgeschichtlichen Apparates des Germanischen Seminars der Universität Berlin*. Edited by Gustaf Kossinna. Hannover: Riemschneider, 1909.

———. "Die Neuaufstellung der vorgeschichtlichen Abteilung in den städtischen Sammlungen zu Heidelberg." *Museumskunde* 16 (1921): 101–12.

Walter, Eva. *Mit Kindern unterwegs: Schwäbische Alb*. Bietigheim-Bissingen: Fleischhauer und Spohn, 2007.

Weber, F. "Mitteilungen aus dem Vereins-Archiv: Aeltere Fundnachrichten aus Oberbayern." *Altbayerische Monatsschrift* 2 (1900): 124–29.

Wegner, Günter. "Auf vielen und zwischen manchen Stühlen: Bemerkungen zu den Auseinandersetzungen zwischen Karl Hermann Jacob-Friesen und Hans Reinerth." In Leube, *Prähistorie und Nationalsozialismus*, 397–417.

Wehrberger, Kurt. "Ausgrabungen und archäologische Bestände des Vereins für Kunst und Altertum in Ulm und Oberschwaben." In *Der Geschichte treuer Hüter . . . : Die Sammlungen des Vereins für Kunst und Altertum in Ulm und Oberschwaben; Festschrift zum 150jährigen Bestehen des Vereins*, 62–91. Ulm: Ulmer Museum, 1991.

Weinland, David F. *Kuning Hartfest: Ein Lebensbild aus der Geschichte unserer deutschen Ahnen, als sie noch Wuodan und Duonar opferten*. 5th ed. Berlin: Neufeld und Henius, n.d.

———. *Rulaman: Erzählung aus der Zeit des Höhlenmenschen und des Höhlenbären*. 6th ed. Leipzig: Spamer, 1906.

Weinland, Martina, and Kurt Winkler, eds. *. . . schaut auf diese Stadt! Die Geschichte Berlins*. Berlin: Museumspädagogischer Dienst Berlin, 1999.

Weiss, Gerhard. "Panem et Circenses: Berlin Anniversaries as Political Happenings." In *Berlin: Culture and Metropolis*, edited by Charles Haxthausen and Heidrun Suhr, 243–52. Minneapolis: University of Minnesota Press, 1990.

Wells, Peter S. *The Battle That Stopped Rome: Emperor Augustus, Arminius, and the Slaughter of the Legions in the Teutoburg Forest*. New York: Norton, 2003.

———. *Beyond Celts, Germans, and Scythians: Archaeology and Identity in Iron Age Europe*. London: Duckworth, 2001.

Weyl, Louis. *Die Sammlung vaterländischer Alterthümer*. Vol. 6 of *Der Führer durch die Kunstsammlungen Berlins*. Berlin: Oehmigke, 1842.

White, Hayden. *The Content of the Form: Narrative Discourse and Historical Representation*. Baltimore: Johns Hopkins University Press, 1987.

————. *Metahistory: The Historical Imagination in Nineteenth-Century Europe*. Baltimore: Johns Hopkins University Press, 1973.

Wiesend, Georg. "Archäologische Funde und Denkmale in den Landgerichtsbezirken Titmaning, Laufen und Burghausen." *Oberbayerisches Archiv für die vaterländische Geschichte* 11 (1850–51): 3–54.

Williamson, George. *The Longing for Myth in Germany: Religion and Aesthetic Culture from Romanticism to Nietzsche*. Chicago: University of Chicago Press, 2004.

Wiwjorra, Ingo. "Die deutsche Vorgeschichtsforschung und ihr Verhältnis zu Nationalismus und Rassismus." In Puschner, Schmitz, and Ulbricht, *Handbuch zur "Völkischen Bewegung," 1871–1918*, 186–207.

————. "'Ex oriente lux'—'Ex septentrione lux': Über den Widerstreit zweier Identitätsmythen." In Leube, *Prähistorie und Nationalsozialismus*, 73–106.

————. "German Archaeology and Its Relation to Nationalism and Racism." In Diaz-Andreu and Champion, *Nationalism and Archaeology in Europe*, 164–88.

————. *Der Germanenmythos: Konstruktion einer Weltanschauung in der Altertumsforschung des 19. Jahrhunderts*. Darmstadt: Wissenschaftliche Buchgesellschaft, 2006.

Zehme, Arnold. "Zur Einführung in die deutschen Altertümer im deutschen Unterricht, besonders der Tertia." *Zeitschrift für den deutschen Unterricht* 10 (1896): 29–41.

Zimmerman, Andrew. *Anthropology and Antihumanism in Imperial Germany*. Chicago: University of Chicago Press, 2001.

————. "Anti-Semitism as Skill: Rudolf Virchow's *Schulstatistik* and the Racial Composition of Germany." *Central European History* 32 (1999): 409–29.

Ziolkowski, Theodore. *Clio the Romantic Muse: Historicizing the Faculties in Germany*. Ithaca, NY: Cornell University Press, 2004.

Zittel, Karl Alfred von. *Rückblick auf die Gründung und die Entwickelung der k. bayerischen Akademie der Wissenschaften im 19. Jahrhundert*. Munich: Verlag der k. b. Akademie, 1899.

"Zur Erinnerung an Albert Voß." *Jahrbuch der königlich preußischen Kunstsammlungen* 28 (1907): i–iv.

Index

Page numbers in italics refer to illustrations.

archaeology (*cont.*)
between, 151; as historical practice, 41, 103; as historical tool, 33, 100; importance of, 187; inventories, compiling of, 64–65; lake dwellings, 118, 121; local landscapes, connection between, 189; local pride, generating of, 8; merry spirit of, 169; methods and record keeping, improvement of, 61; nationalism, influence on, 5, 254; Nazi ideology, 11, 253, 271; in occupied Poland, 272; political use of, 4; and prehistory, 61, 117–18, 156; as process of interpretation, 279–80; professionalization of, 56, 284; as regional, 154; in schools, 179–80, 182, 243–44, 253, 259–62, 267; as science, 178–79, 227, 241–42; settlement archaeology, 234–36, 242–44, 249, 267; territory and history, link between, 26; *Volk*, insights into, 242. *See also* domestic archaeology
Archäologische Staatssammlung (Bavarian state archaeological collection) (Munich), 38–39, 185, 281. *See also* Prähistorische Sammlung; Prähistorische Staatssammlung; Vor- und Frühgeschichtliche Staatssammlung
Archiv für Anthropologie (Archive for anthropology), 122, 128; *Correspondenz-Blatt,* supplement to, 129
Aretin, Karl Maria von, 105
Ariovistus, 32
Arminius, 19–20, 115, 187–88, 196, 199, 230, 300n1 (part 2). *See also* Hermannsdenkmal
Arndt, Ernst Moritz, 43, 104
artifacts, 23, 41, 44, 49–50, 55–56, 72, 75–76, 79, 160, 174, 178, 183, 192, 217, 219, 242, 266; historical associations, 66
Aryans, 211, 221–23
Atlas vorgeschichtlicher Befestigungen in Niedersachsen (Atlas of prehistoric fortifications in Lower Saxony) (Oppermann and Schuchhardt), 188
Attila the Hun, 34, 109
Aubrey, John, 26
Aufseß, Hans von, 74–75
Austria, 88, 90, 92–93, 96, 197, 230
Austrian Empire, 86, 90, 214
Austro-Prussian War, 164. *See also* Wars of Unification

Bab, Alfred, 268–69
Baden, 64, 176
Baiovari, 80, 89, 92, 95, 114
Baltische Studien (journal), 123
Bamberg, 55
Bandel, Ernst von, 187
Bastian, Adolf, 140, 158
Battle of the Teutoburg Forest, 187–88
Bavaria, 28, 34, 36, 39, 44, 46–47, 49–52, 55, 64, 67, 80, 88, 90, 93, 95–96, 103–4, 107–8, 114, 118, 130–31, 175, 179–80, 182, 185
Bavarian Academy of Sciences, 28, 38, 49–50, 175–76
Bayerisches Nationalmuseum (Bavarian national museum, BNM), 103–7, 112, 155, 175, 183–85, 281, 299n69; Bavarian identity, strengthening of, 174
Bekmann, Johann Christoph, 29, 39, 49, 125, 285; *Historische Beschreibung der Chur und Mark Brandenburg* (Historical description of the Electorate and Mark of Brandenburg), 26–27, *28*
Bellermann, Ferdinand, *113*
Beltz, Robert, 176
Berlin, 7, 23, 39, 44, 49, 56–57, 70, 136, 155, 159, 166–68, 184, 238, 275, 281, 283–84; prehistory, study of in, 132; Slavic past of, 270
Berliner Gesellschaft für Anthropologie, Ethnologie and Urgeschichte (BGfAEU), 128–29, 132, 158, 160, 166, 168, 177, 230
Berliner Lehrerverein (Berlin association of teachers), 243
Bersu, Gerhard, 257
Beschreibung entdeckter Alterthümer in Deutschland (A description of antiquities discovered in Germany) (Hummel), 40–41
Bibliothek der deutschen Alterthümer (Library of German antiquities) (Hummel), 40
Bilder aus der deutschen Vergangenheit (Scenes from the German past) (Freytag), 204–6, 208–9, 212–13
Bilder aus der märkischen Vorzeit (Scenes from Brandenburg's prehistory) (Kiekebusch), 266
Bismarck, Otto von, 123, 125, 204
Black Forest, 97–99, 146